Chemical Kinetics and Reaction Dynamics

Chemical Kinetics and Reaction Dynamics

Dr. Hirdayesh Kumar Vatsa

RANDOM PUBLICATIONS
NEW DELHI (INDIA)

Chemical Kinetics and Reaction Dynamics

ISBN 978-93-5111-622-6

Published in 2015 in India by

RANDOM PUBLICATIONS

4376-A/4B, Gali Murari Lal, Ansari Road
New Delhi-110 002
Phone : +9111-43580356, 011-23289044, 011-43142548
e-mail: sales@randompublications.com,
info@randompublications.com, randomexports@gmail.com

Reprinted 2022

Type Setting by : Friends Media, Delhi-110089
Digitally Printed at : Replika Press Pvt. Ltd.

Preface

Chemical kinetics deals with the experimental determination of reaction rates from which rare laws and rate constants are derived. Relatively, simple rate laws exist for zero order reactions first order reactions and second order reactions and can be derived for others.

In chemical kinetics, the time variable is introduced and rate of change of concentration of reactants or products with respect to time is followed. The chemical kinetics is thus, concerned with the quantitative determination of rate of chemical reactions and of the factors upon which the rates depend.

The base methods and responses during a chemical reaction are determined by the chemical itself. No matter what conditions are present, some materials will not react with other chemicals or with environmental factors. This forms the baseline used by chemical kinetics. Once a reaction is understood in a controlled environment, changes to the reaction can be directly attributed to changes in the conditions of the experiment.

Chemical kinetics encompasses the study of reaction rates, how these rates change due to conditions, and the molecular level activity that occurs in a reaction. The four common variables that influence reaction rate are the concentration of the reactant, temperature, catalyst, and surface area. For a reaction to occur, the molecules must collide and more collisions occur with increased thermal energy to greater number of molecules present.

This book clarifies complex aspects of the more theoretical areas of chemical kinetics and the significance of the complex equations which so often baffle students.

I would like to thank my team for standing beside me throughout my career and writing this book. My special thanks go to "Random Publications" who have published the book.

– Dr. Hirdayesh Kumar Vatsa

Contents

Preface .... *v-vi*

1. **Chemical Kinetics** **1**

Introduction 1
Chemical Kinetics, a Clock Reaction 3
Reaction Rates of Chemical Kinetics 7
Precision of Kinetic Measurements 16

2. **Reaction Rates and Mechanisms** **29**

Reaction Rates 29
Rate of Reaction in Molecules Combine 30
Definition of Rate of Chemical Reaction 32
Evaluation of Reaction Rate Forms in Stirred Tank Reactors 34
Surface Area on Reaction Rates 50

3. **Chemical Reactions of Rate Laws** **55**

Chemical Reactions and Rates 55
Temperature Affect a Chemical Reaction 57
Rate Laws 61
Fundamentals of Rate Laws 61
Common Types of Rate Laws 68
Rate Laws and Reaction Order 71
The rates of Chemical Reactions 72

4. **Reaction Rates and Collision Theory** **84**

Measuring Rates of Reaction 85
Molecular Collisions 88
Kinetic Molecular Theory 94

5. **Reaction Mechanisms and the Rate-Law Expression** **123**

Introduction 123
Properties of Mechanisms 123
Description of a Reaction Mechanism 129

Elementary Reaction Steps - Molecularity 132
Factors Affecting Reaction Rate; the Rate Law 134

6. Elementary Reactions and Reaction Mechanisms 148

Elementary Processes and Reaction Mechanism 148
Reaction Mechanism in Chemical Change 150
The Reaction Conditions of Mechanism 153
Order of Reaction, and Rate Constant 157

7. Chemical Equilibrium 161

Introduction 161
Principles of Chemical Equilibrium 161
Le Chatelier's Principle 164
Equilibrium Constant 167
Predicting Equilibrium Concentrations 172
Spreadsheets in Equilibrium Calculations 173

8 Enzymes Kinetics 202

Enzyme 202
General Enzymatic Kinetics 205
Enzyme Inhibition 207
Michaelis-Menten Kinetics of a Two-Substrate Reaction 210
Limiting Kinetics of Enzyme-Catalysed Reactions 212
Kinetic Analysis 214
Enzyme-Substrate Interactions 218

9. Chemical Potential Energy 246

Chemical Potential 246
Potential Energy 247
Kinetic Energy of Systems 248
Kinetic Energy Correction 250
Principle of Work and Energy for a Rigid Body 252
Relativistic Kinetic Energy of Rigid Bodies 254
Equation of Motion for a Rigid Body 257

10. Organic Reaction and Organic Chemistry 259

Functional Groups and Organic Reactions 259
Substitution Reactions 262
Organic Chemistry 271
The Nature of Organic Chemistry 277
Organic Families and Their Functional Groups 279

11. Chemical Dynamics 293

Complex Reactions: Parallel Processes, Consecutive Processes, and Reversible Processes 297

***Bibliography* 303**

***Index* 305**

1

Chemical Kinetics

INTRODUCTION

Chemical kinetics is the study of chemical reactions and how they change based on varying conditions. The primary determination of a chemical reaction's effect is mostly outside the substance's actual makeup. Chemical kinetics studies these conditions to determine how and why a reaction occurs as it does. The primary applied use of the science is creating efficient, non-destructive and non-poisoning reactions for use in manufacturing, materials processing and the pharmaceutical industries. Even so, the majority of this field is pure science and research.

The base methods and responses during a chemical reaction are determined by the chemical itself. No matter what conditions are present, some materials will not react with other chemicals or with environmental factors. This forms the baseline used by chemical kinetics. Once a reaction is understood in a controlled environment, changes to the reaction can be directly attributed to changes in the conditions of the experiment.

Apart from what the initial substance actually is, the state and the concentration of the substance will play an important factor in chemical kinetics. These two factors are directly related to the condition of the initial material, rather than environment. The state of the material refers to whether it is a solid, liquid or gas. The concentration is the amount of reactant when compared to another substance or volume. This may be expressed as a 'parts per 1,000 of X chemical' or as a percentage of space used in the experimental area.

The concentration of material is of great importance to chemical kinetics for two reasons. First, the larger the amount of reactants, the faster the reaction will occur. This was one of the first major breakthroughs in this field and is often considered the birth point of the discipline. The second major effect is in regards to chemical interaction. If the concentration of two reactants is very low within a third non-reactive substance, then it will take a noticeable amount of time for the two reactants to actually encounter one another.

In addition to the physical properties of the material, the environment is also a big part of chemical kinetics. Environmental factors generally come down to temperature and pressure. As the temperature increases, the reactants absorb thermal energy and move faster. This allows them to interact with each other more frequently, which speeds up the reaction. Likewise, higher pressure makes more activities happen in a smaller space, increasing the likelihood that the reactants encounter one another.

The last major part of chemical kinetics is the use of catalysts. These 'third-party' chemicals change the way in which the reaction occurs without changing the end result of the reaction. A catalyst will interact with one or more reactants and change the way they behave. This typically means they increase in energy, move faster, interact more and increase the overall reaction speed. In order to be a catalyst and not a reactant, the catalysing material must remain unchanged during the reaction.

Chemical kinetics is the study and discussion of chemical reactions with respect to *reaction rates, the effect of various variables, re-arrangement of atoms, formation of intermediates etc*. There are many topics to be discussed, and each of these topics is a tool for the study of chemical reactions. By the way, the study of motion is called kinetics, from Greek *Kinesis*, meaning movement.

At the macroscopic level, we are interested in amounts reacted, formed, and the rates of their formation. At the molecular or microscopic level, the following considerations must also be made in the discusion of chemical reaction mechanism.

- Molecules or atoms of reactants must collide with each other in chemical reactions.
- The molecules must have sufficient energy (discussed in terms of activation energy) to initiate the reaction.
- In some cases, the orientation of the molecules during the collision must also be considered.

REACTION RATES

Chemical reaction rates are the rates of change in concentrations or amounts of either reactants or products. For changes in amounts, the units can be one of mol/s, g/s, lb/s, kg/day etc. For changes in concentrations, the units can be one of mol/(L s), g/(L s), per cent/s etc.

With respect to reaction rates, we may deal with *average rates, instantaneous rates, or initial rates* depending on the experimental conditions.

Thermodynamics and kinetics are two factors that affect reaction rates. The study of energy gained or released in chemical reactions is called thermodynamics, and such energy data are called thermodynamic data. However, thermodynamic data have no direct correlation with reaction rates, for which the kinetic factor is perhaps more important. For example, at room temperature (a wide range of temperatures), thermodynamic data indicates that diamond shall convert to graphite, but in reality, the conversion rate is so slow that most people think that *diamond is forever*.

DIFFERENT WAYS OF EXPRESSING THE RATE OF REACTION

There is usually more than one way to measure the rate of a reaction. We can study the decomposition of hydrogen iodide, for example, by measuring the rate at which either H_2 or I_2 is formed in the following reaction or the rate at which HI is consumed.

$$2\,HI(g) \rightleftharpoons H_2(g) + I_2(g)$$

Experimentally we find that the rate at which I_2 is formed is proportional to the square of the HI concentration at any moment in time.

$$\frac{d(I_2)}{dt} = k(HI)^2$$

What would happen if we studied the rate at which H_2 is formed? The balanced equation suggests that H_2 and I_2 must be formed at exactly the same rate.

$$\frac{d(H_2)}{dt} = \frac{d(I_2)}{dt}$$

What would happen, however, if we studied the rate at which HI is consumed in this reaction? Because HI is consumed, the change in its concentration must be a negative number. By convention, the rate of a reaction is always reported as a positive number. We therefore have to change the sign before reporting the rate of reaction for a reactant that is consumed in the reaction.

$$\frac{-d(HI)}{dt} = k'(HI)^2$$

The negative signs does two things. Mathematically, it converts a negative change in the concentration of HI into a positive rate. Physically, it reminds us that the concentration of the reactant decreases with time.

What is the relationship between the rate of reaction obtained by monitoring the formation of H_2 or I_2 and the rate obtained by watching HI disappear? The stoichiometry of the reaction says that two HI molecules are consumed for every molecule of H_2 or I_2produced. This means that the rate of decomposition of HI is twice as fast as the rate at which H_2 and I_2 are formed. We can translate this relationship into a mathematical equation as follows.

$$-\frac{d(HI)}{dt} = 2\left[\frac{d(H_2)}{dt}\right] = 2\left[\frac{d(I_2)}{dt}\right]$$

As a result, the rate constant obtained from studying the rate at which H_2 and I_2 are formed in this reaction (*k*) is not the same as the rate constant obtained by monitoring the rate at which HI is consumed (*k'*)

CHEMICAL KINETICS, A CLOCK REACTION

Chemical kinetics encompasses the study of reaction rates, how these rates change due to conditions, and the molecular level activity that occurs in a reaction. The four common variables that influence reaction rate are the concentration of the reactant, temperature, catalyst, and surface area. For a reaction to occur, the molecules must collide and more collisions occur with increased thermal energy to greater number of molecules present.

To further the understanding of chemical kinetics and the reaction rates dependence on concentration, temperature, and the presence of a catalyst, this experiment focused on the primary reaction between iodide, bromate, and hydrogen ions and determining the reaction's rate law:

$$6I^-(aq) + BrO_3^-(aq) + 6H^+(aq) \rightarrow 3I_2(aq) + Br^-(aq) + 3H_2O(1)$$

$$\text{rate} = k[I^-]^x[BrO_3^-]^y[H^+]^z$$

In this reaction, there are three reactants and different concentrations of each reactant were used in the five different mixtures to obtain the orders, x, y and z. Hence, the rate constant k can then be calculated. It is expected that the reaction occurs more quickly in the presence of greater concentration of reactants. In order to make sure that the same small amount of I_2 product formed for each mixture reaction, sodium thiosulfate was used to create a clock reaction that occurred faster than the primary reaction:

$$I_2(aq) + 2S_2O_3^{2-}(aq) \rightarrow 2I^-(aq) + S_4O_6^{2-}(aq)$$

The addition of a set amount of sodium thiosulfate consumed a set amount of I_2. Therefore, the blue color, which is created by the interaction between I_2 and starch, formed once the thiosulate was fully consumed. The relationship between bromate and thiosulfate was formed:

$$\frac{3\text{ mol } I_2}{1\text{ mol } BrO_3^-} \times \frac{2\text{ mol } S_2O_3^{2-}}{1\text{ mol } I_2} \text{ accounts for } \frac{\Delta [BrO_3^-]}{\Delta t} = \frac{1}{6}\frac{\Delta [S_2O_3^{2-}]}{\Delta t}$$

The four different temperatures settings were used to show the relationship between temperature and the reaction rate. Using the Arrhenius equation:

$$\ln k = \frac{-E_a}{R}\left(\frac{1}{T}\right) + \ln A\text{; A is a constant; R = 8.314 J/molK}$$

the activation energy Ea can be calculated and it is expected that reactions occurring at a higher temperature will have an increased average kinetic energy of the molecules, allowing more molecules to overcome the activation energy barrier. Hence, the increase in temperature speeds up the reaction rate. In addition, a decrease in temperature slows down the reaction. Lastly, the use of ammonium molybdate, which is a catalyst to the reaction, will lower the activation energy and reaction occurs in a shorter time span.

Experimental Reaction

The contents in reaction flask 2 were poured into the reaction flask and the time was taken to see how long it would take for the solution to turn blue. The mixture was continuously swirled until it turned blue. Although it was recommended to use three drops of starch, six drops of starch was used for each trial. The final temperature was also recorded for each trial.

Table. Volume of Reagents for Mixtures 1 to 5

Mix	Reaction Flask 1(250 mL) 0.010 M KI	0.0010 M $Na_2S_2O_3$	H_2O
1	10 mL	10 mL	10 mL
2	20 mL	10 mL	0 mL
3	10 mL	10 mL	0 mL
4	10 mL	10 mL	0 mL
5	8 mL	10 mL	12 mL
	Reaction Flask 2 (125mL) 0.040 M KBrO3	**0.10 M HC1**	**Starch**
1.	10 mL	10 mL	3 drops
2.	10 mL	10 mL	3 drops
3.	20 mL	10 mL	3 drops
4.	10 mL	20 mL	3 drops
5.	5 mL	15 mL	3 drops

For the second part, mixture 1 reaction was completed in the four different temperature settings, 40, 20, 10 and 0 C. To ensure constant temperature from the beginning to end of the reaction, the flasks were kept in the necessary temperature water or ice baths. Time was also taken to see how long it would take for the solution to turn blue.

For the third part, mixture 1 was used and the addition of ammonium molybdate, the catalyst. The catalyst's function is to lower the activation energy and create new elementary steps to the reaction and transition states. Overall, it speeds up the reaction. One drop of 0.5 M $(NH_4)_2MoO_4$ was added to the reagents in reaction flask 2, which was later mixed in with reaction flask 1. The reaction times for with and without the catalyst were both recorded.

Reaction Time

The reaction time varies with the reactant concentrations of iodide, bromate, and hydrogen ions for the different mixtures. As the iodide ion concentration doubles in mix 2, the time of the reaction halves, when compared to mix 1. As the bromate ion concentration doubles in mix 3, the time of the reaction decreases, when compared to mix 1. As the hydrogen ion concentration doubles in mix 4, there is a three fold decrease in reaction time, when compared to mix 1. In mix 5, the reaction has the least reactant concentration and it took the longest for the solution to turn blue.

Table. Reaction time and rates for Mixtures 1 to 5

Mix	Time (s)	Rate $-\Delta[BrO_3^-]/\Delta t$ $= 3.33 \times 10^{-5}$ (mol/L)/Δt	Reactant concentration (mol/L) $[I^-]$	$[BrO_3^-]$	$[H^+]$	Temp (°C)
1	160	2.083E-7	0.0020	0.008	0.002	20.8-21
2	82	4.065E-7	0.0040	0.008	0.002	20.8-21
3	99	3.367E-7	0.0020	0.016	0.002	20.8-21
4	46	7.246E-7	0.0020	0.008	0.004	20.8-21
5	210	1.587E-7	0.0016	0.004	0.003	20.8-21

From chemical kinetics, a higher reactant concentration results in more molecules likely to collide and interact with each other and allowing the reaction to occur more quickly. Therefore, mixtures 2, 3, and 4 have greater number of collisions between the sodium thiosulfate and iodine and allowing the clock reaction to occur quickly and allowing the iodine to react with the starch to turn the solution blue.

When the reactants for mix 1 were heated to 40C and later mixed, it took 60 seconds for the reaction to occur. At higher temperatures, the increased thermal energy allows the molecules to collide more often and react. In contrast, the colder temperatures of 10C and 1.8C resulted in longer reaction times, 406 and 660 seconds, respectively. Hence, the trend shows that the decrease in temperature results in a longer reaction time and smaller rate value. According to the lecture notes, a 10 Celsius increases correlates with a doubling or tripling of the reaction rate. When comparing the reaction rates for 10 and 1.8 C, there is a 1.6 fold difference and when comparing 10C to 20C, there is a 2.5 fold change in the rate. By comparing these results to a theoretical concept, there is variability in the change of rate as the temperature increases in 10 C increments.

Table. Reaction and rates for Mix 1 at 40, 20, 10, 1.8 C

Mixture	Time (s)	Rate = $-\Delta[BrO_3^-]/\Delta t$ (mol/L-s)	Temp (°C)	Calculated k ($1/M^3s$)
1A	60	5.56E-7	40	8680.56
1B	160	2.08E-7	20.8	3255.21
1C	406	8.21E-8	10	1282.84
1D	660	5.05E-8	1.8	789.141

The trend line for the ln k vs 1/T graph shows that the slope is -5526 K and the activation energy derived from this trend line is 46 kJ/mol. The calculated slope for these points is -5446 K and the activation energy Ea is 45 kJ/mol.

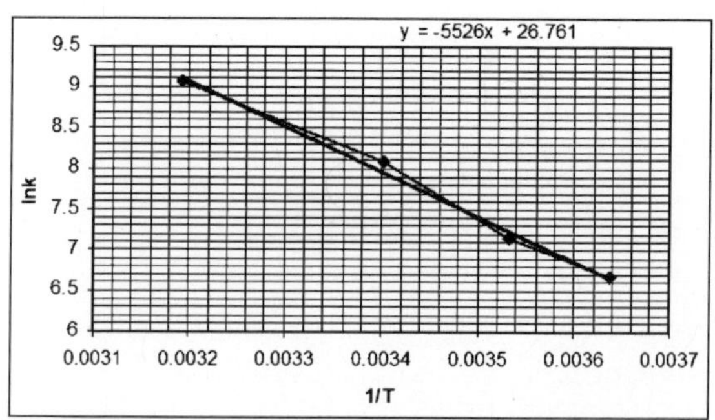

Fig. Graph lnk vs 1/T (y=-5526x +26.761)

Since the rate constant k increases with temperature, the rate of reaction will increase at any give reactant concentration. This clock reaction experiment shows an Arreneius behaviour and the high Ea indicates a strong temperature dependence of the rate constant. When comparing the calculated slope and Ea to the trend line slope and derived Ea, the values are pretty close. From online literature sources, the Ea for an iodine clock reaction is 53kJ/mol, which is higher than what was calculated in the experiment. Since only four temperature settings were used, a more accurate Ea could be derived by conducting the experiment with a greater range of temperature settings.

When introducing the homogenous catalyst ammonium molybdate into the reaction mix 1, the reaction takes 15 seconds. The activation energy of the catalyzed reaction is smaller than the Ea of an uncatalyzed reaction. Since the function of the catalyst is to lower the activation energy, the reactions moves forward a lot quicker, in comparison to the 174 seconds for an uncatalyzed reaction to occur. Although the catalyst doesn't get consumed in the reaction, it creates new transition states and elementary steps to allow the reaction to precede a lot faster. Hence, the catalyst lowers the Ea for the sodium thiosulfate and iodine reaction and this experiment demonstrated the function of a catalyst in chemical kinetics.

REACTION RATES OF CHEMICAL KINETICS

Chemical Kinetics is *the study of reaction rates*; that is, how fast a given reaction does proceeds. It is a measure of the change of the concentration of reactants (or products) as a function of time.

Reaction rates provide information regarding how fast a chemical process occurs as well as the mechanism by which a reaction occurs at molecular level.

MEASURE OF REACTION RATES

The rate of reaction is a measure of the change in concentration of reactants or products over time. Rate can be measured at the beginning of the reaction, which is called the *initial rate*, at any point in time while the reaction is in progress, called *instantaneous rate*, or over an interval of time, which is the *average rate*.

In the *initial rate* the change in concentration of a reactant or product as a function of time is measured within minutes (or seconds) the reaction starts. Determination of the dependence of *initial rate* to the concentrations of reactants allows one to derive the *rate law* for the reaction. A *rate law* is a mathematical equation that shows the dependence of reaction rate to the molar concentrations of reactants at constant temperature.

The *instantaneous rate* is calculated from the slope of a *tangent* drawn at any points on the *graph of concentrations versus time*. The slope of tangent taken at the initial point of the graph is assumed to be equal to its *initial rate*.

The *average rate* of reaction is obtained by dividing the change in concentration of a reactant or product that occurs over a longer period of time by the time interval that the change occurs. For example, consider the following reaction at 300°C:

$$2\,NO_2(g) \rightarrow 2\,NO(g) + O_2(g)$$

If in the first 150 seconds, the concentration of NO_2 has decreased from 0.0100 mol/L to 0.0055 mol/L, the average rate for the disappearance of NO_2 for first 150 s under this condition is, $\frac{-\Delta[NO_2]}{\Delta t}$

$$= \frac{-(0.0055\ \text{mol/L} - 0.0100\ \text{mol/L})}{150\ \text{s}} = \frac{0.0045\ \text{mol/L}}{150\ \text{s}} = 3.0 \times 10^{-5}\ \text{mol/(L.s)}$$

Suppose that in the next 150 seconds the concentration of NO_2 decreases from 0.0055 mol/L to 0.0038 mol/L. The average rate during the second 150 s period is, $\frac{-\Delta[NO_2]}{\Delta t}$

$$= \frac{-(0.0038\ \text{mol/L} - 0.0055\ \text{mol/L})}{150\ \text{s}} = \frac{0.0017\ \text{mol/L}}{150\ \text{s}} = 1.1 \times 10^{-5}\ \text{mol/(L.s)}$$

The *average rate* is the *mean* of several *instantaneous rates* taken over a period of time.

Note that *instantaneous rates* of a reaction decreases as time progresses as the concentrations of reactants decrease.

Then the average rate taken over a longer period would have a smaller value compared to those taken over a shorter period after the reaction begins.

The rate of reaction may be determined by measuring the rate of disappearance (decrease in concentration) of one of the reactants or the rate of formation (increase in concentration) of a product. For example, in the reaction:

$$2\,NO_2(g) \rightarrow 2\,NO(g) + O_2(g)$$

Various expression of rate can be derived, such as,

$$\text{Rate}_1 = \frac{-\Delta[NO_2]}{\Delta t}; \ \text{Rate}_2 = \frac{\Delta[NO]}{\Delta t}; \ \text{or Rate}_3 = \frac{\Delta[O_2]}{\Delta t}$$

The stoichiometric relationships between the different rates are as follows:

$$\frac{-\Delta[NO_2]}{\Delta t} = \frac{\Delta[NO]}{\Delta t} = \frac{2\Delta[O_2]}{\Delta t}$$

That is, in this reaction the rate of appearance of NO is the same as the rate of disappearance of NO_2, but double the rate of appearance of O_2.

RATE LAWS

Consider a general reaction: A + B → C

The rate law is expressed as, Rate = $k[A]^x[B]^y$,

where k is the rate constant; x and y are the *rate orders* with respect to individual reactants. The rate orders x and y are not related to the reaction coefficients. Their values must be experimentally determined from a set of kinetic data. For example, the rate law for reaction:

$$2\,N_2O_5(g) \rightarrow 4\,NO(g) + O_2(g), \text{ is Rate} = k[N_2O_5]$$

That is, the rate of the decomposition of N_2O_5 is *first order* with respect to $[N_2O_5]$, which means that the rate will double if $[N_2O_5]$ is doubled; it will triple if $[N_2O_5]$ is tripled.

For the reaction: $2\ NO_2(g) \rightarrow 2NO(g) + O_2(g)$, the rate law is Rate = $k[NO_2]^2$.

That is, the rate of the decomposition of NO_2 is *second order* with respect to $[NO_2]$. Therefore, the rate will quadruple if $[NO_2]$ is doubled.

The reaction: $2\ HI(g) \rightarrow H_2(g) + I_2(g)$ is found to be a zero order reaction in the presence of gold catalyst. For this reaction, the rate is independent of [HI].

DETERMINATION OF THE RATE LAW

Method of Initial Rates

For many chemical reactions, the expression of rate law contains only the concentrations of reactants, while the concentrations of products do not appear. However, for some reactions that are reversible, the concentration of products may be part of the rate law. To simplify the rate law and avoid the complication that might be contributed by products in reversible reactions, the measurement of rate for rate law determination is normally done at the moment the reactants are mixed.

At this point, we can also assume that the effect of product concentration is considered negligible. Therefore, the rates that are measured are the *initial rates*.

The Graphical Method for determining Zero-order, First-order and Second-order Rates

Consider a reaction: R $\rightarrow$ Products.

- If the reaction is *zero-order* with respect to R, Rate $= \dfrac{-\Delta[R]}{\Delta t} = k$

This gives D[R] = $-k\Delta t$, and $[R]_t = [R]_0 - kt$; a plot of [R] *versus* t yields a straight line with slope = $-k$

- If the rate is *first-order* in [R], Rate $= \dfrac{-\Delta[R]}{\Delta t} = k[R]$, which also

gives $\dfrac{\Delta[R]}{[R]} = -k\text{D}t$,

Integrating this equation from time t = 0 to t = *t*, yields the expression: $ln[R]_t = ln[R]_o - kt$; or $ln([R]_t/[R]_o) = -kt$; If the common log is used, the integrated expression for first rate would be:

$\log[R]_t = \log[R]_o - kt/2.30$; where, $[R]_o$ = concentration at t = 0, and $[R]_t$ = concentration at time *t*.

Plotting $ln[R]_t$ *versus* t yields a straight line with slope = $-k$; (slope = $-k/2.30$ if $\log[R]_t$ is plotted).

- If the rate is *second-order* with respect to R, Rate $= \frac{-\Delta[R]}{\Delta t} = k[R]^2$

  or $\frac{\Delta[R]}{[R]^2} = k\Delta t$;

  Integrating the equation from time t = 0 to t = *t* yields the

  expression: $\frac{1}{[R]_t} = kt + \frac{1}{[R]_0}$

For second order rate, plotting $1/[R]_t$ *versus* t yields a straight line with the slope = *k* (the *rate constant*)

Half-Life of Reactions

The *half-life* ($t_{1/2}$) of a reaction is the time it takes for the reactant concentration to decrease to one-half its initial value.

For *zero-order* reaction, $t_{1/2} = [R]_0/2k$

For *first-order* reaction, $t_{1/2} = 0.693/k$

For *second-order* reactions, $t_{1/2} = 1/(k[R]_o)$

Reactions with More Than One Reactant

For reactions that involve more than one reactant, the integrated rate law is determined for one reactant at a time. This is done by having the concentration of the limiting reactant (which rate order to be determined) considerably lower than that of the other reactant. Under this condition, the time dependent change in concentration is significant only for the limiting reactant, while the other reactant is assumed approximately constant.

For example, the reaction: $2NO(g) + O_2(g) \rightarrow 2NO_2(g)$ has Rate = $k[NO]^2[O_2]$

The reaction can be carried out using $[NO]_0 = 1.0 \times 10^{-4}$ mol/L and $[O_2]$ = 0.10 mol/L. As the reaction progress, [NO] decreases significantly, but the decrease in $[O_2]$ is negligible. Under this condition, the integrated rate law equation can be derived in term of [NO]. Then condition can be reversed and the integrated rate law equation is derived in term of $[O_2]$.

For this reaction, a straight line with positive slope is obtained for the plot of 1/[NO] *versus* time and for $ln[O_2]$ *versus* time, which yield a rate law: Rate = $k[NO]^2[O_2]$.

REACTION MECHANISM

A *reaction mechanism* is the detailed picture of how a reaction occurs at molecular level. It consists of *a set of proposed elementary steps* involving molecular species – reactants as well as reaction *intermediates*. A reaction mechanism explains how a given reaction might occur at molecular level and from which a rate law can be derived, which must agree with the one determined experimentally. If the mechanism consists of more than one *elementary step*, the sum of these *steps* must be equal to the overall balanced equation for the reaction.

For example, the reaction: 2A + B → C + D may involve the following *elementary steps*:

- *Step-1:* A + B → X;
- *Step-2:* X + A → Y;
- *Step-3:* Y → C + D
- *Overall reaction:* 2A + B → C + D;

Elementary Steps and Molecularity

Elementary steps or *elementary reactions* are *simple* steps that together make up the *reaction mechanism* for a given reaction. Each *elementary reaction* describes *individual molecular event*, such as two particles combining or a particle (simple or complex) decomposing.

An *elementary reaction* is characterized by its *molecularity*, which is the number of particles/molecules involved in the formation of *transition-state complex*. Thus, an elementary reaction may be characterized as *unimolecular*, *bimolecular*, or *termolecular*. Unimolecular and bimolecular reactions are considered the most common elementary processes.

Rate Law for Elementary Reaction Steps:

Elementary Reactions	Molecularity	Rate Law
A → product	Unimolecular	Rate = k[A]
2A → product	Bimolecular	Rate = $k[A]^2$
A + B → product	Bimolecular	Rate = k[A][B]
2A + B → product	Termolecular	Rate = $k[A]^2[B]$

For example, in the reaction: $2N_2O_5(g) \rightarrow 4NO_2(g) + O_2(g)$, the proposed mechanism is:

Step-1: $N_2O_5 \rightarrow NO_2 + NO_3$ (*Unimolecular*)
Step-2: $NO_2 + NO_3 \rightarrow 2NO_2 + O$ (*Bimolecular*)
Step-3: $N_2O_5 + O\ 2NO_2 + O_2$ (*Bimolecular*)

Termolecular event is considered very rare; the chances that three particles colliding at the same time, with proper orientation and sufficient energy, are considered extremely small. A possible *termolecular* event is thought to occur during the formation of ozone from oxygen in the outer atmosphere:

$$2\,O_2 + N_2 \rightarrow O_3 + O + N_2^*$$

Deriving Rate Law from the Rate-Determining Step

When a proposed mechanism consists of more than one elementary steps, the one with the slowest rate will determine the overall rate of reaction. It is called *the rate-determining step*, which *step represents the rate law for the overall reaction*.

Consider the reaction: $NO_{2(g)} + CO_{(g)} \rightarrow NO_{(g)} + CO_{2(g)}$

If the reaction follows a *one elementary step* mechanism, the rate law would be: Rate = $k[NO_2][CO]$

However, the experimentally determined rate law is Rate = $k[NO_2]^2$

It is proposed that the reaction must involve at least *two elementary steps*:

$$\text{step}-1: NO_2 + NO_2 \xrightarrow{k_1} NO_3 + NO; \ [\text{slow; rate-determining}]$$

$$\text{step}-2: NO_3 + CO \rightarrow CO_2 + NO; \quad [\text{fast}]$$

The rate law for the rate-determining step: *Rate* = $k_1[NO_2]^2$, which is identical in form to the rate law obtained experimentally, in which $k_1 = k$. The second step, which occurs very fast, does not influence the overall rate.

In fact, if the exponents in the rate law differ from the coefficients in the chemical equation, the mechanism for the reaction normally involves more than one elementary step. Most reactions are believed to involve more than one elementary step.

Note that, unlike the overall rate law which must be determined experimentally, the rate law for an *elementary reaction* can be deduced from its *molecularity*, since each elementary reaction is a single step event. The exponents in the expression of rate law are the same as the equation coefficients. For example, the rate laws for the following elementary reactions are written as indicated:

1. $NO + O_3 \rightarrow NO_2 + O_2$; Rate = $k[NO][O_3]$
2. $N_2O_5 \rightarrow NO_3 + NO_2$; Rate = $k[N_2O_5]$
3. $Br_2 + NO \rightarrow ONBr_2$; Rare = $k[Br_2][NO]$

In addition, certain species such as NO_3 may be formed in on *elementary step*, but immediately consumed in subsequent elementary step; they do not appear in the overall equation. These species are called *reaction intermediates*.

Rate Law Derived from Mechanism with Slow First Elementary Step

Consider the reaction: $2NO_2(g) + F_2(g) \rightarrow 2NO_2F(g)$

The experimental rate law is Rate = $k[NO_2][F_2]$

The proposed mechanism for this reaction consists of the following elementary steps:

Step-1: $NO_2 + F_2 \rightarrow NO_2F + F$; (slow, rate-determining)

Step-2: $NO_2 + F \rightarrow NO_2F$; (fast)

Overall Rxn: $2NO_2 + F_2 \rightarrow 2NO_2F$

Rate law for the rate-determining step: Rate = $k_1[NO_2][F_2] = k[NO_2][F_2]$; $\rightarrow (k = k_1)$

Rate Law Derived from Mechanism *with* Slow Second Elementary Step:

Consider the reaction: $2\,NO(g) + O_2(g) \rightarrow 2\,NO_2(g)$;

The rate law is: Rate = $k[NO]^2[O_2]$

The proposed mechanism composed of the following elementary reactions:
Step-1: $NO + O_2 \rightleftarrows NO_3$; (fast; equilibrium)
Step-2: $NO_3 + NO \rightarrow 2\ NO_2$; (slow, rate-determining)
Overall Rxn: $2NO + O_2 \rightarrow 2\ NO_2$;
Rate law for the *slow, rate-determining step*: Rate = $k_2[NO_3][NO]$

NO_3 is a *reaction intermediate* formed in a fast reversible reaction (step-1) and it concentration quickly reaches a steady state such that, $[NO_3] = K_{eq}[NO][O_2]$, where K_{eq} is the equilibrium constant for reversible step-1. Substituting for $[NO_3]$ in the rate law expression, we obtain,

$$\text{Rate} = k_2K_{eq}[NO]^2[O_2] = k[NO_2]^2[O_2]; \text{ (where } k = k_2K_{eq})$$

Correlating Reaction Mechanism with Rate Law

For any reaction, the proposed *mechanism* must obey three fundamental criteria:

1. *The elementary steps must add up to give the overall equation.*
2. *The elementary steps must be physically reasonable.* That is, the proposed step can occur with reasonable probability; they are either *unimolecular* or *bimolecular*. A *termolecular* elementary step is considered unlikely.
3. *The mechanism must correlate with the rate law.* That is, the rate law derived from *rate-determining step* must agree with the actual rate law obtained experimentally.

For example, the reaction: $2\ N_2O_5(g) \rightarrow 4\ NO_2(g) + O_2(g)$ has Rate = $k[N_2O_5]$.

The proposed mechanism involves the following elementary steps:
Step-1: $N_2O_5 \rightleftarrows NO_3 + NO_2$; (fast; equilibrium)
Step-2: $NO_3 + NO_2 \rightarrow 2NO_2 + O$; (slow)
Step-3: $N_2O_5 + O \rightarrow 2NO_2 + O_2$; (fast)
Overall: $2\ N_2O_5 \rightarrow 2NO_2 + O_2$; Rate = $k[N_2O_5]$

A REPRODUCTION FOR CHEMICAL KINETICS

The rate law of a reaction relates the dependence of reaction rates on the concentrations of reactants. As we know, the rate of reaction also depends on other factors, such as temperature. This dependency can be explained by the *collisional theory* of reaction and the formation of *activated complex*, or *transition state*.

Consider the following reaction: $2BrNO(g) \rightarrow 2NO(g) + Br_2(g)$

For the reaction to occur:

- *Reactant molecules must collide;*
- *Molecular collisions must occur with proper orientations;*
- *Collisions must be energetic and lead to the formation of an activated complex;*

- *The rate of formation of activated complex is the rate determining step;*
- *The activated complex eventually leads to the formation of products;*

1. The rate of formation of *activated complex* is proportional to the *frequency of effective molecular collisions*, which is dependent on reactant concentrations.
2. The *activated complex* or *transition state* is separated from the reactants by an *energy barrier*, called the *activation energy* (E_a). The formation of *transition state complex* depends on the fraction of molecules with sufficient *kinetic energy* that can overcome this *energy barrier*. This fraction is temperature dependent. At low temperature, the *fraction* of molecules with sufficient energy to overcome the *activation energy barrier* and subsequently forms the *transition state complex* will be low. This *fraction* increases with temperature, which also increase the overall reaction rate.
3. High *activation energy* for a reaction also means a lower fraction of molecules with sufficient energy to form the *transition state complex*. Thus, higher E_a leads to a slower rate of reaction.

The dependence of rate on temperature and *activation energy* (E_a) is shown the Arrhenius equation for the rate constant, *k*:

$k = Ae^{-Ea/RT}$, where *A* is called the *Arrhenius frequency factor*.

Taking the natural log, yields the equation:

$$\text{In}(k) = -\left(\frac{E_a}{R}\right)\left(\frac{1}{T}\right) + \text{In}(A).$$

A plot of *ln*(*k*) *versus* 1/T yields a straight line with slope = $-E_a/R$; (R = 8.314 J/mol.K).

The *activation energy* (E_a) can also be calculated using two values of the rate constant *k* according to the expression: $\text{In}(k_2/k_1) = -\frac{E_a}{R}\left(\frac{t}{T_2} - \frac{1}{T_1}\right)$; or $\text{In}(k_1/k_2) - \frac{E_a}{R}\left(\frac{1}{T_2} - \frac{1}{T_1}\right)$;

(T must be in Kelvin)

CATALYSIS INCREASE THE RATE OF REACTIONS

Catalysts are substances that increase the rate of reactions, but do not get used up by the reactions. A catalyst functions by providing an *alternative reaction pathway with lower activation energy*. It increases the reaction rate, but does not affect the reaction enthalpy or the equilibrium position.

Homogeneous Catalysis

These are catalysts that have the same phase as the reactants. For example, nitric oxide (NO) catalyzes the reaction between SO_2 and O_2:

$$2SO_2(g) + O_2(g) \xrightarrow{\text{NO catalyst}} 2SO_3(g)$$

The above reaction, though spontaneously and highly exothermic, is very slow because of the very high activation energy. NO molecules provide an alternative pathway that has a lower E_a.

Step-1: $2NO + O_2 \rightarrow 2NO_2$

Step-2: $2NO_2 + 2SO_2 \rightarrow 2NO + 2SO_3$

Overall: $2\ SO_2(g) + O_2(g) \rightarrow 2SO_3(g)$

Some catalysts may be involved in the reaction kinetics and the concentration appears in the rate law. For example, iodine which catalyzes the *isomerization* of *cis*-2-butene to *trans*-2-butene is included in the rate law expression. The Uncatalyzed reaction:

$$\text{cis-}CH_3-\overset{\overset{H}{|}}{C}=\overset{\overset{H}{|}}{C}-CH_3 \rightarrow \text{trans-}CH_3-\overset{\overset{H}{|}}{C}=\underset{\underset{H}{|}}{C}-CH_3$$

Rate = k[*cis*-2-butene]

Iodine-catalyzed reaction:

$$\text{cis-}CH_3-\overset{\overset{H}{|}}{C}=\overset{\overset{H}{|}}{C}-CH_3 \rightarrow \text{trans-}CH_3-\overset{\overset{H}{|}}{C}=\underset{\underset{H}{|}}{C}-CH_3$$

Rate = k[*cis*-2-butene]$[I_2]^{1/2}$

Heterogeneous Catalysis

These are catalysts which have a different physical state from the reactants. Most reactions involving gases use inert metals or metal oxides as catalysts. These solid catalysts provide surface areas for molecules to interact effectively. The metal surface also facilitates the breaking and formation of bonds. Ni, Pd and Pt are often used in the hydrogenation of vegetable oil to make margarine and "Crisco oil", where a hydrogen molecule is added to each carbon-carbon double bond in the unsaturated fatty acid chains:

$$-\underset{\underset{H}{|}}{C}=\underset{\underset{H}{|}}{C}- + H_2 \xrightarrow[\text{(Ni or Pt)}]{} -\overset{\overset{H}{|}}{\underset{\underset{H}{|}}{C}}-\overset{\overset{H}{|}}{\underset{\underset{H}{|}}{C}}-$$

Heterogeneous catalyses are used in automobile catalytic converters, which consist of Pt, Pd and Rh mixture embedded in ceramic honeycombs. The metals catalyze the following reaction:

$$2NO(g) + 2CO(g) \xrightarrow{(Pt+Pd+Rh)} N_2(g) + 2CO_2(g)$$

Catalytic converters reduce the emission of toxic exhaust gases

Some reactions require specific catalysts. For example, Nickel catalyzes the reaction:

$$CO(g) + 3H_2(g) \xrightarrow{\text{Ni-catalyst}} CH_4(g) + H_2O(g)$$

Whereas ZnO-Cr_2O_3 mixture catalyzes the formation of methanol from the same reactants:

$$CO(g) + 2H_2(g) \xrightarrow{\text{ZnO-}Cr_2O_3} CH_3OH(g)$$

PRECISION OF KINETIC MEASUREMENTS

Random errors can never be completely avoided in kinetic measurements but we can often estimate the maximum possible error in our measurements of concentrations, temperature fluctuations, etc.

In general, if we are interested in obtaining values of a dependent variable f from values of independent variables x_i, $i = 1,2,...$
where,

$$f = f(x_i, x_2, x_3, ...)$$

the relative error in f can be related to the relative errors in x_i by

$$\left(\frac{\Delta f}{f}\right)^2 = \sum_{j=1}^{n} \left(\frac{\partial \ln f}{\partial \ln x_j}\right)^2 \left(\frac{\Delta x_j}{x_j}\right)^2$$

Suppose we are interested in the accuracy of the rate constant which we have obtained from the following expression:

$$k = \frac{-r_A}{C_A C_B}$$

$$\ln k = \ln(-r_A) - \ln C_A - \ln C_B$$

$$\frac{\partial \ln k}{\partial \ln(-r_A)} = 1 \quad \frac{\partial \ln k}{\partial \ln C_A} = -1 \quad \frac{\partial \ln k}{\partial \ln C_B} = -1$$

$$\left(\frac{\Delta k}{k}\right)^2 = \left(\frac{\Delta(-r_A)}{(-r_A)}\right)^2 + \left(\frac{\Delta C_A}{C_A}\right)^2 + \left(\frac{\Delta C_B}{C_B}\right)^2$$

If we measured each concentration with precision better than 2per cent and the precision of the rate measurements is 10per cent we get,

$$\left(\frac{\Delta k}{k}\right)^2 = 0.1^2 + 0.02^2 + 0.02^2 = 0.0108$$

$$\frac{\Delta k}{k} = 0.1039$$

The error in k is 10.4per cent

From the Arrhenius relationship we see,

$$\left(\frac{dk}{k}\right) = \frac{E}{RT}\left(\frac{dT}{T}\right)$$

The larger the error in T and the larger the activation energy, E, the larger the error in k.

We usually estimate E from,

$$\frac{E}{R} = \frac{\ln\,(k_1/k_0)}{\frac{1}{T_0} - \frac{1}{T_1}}$$

$$\ln\frac{E}{R} = \ln\left[\ln\,(k_1/k_0)\right] - \ln\left(\frac{1}{T_0} - \frac{1}{T_1}\right)$$

$$\left(\frac{\Delta E}{E}\right)^2 = \left(\frac{1}{\ln\,k_1/k_0}\right)^2\left[\left(\frac{\Delta k_0}{k_0}\right)^2 + \left(\frac{\Delta k_1}{k_1}\right)^2\right] + \left(\frac{T_1}{T_0 - T_1}\right)^2 + \left(\frac{T_0}{T_0 - T_1}\right)^2\left(\frac{\Delta T_1}{T_1}\right)^2$$

The larger the interval $T_1 – T_0$ the smaller the error in E provided that classical Arrhenius form with E = const holds over such a temperature range.

Useful guidelines can be found in Benson, Foundations of Chemical Kinetics.

To obtain the rate constant with an error of $\pm\,\varepsilon$ it is necessary:

- To measure the concentration with an accuracy,

$$\pm\,\frac{\text{Change in concentration}}{\text{Largest concentration}}\,\frac{\varepsilon}{1.4}\%$$

- To measure time (or equivalent) with accuracy of,

$$\pm\,\frac{\text{Time interval}}{\text{Largest time}}\,\frac{\varepsilon}{1.4}\%$$

- To measure temperature with an accuracy of,

$$\pm\,\frac{\varepsilon}{35}\% = \pm\,\frac{\varepsilon\ T}{3500}\ T \text{ in } K$$

KINETIC PARAMETERS IN MORE COMPLEX REACTION SYSTEMS

So far we have dealt with single reactions only, yet an engineer will almost always encounter a problem involving multiple-mixed reactions. Nevertheless the methods described so far can often be utilized, if adequate amount of experimentation can be obtained, to estimate at least some of the kinetic parameters.

The method of half-time is often utilized to:

- Evaluate the overall order of reaction (single reaction) by using all reactants in stoichiometric ratio;
- Evaluate the order with respect to A when other reactants are used in excess.

The assumption is that the reaction is irreversible and follows n-th order behaviour,

$$-\frac{dC_A}{dt} = k\,C_A^{\alpha} \qquad -\int_{C_{Ao}}^{C_{Ao}/2} \frac{dC_A}{C_A^{\alpha}} = kt_{1/2}$$

$$t_{1/2} = \frac{2^{\alpha-1} - 1}{k(\alpha-1)}\,C_{Ao}^{1-\alpha}$$

$$\underbrace{\log t_{1/2}}_{y} = \log\left\{\frac{2^{\alpha-1} - 1}{k\,(\alpha - 1)}\right\} + (1 - \alpha)\,\underbrace{\log C_{Ao}}_{x}$$

Starting with different C_{Ao}'s and measuring the time that it takes for C_{Ao} to drop to $C_{Ao}/2$ and by plotting $t_{1/2}$ vs. C_{Ao} on a log-log plot one can determine overall order α and rate constant k.

When investigating multiple reactions in batch systems of constant volume we have to monitor as many variables as there are independent reactions in the system. For example, if we are studying the rates in a system of the type

$A + B = D$ – desired product

$2A = U$ – unwanted product

we could monitor C_A and C_D in time. If the experiments are performed in a constant volume batch system and if we assume n-th order rate forms we have to evaluate the parameters $k_{1f}, k_{1b}, k_{2f}, k_{2b}, \alpha, \alpha_2, \beta, \gamma, \delta$, which appear in the following description of the systems component mass balance:

$$-\frac{dC_A}{dt} = k_{1f}\,C_A^{\alpha_1}\,C_B^{\beta} - k_{1b}\,C_D^{\delta} + 2k_{2f}\,C_A^{\alpha_2} - k_{2b}\,C_U^{\gamma}$$

$$\frac{dC_D}{dt} = k_{1f}\,C_A^{\alpha_1}\,C_B^{\beta} - k_{1b}\,C_D^{\delta}$$

We could again perform the experiments at low conversions of A and B so as to measure the initial rates in the region where reverse rates can be neglected. Then:

$$-\frac{dC_A}{dt} = k_{1f}\,C_A^{\alpha_1}\,C_B^{\beta} + 2k_{2f}\,C_A^{\alpha_2}$$

$$\frac{dC_D}{dt} = k_{1f}\,C_A^{\alpha_1}\,C_B^{\beta}$$

By forming the expression for a point yield $y\left(\frac{D}{A}\right)$ we get:

$$y\left(\frac{D}{A}\right) = \frac{dC_D}{-dC_A} = \frac{1}{1 + \frac{2k_{2f}}{k_{1f}}\,C_A^{\alpha_2-\alpha_1}\,C_B^{-\beta}}$$

The stoichiometry of the reaction dictates:

$$C_A = C_{Ao} - \xi_1 - 2\xi_2 \qquad C_D = C_{Do} + \xi_1$$

$$C_B = C_{Bo} - \xi_1 \qquad C_U = C_{Uo} + \xi_2$$

Eliminating the extents ξ_1, ξ_2 in terms of measured concentrations we get:

$$\xi_1 = C_D - C_{Do} = C_D \text{ since } C_{Do} = 0 \text{ was selected}$$

$$\xi_2 = \frac{1}{2}\left(C_{Ao} - C_A - C_D\right)$$

Thus $C_B = C_{Bo} - C_D$

$$\frac{dC_D}{-dC_A} = \frac{1}{1 + \dfrac{2k_{2f}}{k_{1f}} C_A^{\alpha_2-\alpha_1}\left(C_{Bo} - C_D\right)^{-\beta}}$$

If in addition we perform the experiment with large excess of B, then $C_{Bo} >> C_D$

$$\frac{dC_D}{-dC_A} = \frac{1}{1 + \dfrac{2k_{2f}\, C_{Bo}^{-\beta}}{k_{1f}} C_A^{\alpha_2-\alpha_1}}$$

$$\frac{dC_A}{-dC_D} - 1 = \frac{2k_{2f}}{k_{1f}\, C_{Bo}^{\beta}} C_A^{\alpha_2-\alpha_1}$$

$$\log\left(-\frac{dC_A}{dC_D} - 1\right) = \log\left(\frac{2k_{2f}}{k_{1f}\, C_{Bo}^{\beta}}\right) + \left(\alpha_2 - \alpha_1\right)\log C_A$$

From C_A vs. C_D data we can get estimates of $-\dfrac{dC_A}{dC_D}$ as shown before, and by plotting the quantities indicated above on a semi-log plot we get a slope of $\alpha_2 - \alpha_1$.

Working now at moderate levels of C_{Bo} so that $C_{Bo} - C_D \neq C_{Bo}$ we can plot for the new set of experiments,

$$\underbrace{\log\left(\frac{-\dfrac{dC_A}{dC_D} - 1}{C_A^{\alpha_2-\alpha_1}}\right)}_{y} = \log\left(\frac{2k_{2f}}{k_{1f}}\right) - \beta \log \underbrace{\left(C_{Bo} - C_D\right)}_{x}$$

From the slope one can get β and from all points an average for $2k_{2f}/k_{1f}$. Now one can use C_D vs. t data C_A vs. t

$$-\frac{dC_D}{dt} = k_{1f}\, C_A^{\alpha_1}\, C_B^{\beta} - k_{1f}\, C_A^{\alpha_1}\left(C_{Bo} - C_D\right)^{\beta}$$

$$\underbrace{\log\left(\frac{-\dfrac{dC_D}{dt}-1}{\left(C_{Bo}-C_D\right)^b}\right)}_{y} = \log k_{1f} + \alpha_1 \log \underbrace{\left(C_A\right)}_{x}$$

From the slope α_1 is determined and from the data points an average value of k_{1f}. Now since $\alpha_2 - \alpha_1$, and $\dfrac{k_{2f}}{k_{1f}}$ are already known, also k_{2f} and α_2 can be evaluated.

The above served to demonstrate that the differential method of analysis, as illustrated for single reactions, can also be used in more complex reaction schemes. In order to do so one needs to plan the experiments in a proper way and manipulate the data properly.

Similarly for many practical reaction orders of interest the expression for the yield can be integrated and integral analysis can be applied. For example if we guess,

$$\alpha_2 = 1 = \alpha_1 = 1 \quad \beta = 1$$

$$-\frac{dC_D}{dC_A} = \frac{1}{1+\dfrac{2k_{2f}}{k_{1f}}C_B^{-1}} = \frac{1}{1+\dfrac{2k_{2f}}{k_{1f}}\left(C_{Bo}-C_D\right)^{-1}}$$

$$\int_o^{C_D}\left[1+\frac{2k_{2f}}{k_{1f}}\left(C_{Bo}-C_D\right)^{-1}\right]dC_D = -\int_{C_{Ao}}^{C_A} dC_A$$

$$C_D + \frac{2k_{2f}}{k_{1f}}\ln\left(\frac{C_{Bo}}{C_{Bo}-C_D}\right) = C_{Ao} - C_A$$

Thus, if the hypothesized orders $\alpha_2 = \alpha_1 = 1 = \beta$ are correct, then a plot of,

$$\underbrace{\ln\left(\frac{C_{Bo}}{\left(C_{Bo}-C_D\right)}\right)}_{y} = \frac{k_{1f}}{2k_{2f}}\underbrace{\left(C_{Ao}-C_A-C_D\right)}_{x}$$

should yield a straight line with a slope $\dfrac{k_{1f}}{2k_{2f}}$.

If we guessed $\alpha_2 = 2$, $\alpha_1 = \beta = 1$ we would have to integrate:

$$-\frac{dC_D}{dC_A} = \frac{1}{1+\dfrac{2k_{2f}}{k_{1f}}C_A\left(C_{Bo}-C_D\right)^{-1}}$$

which upon integration yields,

$$\left[\frac{C_{Bo} - C_D}{C_{Bo}}\right]^{-\frac{2k_{2f}}{k_{1f}}} = \frac{C_{Ao} - \dfrac{k_{1f}}{k_{1f} - 2k_{2f}} C_{Bo}}{C_A - \dfrac{k_{1f}}{k_{1f} - 2k_{2f}} (C_{Bo} - C_D)}$$

Now it is difficult to rewrite the result in a form which would yield a straight line plot in a properly selected coordinate system. Clearly when dealing with multiple reactions a point is reached very fast when conventional methods of integral analysis become tedious and often impossible to use. In such a situation one must perform a sufficient number of experiments so that some statistical parameter estimation methods can be used. The initial estimates may frequently be obtained by differential analysis as presented above.

PARAMETERS IN SYSTEMS OF MONOMOLECULAR REACTIONS

In many complex reaction schemes there often is some indication that most or even all reactions can be considered to be first order. (The same assumption is frequently arbitrarily made, since this simplifies tremendously the treatment of the system, and this is to be avoided). For M independent first order reversible reactions between S species in a batch system of constant volume one can write:

$$\frac{dC_1}{dt} = \left(-\sum_{j=2}^{S} k_{j1}\right) C_1 + k_{12} C_2 + k_{13} C_3 + \ldots + k_{1s} C_s$$

$$\frac{dC_2}{dt} = k_{21} C_1 \left(-\sum_{\substack{j=1 \\ j\neq 2}}^{S} k_{j2}\right) C_2 + k_{23} C_3 + \ldots + k_{2s} C_s$$

$$\frac{dC_s}{dt} = k_{s1} C_1 + k_{s2} C_2 + k_{s3} C_3 + \ldots + \left(-\sum_{\substack{j=1 \\ j\neq s}}^{S} k_{sj}\right) C_s$$

where k_{ji} is the rate constant for the rate for formation of component j in reaction $A_i \rightarrow A_j$. For example, k_{12} is the rate constant for the rate of formation of species 1 in reaction $A_2 \rightarrow A_1$. The reversible rate is viewed for each species as rate of formation - rate of disappearance. Thus, to obtain the accumulation of say species 1, $\frac{dC_1}{dt}$, we have to sum over the rates of formation of 1 in all reactions where 1 appears. Then $\sum_{j=2}^{S} k_{j1}$ is the sum of the rate constants for rates of formation of species j from species 1 summed up over all species j. This properly appears with the negative sign since it is equal to disappearance of species 1. We could arbitrarily define $k_{11} = -\sum_{j=2}^{S} k_{j1}$ or in general,

$$k_{ii} = -\sum_{\substack{j=1 \\ j \neq i}}^{S} k_{ji} .$$

The above scheme considers the possibility that every species in the system can produce all the others or be produced from them. For example:

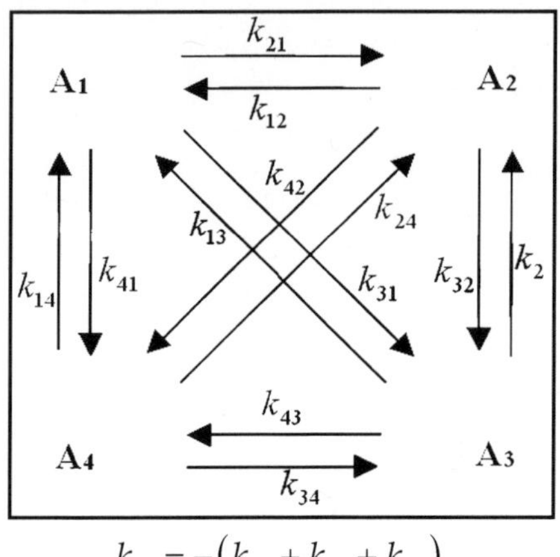

$$k_{11} = -(k_{21} + k_{31} + k_{41})$$
$$k_{22} = -(k_{12} + k_{32} + k_{42})$$
$$k_{33} = -(k_{13} + k_{23} + k_{43})$$
$$k_{44} = -(k_{14} + k_{24} + k_{34})$$

Naturally the above form is the most general one and in specific situations many constants can be a priori set to be zero. The problem in matrix form is:

$$\frac{d\underline{C}}{dt} = \underline{\underline{K}}\,\underline{C}; \quad t = 0 \quad \underline{C} = \underline{C_o}$$

where,

$$\underline{C} = \begin{bmatrix} C_1 \\ C_2 \\ \cdot \\ \cdot \\ C_s \end{bmatrix}$$

$$\underline{\underline{K}} = \begin{bmatrix} k_{11} & k_{12}\ldots.. & k_{1s} \\ k_{21} & k_{22}\ldots.. & k_{2s} \\ \ldots.. & \ldots.. & \ldots.. \\ k_{s1} & k_{s2}\ldots.. & k_{ss} \end{bmatrix}$$

with,

$$k_{ii} = -\sum_{\substack{j=1 \\ j \neq i}}^{S} k_{ji}$$

Most often in order to deal with unit vectors every row is divided by C_{ToT}, total molar concentration, and the equations are written in terms of mole fractions:

$$y_j = \frac{C_j}{C_{ToT}}$$

K is (s × s) matrix of rate constants.

$$\frac{d\underline{y}}{dt} = \underline{\underline{K}}\,\underline{y}; \quad t = 0 \quad \underline{y} = \underline{y}_o$$

The problem is as follows:

From a continuous record of $\underline{y}$ vs. t (or from a set of N discrete measurements of $\underline{y}$ at $t_1, l = 1, 2......N$ where $N >> S$) determine the constants comprising the matrix $\underline{\underline{K}}$.

Basically the theory states that since $\underline{\underline{K}}$ is an (s × s) positive definite matrix it must have S real eigenvalues λ_m, m, = 0, 1, 2 .....s – 1. To each eigenvalue corresponds a particular eigenvector $\underline{x}_m$ so that the following matrix equation is satisfied:

$$\underline{\underline{K}}\,\underline{\underline{X}} = \underline{\underline{X}}\,\underline{\underline{\Lambda}} \quad \text{or} \quad \underline{\underline{K}}\,\underline{x}_m = -\,\underline{x}_m$$

where,

$$\underline{\underline{X}} = \left[\underline{x}_0\ \underline{x}_1\ \underline{x}_2 .....\underline{x}_{s-1}\right] = \begin{bmatrix} x_{01} & x_{11} & x_{s-1,\,1} \\ x_{02} & x_{12} & x_{s-1,\,2} \\ \cdot & \cdot & \cdot \\ \cdot & \cdot & \cdot \\ x_{05} & x_{15} & x_{s-1,\,s} \end{bmatrix}$$

$$-\,\underline{\underline{\Lambda}} = \begin{bmatrix} \lambda_o & 0 & 0 & \cdot & \cdot & 0 \\ 0 & \lambda_1 & 0 & \cdot & \cdot & 0 \\ 0 & 0 & \lambda_2 & \cdot & \cdot & 0 \\ \cdot & \cdot & \cdot & \cdot & \cdot & \cdot \\ 0 & 0 & 0 & \cdot & \cdot & \lambda_{s-1} \end{bmatrix} = \text{diagonal matrix}$$

Thus the matrix $\underline{\underline{K}}$ maps every eigenvector $\underline{x}_m$ onto itself augmented by a "stretching" factor λ. Eigenvectors $\underline{x}_m$ establish the characteristic directions for the matrix $\underline{\underline{K}}$, eigenvalues λ_m give the scale factors in these directions.

If one selects now,

$$\underline{y} = \underline{\underline{X}}\,\underline{b} \qquad \underline{b} = \begin{bmatrix} -b_0 \\ b_1 \\ b_{s-1} \end{bmatrix}$$

$$\frac{d\underline{y}}{dt} = \underline{\underline{X}}\frac{d\underline{b}}{dt} = \underline{\underline{K}}\underline{y} = \underline{\underline{K}}\underline{\underline{X}}\underline{b}$$

Premultiplying both sides with $\underline{\underline{X}}^{-1}$

$$\underline{\underline{X}}^{-1}\underline{\underline{X}}\frac{d\underline{b}}{dt} = \underline{\underline{X}}^{-1}\underline{\underline{K}}\underline{\underline{X}}\underline{b}$$

and remembering,

$$\underline{\underline{X}}^{-1}\underline{\underline{X}} = \underline{\underline{I}} \quad - \quad \text{Identity matrix}$$

$$\underline{\underline{K}}\underline{\underline{X}} = \underline{\underline{X}}\,\underline{\Lambda}$$

we get,

$$\frac{d\underline{b}}{dt} = \underline{\underline{\Lambda}}\,\underline{b} \qquad t = 0 \qquad \underline{b} = \underline{b}_0 = \underline{\underline{X}}^{-1}\underline{y}_0$$

This is a decoupled matrix equation,

$$\frac{db_0}{dt} = -\lambda_0 b_0 \qquad b_0 = b_{00}\, e^{-\lambda_0 t}$$

$$\frac{db_1}{dt} = -\lambda_1 b_1 \qquad b_1 = b_{10}\, e^{-\lambda_1 t}$$

$$\frac{db_{s-1}}{dt} = -\lambda_{s-1}\, b_{s-1} \qquad b_{s-1} = b_{s-10}\, e^{-\lambda_{s-1} t}$$

The above matrix manipulation shows that in principle one can find a set of S fictitious components $\underline{B}$ so that the equations for quantities of $\underline{B}$, $\underline{b}$, are uncoupled and can be solved one by one. Furthermore at equilibrium:

$$\frac{d\underline{y}_{eq}}{dt} = \underline{0} = \underline{\underline{K}}\,\underline{y}_{eq}$$

Remember $-\lambda_m\, \underline{x}_m = \underline{\underline{K}}\,\underline{x}_m, \quad m = 0, 1, 2, \ldots\ldots s - 1$

Thus the first eigenvector can be selected as the equilibrium composition:

$$\underline{x}_0 = \underline{y}_{eq} \qquad \text{and} \qquad b_0 = b_{00} = const = 1$$

and the first eigenvalue is zero $\lambda_0 = 0$

Thus,

$$\underline{y} = \underline{\underline{x}}\,\underline{b} = b_0\,\underline{x}_0 + b_1\,\underline{x}_1 + b_2\,\underline{x}_2 + b_3\,\underline{x}_3\ldots\ldots$$

$$\underline{y} - \underline{y}_{eq} = b_1\,\underline{x}_1 + b_2\,\underline{x}_2 + b_3\,\underline{x}_3\ldots\ldots + b_{s-1}\,\underline{x}_{s-1}$$

$$= b_{10}\, e^{-\lambda_1 t}\,\underline{x}_1 + b_{20}\, e^{-\lambda_2 t}\,\underline{x}_2 + \ldots\ldots + b_{s-1}\, e^{-\lambda_{s+1} t}\,\underline{x}_{s-1}$$

This describes the decay of deviations from equilibrium. Each eigenvector $\underline{x}_m$ is a direction in space, and the right hand side of the equation represents all contributions that make up the reaction paths.

A good choice of initial compositions $\underline{y}_0$, so that say $b_{10} \neq 0$ but $b_{20} = b_{30} = b_{40} = \ldots\ldots.0$, would yield a straight line reaction path in the direction of $\underline{x}_1$.

Thus when one knows the rate constants k_{ji} one can compute all λ_m's and corresponding $\underline{x}_m$'s, find the $\underline{b}$ vector and determine all the straight line reactions paths.

However, this theory is helpful even when k_{ji} are not known and are sought since it provides the guidelines for searching for straight line reaction paths. Once these are experimentally established semi log plots of $\frac{b_i}{b_{io}}$ vs. t give all the eigenvalues λ_m.

$$\ln \frac{b_1}{b_{10}} = -\lambda_1 t$$

$$\ln \frac{b_2}{b_{20}} = -\lambda_2 t$$

From the eigenvalues λ_m, by simple matrix manipulation, k_{ji} are determined.

Since y_m, $m = 1,2....s$ are measured as a function of time and b_m, $m = 0, 1, 2....s - 1$ are obtainable from y_m's only if k_{ji}'s are known, which they are not, this means that one has to use a trial and error method and search for suitable combinations of y_{io}'s giving the desired straight line reaction paths from which λ_n and b_is can be determined leading to evaluation of k_{ji}'s.

One of the best examples is the classical analysis of the isomerization of butane on alumina which was studied extensively by Haag and Pines and Lago and Haag and was used by Wei and Prater to illustrate the power of their method.

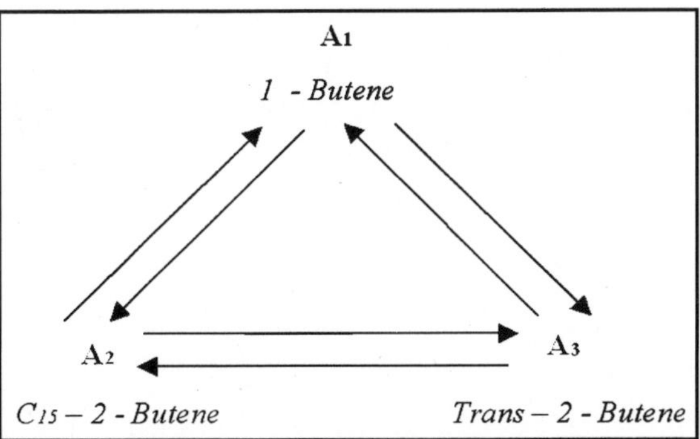

Let us describe now a set of experiments and the use of Wei-Prater technique. We will follow the reaction path for every experiment on the triangular composition diagram. In the first experiment (path 1) start with pure A_1 and let the system reach equilibrium. The equilibrium composition of point e establishes the zeroth eigenvector $\underline{x}_0$ for $\lambda_0 = 0$. Since path 1 is not straight we know that pure A_1 is not a component of b.

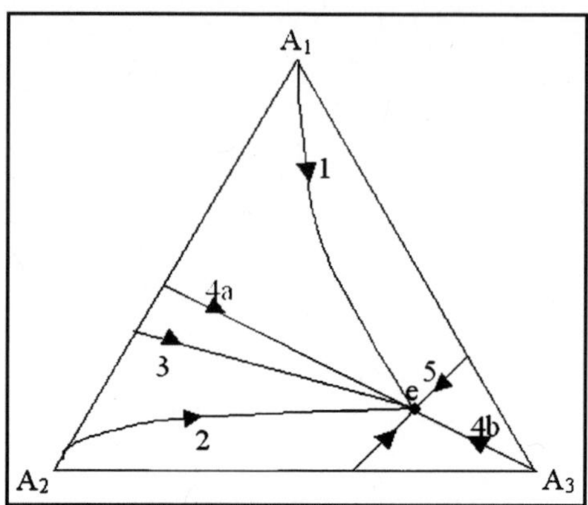

We can now perform an experiment starting with pure A_2 (path 2). Since this does not yield a straight line we know that pure A_2 is not a component of $\underline{b}$ vector. However, Wei and Prater method shows how to use the data from path 2 close to equilibrium in order to extrapolate to the side of the triangle and select a new starting composition for run 3. Run 3 (path 3) does not give a straight line but with one more correction leads to starting point for run 4. Path 4a is a straight line, thus, the composition on the side of the triangle determines b_{10} and λ_1 can be found from $\ln \frac{b_1}{b_{10}} = -\lambda_1 t$. However, a run 4b from pure A_3 to verify the straight line path is necessary. Then matrix algebra allows the prediction of the last straight line, path 5, and calculation of all λ's and b_i's and k_{ji}'s by matrix manipulation as shown in Boudart's book. Straight reaction path 5 can naturally be verified by experiments.

Certainly a powerful method to cut down on experimental work but restricted to first order reactions.

Let $\bar{C} = \left[C_1, C_2, C_3 \ldots C_N\right]$ be a row vector of concentration of all the species and let these concentrations be determined at equal time increments Δt. Then,

$$\bar{C}(\Delta t) = \bar{C}(0)\, e^{\underline{\underline{K}}\,\Delta t}$$
$$\bar{C}(2\Delta t) = \bar{C}(\Delta t)\, e^{\underline{\underline{K}}\,\Delta t} = \bar{C}(0)\, e^{2\underline{\underline{K}}\,\Delta t}$$
$$\bar{C}(m\,\Delta t) = \bar{C}((m-1)\,\Delta t)\, e^{2\underline{\underline{K}}\,\Delta t}$$

or in matrix notation,

$$\begin{bmatrix} C(\Delta t) \\ C(2\,\Delta t) \\ C(m\,\Delta t) \end{bmatrix} = \begin{bmatrix} C(0) \\ C(\Delta t) \\ C((m-1)\,\Delta t) \end{bmatrix} \left[e^{\underline{\underline{K}}\,\Delta t}\right]$$

$$\left[e^{\underline{\underline{K}}\,\Delta t}\right] = \begin{bmatrix} C(0) \\ C(\Delta t) \\ C((m-1)\,\Delta t) \end{bmatrix} \begin{bmatrix} C(\Delta t) \\ C(2\,\Delta t) \\ C(m\,\Delta t) \end{bmatrix}$$

$$\underline{\underline{K}} = \frac{1}{\Delta t} \ln \left\{ \begin{bmatrix} C(0) \\ C(\Delta t) \\ C((m-1)\,\Delta t) \end{bmatrix}^{-1} \begin{bmatrix} C(\Delta t) \\ C(2\,\Delta t) \\ C(m\,\Delta t) \end{bmatrix} \right\}$$

All of the above computations (matrix inversion, logarithm of a matrix *etc.*) are readily programmable or available on modern computers. Thus the above computations yield the matrix of rate constants directly.

N-th Order Reactions

A more general procedure valid for a set of n-th order reactions of known order for evaluation of the rate constants. In a batch constant volume system the equations for materials balance for all components can be cast in the following form:

$$\frac{dC_j}{dt} = \sum_{i=1}^{M} k_i\, r_{ji}\, (\underline{C});\; j = 1, 2, ... S$$

S – total number of components

M – total number of reactions i.e. number of tems in the rate expressions

r_{ji} – concentration dependent term in rate of production (if > 0) or depletion (if with negative sign) of component j in reaction path i

k_i – rate constant for path i

$\underline{C}$ – vector of S concentrations

The derivatives $\frac{dC_j}{dt}$ cannot be measured directly but the differences in concentrations at various times can. By integrating the above equations one gets:

$$\underbrace{C_j(t_n) - C_j(t_o)}_{\text{directly from measured data}} = \sum_{i=1}^{M} k_i \underbrace{\int_{t_o}^{t_n} r_{ji}\,(\underline{C}\,(t))\,dt}_{\text{by integration (since concentration dependence of the rates is known) of measured data}}$$

$$C_j(t_n) - C_j(t_o) = \sum_{i=1}^{M} k_i X_{nji}$$

The standard least squares method can now be used to minimize the following expression,

$$S = \sum_{n=1}^{N} \sum_{j=1}^{S} \left[w_{nj}\,(y_{nj} - \hat{y}_{nj})\right]^2$$

n – indicates n-th data point at t_n

N – total number of data points $y_{nj} = C_j(t_n) - C_j(t_o)$

S – total number of species

experimentally measured concentration difference of component j at time t_n

and at time t_o = ⇒ dependent variable $\hat{y}_{nj} = \sum_{i=1}^{M} k_i X_{nji}$

– predicted valued of the dependent variable y_{nj} at point n where X_{nji} are calculated from experimental data and k_i are to be found.

w_{nj} - any desired weighing function

These weights can be picked all to be 1 leading to the usual least square fits, or they can be selected to be proportional to the inverse of measured concentration difference, or to be inversely proportional to the $\sqrt{\text{variance}}$ of concentration values for each species j.

Setting then,

$$\frac{\partial S}{\partial k_i} = 0 \text{ for all } i = 1, 2, \ldots M$$

leads to a set of linear equations from which k_i can be promptly determined.

The above method although quite powerful in many situations also leads in general only to rough estimates of reaction constants.

2

Reaction Rates and Mechanisms

REACTION RATES

Consider the reaction,

$$2\ NO(g) + O_2(g) \rightarrow 2\ NO_2(g).$$

We can specify the rate of this reaction by telling the rate of change of the partial pressures of one the gases. However, it is convenient to convert these pressures into concentrations, so we will write our rates and rate equations in terms of concentrations, where square brackets, [ ], mean concentration in mol/L.

We might try to write the rate variously as,

$$\frac{d[NO_2]}{dt}$$

or as,

$$-\frac{d[O_2]}{dt}$$

but these are not the same because each molecule of O_2 gives two molecules of NO_2. To arrive at an unambiguous definition of reaction rate we define the "reaction velocity," v, as,

$$v = \frac{1}{2}\frac{d[NO_2]}{dt} = -\frac{d[O_2]}{dt} = -\frac{1}{2}\frac{d[NO]}{dt}$$

This is unambiguous. The negative sign tells us that that species is being consumed and the fractions take care of the stoichiometry. Any one of the three derivatives can be used to define the rate of the reaction.

For a general reaction,

$$aA + bB \rightarrow cC + dD,$$

the reaction velocity can be written in a number of different but equivalent ways,

$$v = -\frac{1}{a}\frac{d[A]}{dt} = -\frac{1}{b}\frac{d[B]}{dt} = \frac{1}{c}\frac{d[C]}{dt} = \frac{1}{d}\frac{d[D]}{dt}$$

As in our previous example, the negative signs account for material that is being consumed in the reaction and the positive signs account for material that is being formed in the reaction. The stoichiometry is preserved by dividing the rate of change of concentration of each substance by its stoichiometric coefficient.

RATE OF REACTION (AVERAGE AND INSTANTANEOUS RATE)

We are all familiar with processes in which something changes with time. Rate is usually expresses as the ratio of the amount of change in some quantity to the time required to produce that change.

Rate = (Change in some quantity)/(Time taken for the change) = $\Delta X/\Delta t$

The term ΔX means X_{final} - $X_{initial}$ and Δt is the amount of time elapsed. For example, a car driver starts his journey at 9.0 a.m. with odometer reading x miles. At 11.a.m. he reached his destination. The odometer reading at destination is y miles. The rate of his travel can be calculated as

Rate = $(\Delta(\text{distance}))/(\Delta(\text{time})) = (\text{distance}_{(f)} - \text{distance}_{(in)})/(\text{time}_{(f)} - \text{time}_{(in)})$

$= (y\text{-}x)/(11.0\text{-}9.0) = (y\text{-}x)/2$ miles h^{-1}

The above example indicates that the car has been driven with uniform rate but actually it has been driven sometimes faster and sometimes slower depending upon the conditions of the road. Thus, the overall rate is an average rate and the rate at which the car was moving at any instant, i.e., instantaneous rate was changeable. The rate measured over a long time interval is called average rate and the rate measured for an infinitesimally small time interval is called instantaneous rate.

In a chemical change, reactants and products are involved. As the chemical reaction proceeds, the concentration of the reactants decreases, i.e., products are produced. The rate of reaction (average rate) is defined as the change of concentration of any one of its reactants (or products) per unit time.

Average rate of reaction,

r_{av} = (Change of concentration one of the reactants or products)/(Time taken for the change)

Consider a common hypothetical reaction,

$$A \rightarrow B$$

The average rate of reaction may be expressed either in terms of decrease in concentration of A (reactants) or in terms of increase in concentration of B (product). or Average rate of reaction= (Decrease in concentration of A)/(Time taken) Average rate of reaction= (Increase in concentration of B)/(Time taken)

RATE OF REACTION IN MOLECULES COMBINE

The rate of a reaction is the speed at which areaction happens. If a reaction has a low rate, that means the molecules combine at a slower speed than a reaction with a high rate. Some reactions take hundreds, maybe even thousands, of years while others can happen in less than one second. The rate of reaction depends on the type of molecules that are combining. If you want to think of a very slow reaction, think about how long it took dinosaur bones to become fossils through breakdown.

There is another big idea for rates of reaction called collision theory. The collision theory says that as more collisions in a system occur, there will be more combinations of molecules bouncing into each other. If there are a higher number of collisions in a system, more combinations of molecules can occur. The reaction will go faster and the rate of that reaction will be higher. Even though they are both liquids, think about how slowly molecules move in honey when compared to your soda. There are a lower number of collisions in the honey. Reactions happen - no matter what. Chemicals are always combining or breaking down. The reactions happen over and over, but not always at the same speed. A few things affect the overall speed of the reaction and the number of collisions that can occur.

Concentration

If there is more of a substance in a system, there is a greater chance that molecules will collide and speed up the rate of the reaction. If there is less of something, there will be fewer collisions and the reaction will probably happen at a slower speed. Sometimes when you are in a chemistry lab, you will add one solution to another. When you want the rate of reaction to be slower, you will add only a few drops at a time instead of the entire beaker.

Temperature

When you raise the temperature of a system, the molecules bounce around a lot more because they have more energy. When they bounce around more, they are more likely to collide.

That fact means they are also more likely to combine. When you lower the temperature, the molecules are slower and collide less. That temperature drop lowers the rate of the reaction. Back to the chemistry lab! Sometimes you will mix solutions in ice so that the temperature of the system stays cold and the rate of reaction is slower.

Pressure

Pressure affects the rate of reaction, especially when you look at gases. When you increase the pressure, the molecules have less space in which they can move. That greater density of molecules increases the number of collisions. When you decrease the pressure, molecules don't hit each other as often. The lower pressure decreases the rate of reaction.

MEASURING REACTION RATES

Scientists like to know the rates of reactions. They like to measure different kinds of rates too. Each rate that can be measured tells scientists something different about the reaction. We're going to take a little time to cover a few different measures of reaction rates.

- *Forward Rate:* The rate of the forward reaction when reactants combine to become products.
- *Reverse Rate:* The rate of the reverse reaction when products break apart to become reactants.
- *Net Rate:* The forward rate minus the reverse rate.
- *Average Rate:* The speed of the entire reaction from start to finish.
- *Instantaneous Rate:* The speed of the reaction at one moment in time. Some reactions can happen quickly at the start and then slow down. You have one average rate, but the instantaneous rates can tell you the whole story.

Scientists measure all of these rates by finding out the concentrations of the molecules in themixture. If you find out the concentration of molecules at two different times, you can find out what direction the reaction is moving towards and how fast it is going. Even if the concentrations are equal at the two points of measurement, scientists still learn something. If the concentrations are stable during two measurements, the reaction is at an equilibrium point.

One Step at a Time

There is still more to know about measuring the rates of reactions. Since many reactions happen in several steps, the rate for each step needs to be measured. There will always be one step that happens at the slowest speed. That slowest step is called the rate-limiting step. That rate-limiting step is the one reaction that really determines how fast the overall reaction can happen. If you have six steps in your series of reactions and the third step goes incredibly slow, that is the rate-limiting step.

As far as the overall reaction is concerned, none of the other rates really matter. If you want to speed up the overall reaction, you would focus on that slowest step. Don't forget that if you only speed up one step, another step may become the new rate-limiting step. You should always understand how all of the steps are involved in the overall reaction.

DEFINITION OF RATE OF CHEMICAL REACTION

Any chemical reaction that has time-independent stoichiometry. For example, nitrogen dioxide (NO_2) decomposes in the gas phase at temperatures in the region of 300 °C to give nitric oxide (NO) and oxygen,

$$2NO_2(g) = 2NO(g) + O_2(g)$$

If the progress of this gas-phase reaction is monitored in a closed reaction vessel then concentrations can simply be expressed in terms of mol dm^{-3}. At any time in the decomposition, the rate of *decrease* in the concentration of NO_2

will be directly related to the rates of *increase* in the concentrations of NO and O_2, respectively.

What is the relationship between these quantities?

If we consider $-d[NO_2]/dt$, which is a positive quantity, then,

$$-\frac{d[NO_2]}{dt}=\frac{d[NO]}{dt}=2\frac{d[O_2]}{dt}$$

These relationships are consistent with the fact that according to the stoichiometry, the rate of increase in the concentration of O_2 is 'only equal to one-half of that for NO.

The answer to the above question could equally well have been written in a fractional form, that is,

$$-\frac{1}{2}\frac{d[NO_2]}{dt}=\frac{1}{2}\frac{d[NO]}{dt}=\frac{d[O_2]}{dt}$$

It is this form that is conventionally used to define the rate of reaction and so,

$$J=-\frac{1}{2}\frac{d[NO_2]}{dt}=\frac{1}{2}\frac{d[NO]}{dt}=\frac{d[O_2]}{dt}$$

It is common practice in chemical kinetics (as well as in chemical thermodynamics) to use a chemical reaction written in an 'alphabetical' form to help to express a definition in a general way. Thus, we could write a chemical reaction with *known* stoichiometry as,

$$aA + bB + \ldots = pP + qQ + \ldots$$

where A, B and so on, represent reactants and P, Q and so on, represent products.

The numbers a, b... and *p*, *q*... ensure that the equation is balanced and so are

known as balancing coefficients. In practice they are usually chosen to have their *smallest* possible integer values, and they must be positive. Writing a chemical reaction in this way allows a quantity called the stoichiometric number to be introduced. It is given the symbol ν_Y where the subscript Y represents a given species (reactant or product) in the reaction. The stoichiometric number is then defined so that for

reactant A, $\nu_A = -a$

reactant B, $\nu_B = -b$

product P, $\nu_P = +p$

product Q, $\nu_Q = +q$

Hence, the stoichiometric number for a reactant is always negative, and that for a product is always positive.

What are the stoichiometric numbers of NO_2, NO and O_2?

According to the definition given above,

$$\nu_{NO_2} = -2, \nu_{NO} = +2 \text{ and } \nu_{O_2} = +1$$

We can now make a general definition for the rate of a chemical reaction,

$$J = \frac{1}{\nu_A}\frac{d[A]}{dt} = \frac{1}{\nu_B}\frac{d[B]}{dt} = \frac{1}{\nu_P}\frac{d[P]}{dt} = \frac{1}{\nu_Q}\frac{d[Q]}{dt}$$

Is the term,

$$\frac{1}{\nu_A}\frac{d[A]}{dt}$$

positive or negative in value?

It is positive. Since A is a reactant d[A]/dt is negative. But the stoichiometric number for a reactant is defined to be negative. So, dividing d[A]/dt by n_A is equivalent to dividing one negative quantity by another, and this gives a positive result.

It is important to remember, however, that this definition only holds for a reaction with time-independent stoichiometry. If, for example, intermediates build up to measurable quantities during the course of a reaction then there are no simple relationships between the rates of change of concentrations of reactants and products.

A reactant or a product in a reaction is simply represented by Y (as we have already done in our discussion of the stoichiometric number). It then follows that,

$$J = \frac{1}{\nu_Y}\frac{d[Y]}{dt}$$

Strictly, this definition assumes constant *volume* conditions during the course of a reaction, For solution reactions this is a reasonable approximation. It is also valid for gas-phase reactions carried out in sealed containers.

EVALUATION OF REACTION RATE FORMS IN STIRRED TANK REACTORS

Most of the problems associated with evaluation and determination of proper rate forms from batch data are related to the difficulties in estimating derivatives of concentrations or in fitting cumbersome integral forms to the data.

In contrast, if a perfectly mixed stirred tank reactor shown below is operated at steady state, direct estimates of the rate itself are obtained.

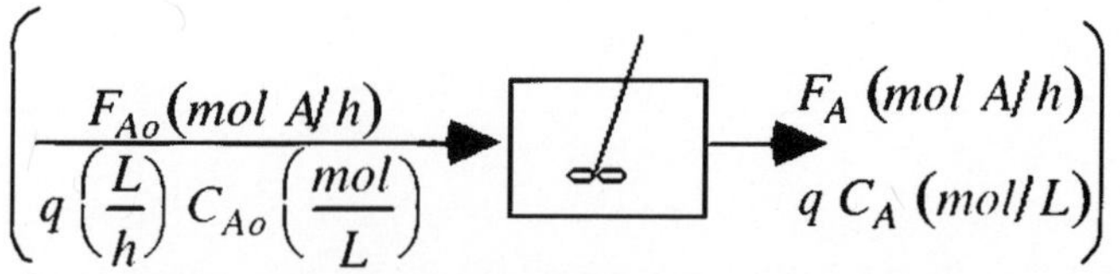

$F_{Ao}(mol\ A/h)$ → [reactor] → $F_A\ (mol\ A/h)$

$q\left(\frac{L}{h}\right)\ C_{Ao}\left(\frac{mol}{L}\right)$ $\qquad$ $q\ C_A\ (mol/L)$

For a reaction A→ products that proceeds without appreciable change in volume of the reaction mixture a mass balance on *A* gives at steady state:

$$(\text{Rate in}) - (\text{Rate out}) - (\text{Rate of Reaction}) = O$$

$$F_{Ao} - F_A - (-R_A)V = 0$$

$$F_{Ao}, F_A - \text{in}(\text{mol}/\text{hr})$$

$$(-R_A)\left(\frac{\text{mol}}{\text{hr lit}}\right)V(L) - \text{reaction mixture in the reactor}$$

$$F_{Ao} = qC_{Ao} \qquad F_A = qC_A \; q\left(\frac{L}{hr}\right)$$

$$-R_A = \frac{F_{Ao} - F_A}{V} = \frac{C_{Ao} - C_A}{(V/q)}$$

By varying $\tau = \frac{V}{q}$ we can readily obtain a set of data $-R_A$ vs C_A and plot directly,

$$\underbrace{\log(-R_A)}_{Y} = \log k + \alpha \underbrace{\log C_A}_{X} \text{ as Y vs X}$$

Let us in addition assume that every run was performed isothermally, but at a different temperature, and that the rate can be expressed by an n-th order form. Now we have generated a set of data - rA vs T vs CA from which we have to extract α, k_o, E since it was assumed that the rate is of an n-th order form:

$$-R_A = k_o e^{-E/RT} C_A{}^{\alpha}$$

$$\underbrace{\log(-R_A)}_{Y} = \log k_o - \frac{E}{2.3R}\underbrace{\frac{}{T}}_{X_1} + \alpha \underbrace{\log C_A}_{X_2}$$

Say we had N-measurements. Then,

$$Y_i = A + B\,X_{1i} + CX_{2i} \qquad ; i = 1, 2 \ldots N$$

$$X_{1i} = \frac{1}{T_i} \qquad ; X_{2i} = \log C_{Ai}$$

$$Y_i = \log(-R_A) \text{ predicted at } i \qquad ; A = \log k_o, B = -\frac{E}{2.3R}, C = \alpha$$

X_{1i}, X_{2i}, Y_i are known at every data point. We want to minimize the sum of the squares of the errors E = $y_i - Y_i$ where y_i = log (– R_{Ai}) measured at point i.

$$S = \sum_{i=1}^{N}\left[w_i\left(y_i - Y_i\right)\right]^2$$

$$\frac{\partial S}{\partial A} = \frac{\partial S}{\partial B} = \frac{\partial S}{\partial C} = 0$$

which yields the three linear equations for best estimates of A,B,C (when wi = 1)

$$\sum_{i=1}^{N} y_i = AN + B\sum_{i=1}^{N} X_{1i} + C\sum_{i=1}^{N} X_{2i}$$

$$\sum_{i=1}^{N} y_i = A\sum_{i=1}^{N} X_{1i} + B\sum_{i=1}^{N} X_{1i}^2 + C\sum_{i=1}^{N} X_{1i}X_{2i}$$

$$\sum_{i=1}^{N} X_{2i}y_i = A\sum_{i=1}^{N} X_{2i} + B\sum_{i=1}^{N} X_{1i}X_{2i} + C\sum_{i=1}^{N} X_{2i}^2$$

Thus, multiple linear regression can be used successfully to determine all three parameters but at some loss of statistical rigour due to the logarithmic transformation performed. Often multiple linear regression is adequate to determine the parameters with satisfactory accuracy. When further refinements are needed, one should use the results of linear regression analysis as the starting estimates of the parameters and proceed to refine them by nonlinear regression. Suppose we wanted to find now the values of parameters that minimize the sum of the squares of errors in the original space. Now S is defined by:

$$S = \sum_{i=1}^{N}\left[\left(y_i - Ae^{BX1i}X_{2i}^{C}\right)w_i\right]^2$$

All of the techniques will rely on some sort of iteration scheme. Briefly if

$\underline{P} = \begin{pmatrix} A \\ B \\ C \end{pmatrix}$ is the vector of parameters and $\underline{P}^k$ is the k-th guess for parameters, $\underline{P}^{k+1}$ improved guess given by,

$$\underline{P}^{k+1} = \underline{P}^k - \lambda\,\underline{R}\,\underline{G}$$

where,

$$\underline{G} = \left[\frac{\partial S}{\partial A}\frac{\partial S}{\partial B}\frac{\partial S}{\partial C}\right]_{\underline{P}^k}^{T}$$

is the gradient vector in the direction of increasing (decreasing S), $\underline{R}$ is a matrix, the components of which determine whether in search of the minimum of S we will move directly in the gradient direction or at some angle to the gradient direction, λ is a scalar regulating the size of the step taken in improving the parameter value.

For example if $\underline{R} = \underline{L}$ is the identity matrix

$$\underline{R} = \begin{bmatrix} 1 & 0 & 0 \\ 0 & 1 & 0 \\ 0 & 0 & 1 \end{bmatrix}$$

$$\underline{P}^{k+1} = \underline{P}^k - \lambda\underline{G}$$

and we have the steepest descent method.

If, however, $\underline{R}$ is related to the Hessian matrix,

$$\underline{R} = \begin{pmatrix} \frac{\partial^2 S}{\partial A^2} & \frac{\partial^2 S}{\partial A \partial B} & \frac{\partial^2 S}{\partial A \partial C} \\ \frac{\partial^2 S}{\partial A \partial B} & \frac{\partial^2 S}{\partial B^2} & \frac{\partial^2 S}{\partial B \partial C} \\ \frac{\partial^2 S}{\partial A \partial C} & \frac{\partial^2 S}{\partial B \partial C} & \frac{\partial^2 S}{\partial C^2} \end{pmatrix}^{-1}$$

one has the Newton-Raphson Method, and with approximations to $\underline{R}$ the Gaus-Newton Method. Some available computer programmes would require explicit expressions for the partial derivatives $\frac{\partial S}{\partial A}, \frac{\partial^2 S}{\partial A \partial B}$, etc. to evaluate $\underline{G}$ and $\underline{R}$. Others estimate them by finite differences.

KINETIC FORMS

The integrated most commonly encountered kinetic forms in batch reactors. Only the frequently encountered reaction rates for single reactions are integrated and presented. Expressions for some simpler multiple reactions can also be readily found and are given in almost all textbooks (Carberry, Levenspiel, *etc.*).

We have seen that reaction rates are functions of temperature and composition and usually are much more sensitive to changes in temperature than in composition (i.e the rate changes much more dramatically with temperature than with composition).

Consider an n-th order irreversible reaction,

$$-R_A = k_o e^{-E/RT} C_A^{\ n}$$

with E = 20,000 cal; n = 2.

At constant temperature doubling the concentration would increase the rate by a factor of 4. However, doubling the temperature from 300K to 600K would increase the rate at constant concentration by a factor of 17 × 106, i.e, seventeen million times!

An increase of 27o i.e of 1.09 times the original temperature (less than 10per cent) would increase the rate by a factor of 4.

Thus, for single irreversible reactions we should choose the highest allowable temperatures in order to get the highest rates.

For a reversible reaction the problem is slightly more complicated. For example, for a first order reversible reaction we have,

$$-R_A = k_{10}e^{-E_1/RT}C_A - k_{20}e^{-E_2/RT}Cp$$

for a reaction A = P. At equilibrium - *RA* = 0.

$$\frac{k_{10}}{k_{20}}e^{\frac{-(E_1 - E_2)}{RT}} = \left(\frac{C_p}{C_A}\right)_{eq} = K_c$$

$$C_p = C_{A_o}x_A \qquad C_A = C_{A_o}(1 - x_A)$$

$$\frac{x_{A_o}}{1 - x_{A_o}} = K_c \qquad x_{A_o} = \frac{K_c}{1 + K_c}$$

Now, if we raise the concentration of A, the net rate forward will increase while the equilibrium remains unaffected.

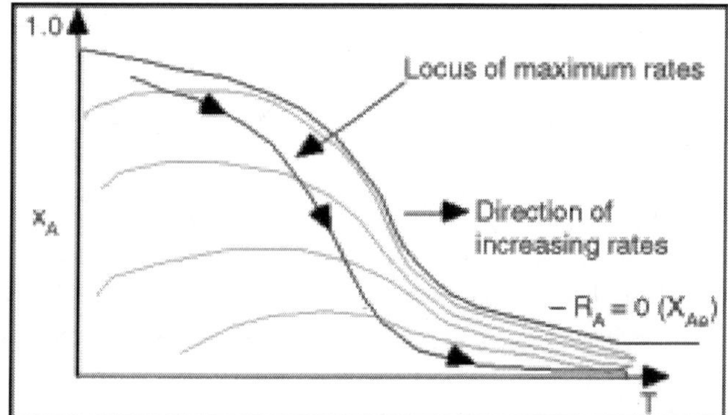

Fig. Conversion vs Temperature for an Exothermic Reaction

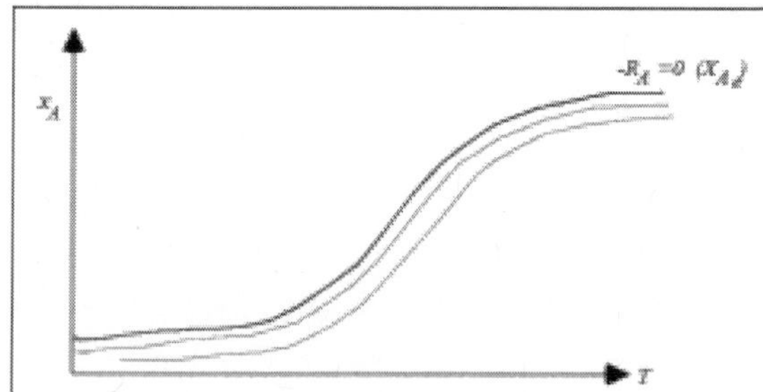

Fig. Conversion vs Temperature for an Endothermic Reaction

If we raise the temperature, the rate will increase both in forward and reverse direction, the portion of the rate with the higher activation energy will increase more and the equilibrium will be affected. In the above example $\Delta H = E_1 - E_2$. $K_c = K_P (RT)^{-\Sigma \upsilon j}$ For this example,

$$\Sigma \upsilon j = O$$

and thus,

$$\frac{d\ell n K_c}{dT} = \frac{\Delta H}{RT^2}$$

$$\frac{dX_{A_o}}{dT} = \frac{\frac{dK_c}{dT}(1+K_c) - \frac{dK_c}{dT}}{(1+K_c)^2} = \frac{K_c}{(1+K_c)^2}\frac{dK_c}{dT}$$

$$= \frac{K_c^2}{(1+K_c)^2}\frac{d\ell n K_c}{dT} = \frac{K_c^2}{(1+K_c)^2}\frac{\Delta H}{RT^2}$$

The sign of the above derivative syn $\frac{dx_{A_e}}{dT}$ depends exclusively on the sign of ΔH since all other terms are positive. Thus $\frac{dx_{A_e}}{dT} > 0$ for $\Delta H > 0$ i.e endothermic reactions, and x_{Ae} goes up as temperature rises $x_{A_e} \uparrow$ as $T \downarrow$. $\frac{dx_{A_e}}{dT} > 0$ for $\Delta H > 0$ i.e exothermic reactions, and the equilibrium conversion drops as the temperature rises $x_{A_e} \uparrow$ as $T \downarrow$.

This conclusion can be generalized and it holds for all types of reversible reactions. Thus, in principle for every reversible reaction we could plot on a conversion versus temperature plot a locus for the equilibrium conversion as function of temperature and loci of constant rates. The locus of constant rates is obtained by connecting the points in the x_A vs T plane which have the same rates.

For endothermic reactions no peculiar behaviour is observed. The higher the temperature at fixed conversion the higher the rate, i.e if we travel along a line parallel to T-axis from left to right we will cross the lines of constant rates moving from lower to higher rates. At constant temperature the lower the conversion (the higher the reactant concentration) the higher the rate i.e, when moving from the top to the bottom along a line parallel to x_A - axis we will constantly get into regions of higher rates.

For exothermic reactions, we can quickly observe that at constant temperature T the lower the conversion the higher the rate. However, when we look at lines of constant conversion x_A we see that if we move from left to right we go through a region of increasing rates, hit a maximum rate and go then through a region of decreasing rates. Thus, we can plot on the x_A vs T diagram a locus of maximum rates. For the given example of first order reversible exothermic reaction this locus would be obtained as follows:

$$-R_A = k_{10}e^{-E_1/RT} C_{A_o}(1-x_A) - k_{20}e^{-E_2/RT} C_{A_o}x_A$$

$$d(-R_A)=\left[\frac{E_1}{RT^2}k_{10}e^{-E_1/RT}C_{A_o}(1-x_A)-\frac{E_2}{RT^2}k_{20}e^{-E_2/RT}C_{A_o}x_A\right]dT$$
$$+\left[-k_{10}e^{-E_1/RT}C_{A_o}-k_{20}e^{-E_2/RT}C_{A_o}\right]dx_A$$

Now for the loci of constant rates $-RA$ = const, $d\,(-RA\,) = 0$ and,

$$\left.\frac{dx_A}{dT}\right|_{-R_A-\text{const}}=\frac{\frac{E_1}{RT^2}k_1(1-x_A)-\frac{E_2}{RT^2}k_2x_A}{k_1+k_2}$$

If we want to identify the points at which the loci of constant rates go through maxima in the x_A vs T plot then at those points we must have:

$$\left.\frac{dx_A}{dT}\right|_{-R_A=\text{const}}=0$$

which leads to:

$$E_1k_1(1-x_A)-E_2k_2x_A=0$$

$$E_1k_{10}e^{-E_1/RT}x_A\left[\frac{E_1k_{10}(1-x_A)}{E_2k_{20}}e^{\frac{-E_1-E_2}{RT}}-1\right]=0$$

$$\frac{E_1}{E_2}\frac{k_{20}}{k_{20}}\frac{1-x_A}{x_A}=e\frac{E_1-E_2}{RT}$$

$$T_m=\frac{E_1-E_2}{R\ell n\left\{\frac{E_1k_{10}(1-x_A)}{E_2k_{20}x_A}\right\}}=\frac{E_2-E_1}{R\ell n\left\{\frac{E_2k_{20}x_A}{E_1k_{10}(1-x_A)}\right\}}$$

This is the locus of maximum rates. In other words, pick a set of values for $0 < x_A < x_{Ae}$ (T_{min}) and from the above formula calculate the corresponding T. Every pair $x_A - T$ represents a point on the locus of maximum rates.

By solving for conversion one can express the locus of maximum rates as,

$$x_A=\frac{1}{\frac{E_2k_{20}}{E_1k_{10}}\exp\frac{E_1-E_2}{RT}+1}$$

Thus, one can pick a set of values for temperature, T, in the region of interest, and from the above formula calculate the corresponding x_A. Every pair $x_A - T$ is a point on the locus of maximum rates on our graph (dashed line).

The above expressions were developed for a first order reversible reaction only. Nevertheless the procedure can be readily generalized for any rate form,

$$dr = \frac{\partial r}{\partial T}dT + \frac{\partial r}{\partial x_A}dx_A = 0$$

$$\left.\frac{dx_A}{dT}\right|_{r=const} = -\frac{\dfrac{\partial r}{\partial T}}{\dfrac{\partial r}{\partial x_A}} = 0$$

The locus of maximum rates can thus always be obtained by setting,

$$\frac{\partial r}{\partial T} = 0$$

The above discussion leads to two basic conclusions. In the case of reversible endothermic reactions, the higher the temperature the higher the rate and the more favorable the equilibrium. For reversible exothermic reactions, the higher the temperature the less favorable the equilibrium. Thus, in a batch system one should move along the locus of maximum rates by starting at high temperature and then as conversion increases, lower the temperature accordingly in such a manner that at every temperature T one stays at the maximum of the rate loci. Towards the end of reaction low temperature will allow favorable equilibrium while the rates will still stay as high as possible at that temperature.

Besides being affected by temperature, the equilibrium composition is also affected by initial composition. For example for a reaction aA = pP,

$$K_c = \frac{\left(C_{A_o}x_{A_e}\right)^p}{C_{A_o}^a\left(1-x_{A_e}\right)^a} = C_{A_o}{}^{p-a}\frac{x_{A_e^p}}{\left(1-x_{A_e}\right)^a} = \text{const at fixed T}$$

Differentiating the above with respect to C_{Ao},

$$0 = (p-1)C_{A_o}{}^{p-a-1}\frac{x_{A_o}^p}{\left(1-x_{A_e}\right)^a}$$

$$+\, C_{A_o}{}^{p-a}\frac{px_{A_o}{}^{p-1}\left(1-x_{A_e}\right)^a + a\left(1-x_{A_e}\right)^{a-1}x_{A_e}^p}{\left(1-x_{A_e}\right)^{2a}}\frac{dx_{A_e}}{dC_{A_o}}$$

Thus, the sign of $\dfrac{dx_{A_e}}{dC_{A_o}}$ depends only on *p-a*,

$$\operatorname{sgn}\frac{dx_{A_e}}{dC_{A_o}} = -(p-a)$$

This can readily be generalized to:

$$\text{sgn}\frac{dx_{A_e}}{dC_{A_o}} = -\sum_{j=1}^{S} \upsilon_j$$

If the reaction proceeds with an increase in the number of moles then $\sum \upsilon_j > 0$ and the equilibrium conversion decreases with increased initial concentration i.e $C_{Ao} \uparrow x_{Ae} \downarrow$. If the reaction proceeds with the decrease in the number of moles ($\sum \upsilon_j < 0$) and equilibrium conversion increases with increasing C_{Ao} i.e. $C_{Ao} \uparrow x_{Ae} \uparrow$.

This of course assumes ideal mixtures in which the activity or fugacity coefficients will not change with composition.

For example, consider A = 2P, a = 1, p = 2 at the temperature at which $K_c = 1\left(\frac{\text{mol}}{\text{lit}}\right)$. If we start with $C_{A_o} = 1\left(\frac{\text{mol}}{\text{lit}}\right)$ the attainable equilibrium conversion is obtainable from the expression for the equilibrium constant:

$$C_{A_o}\frac{x_{A_e}^2}{1-x_{A_e}} = K_c \text{ and } x_{A_e} = \frac{-\frac{K_c}{C_{A_o}} + \sqrt{\frac{K_c^2}{C_{A_o}{}^2} + 4\frac{K_c}{C_{AO}}}}{2}$$

$$x_{A_e} = \frac{1}{2}\left\{-1+\sqrt{1+4}\right\} = 0.618$$

If at the same temperature we started with,

$$C_{A_o} = 2\left(\frac{\text{mol}}{\text{lit}}\right) \; ; \; x_{A_e} = \frac{1}{2}\left\{-\frac{1}{2} + \sqrt{\frac{1}{4} + 4 \, x \frac{1}{2}}\right\} = 0.5$$

Since $\sum \upsilon_j = 2\text{–}1 = 1 > 0$ as $C_{Ao} \uparrow x_{Ae} \downarrow$ as stated above.

The effect of total pressure on the equilibrium constant is,

$$\frac{d\ell n \, K_P}{dP} = -\frac{\Delta V}{RT}$$

Thus,

$$\text{sgn}\frac{dx_{A_e}}{dP} = -\Delta V$$

For a reaction that is accompanied by an increase in volume ($\Delta V > 0$) an increase in pressure on the system will reduce the equilibrium conversion, i.e $P\uparrow x_{Ae}\downarrow$.

For a reaction that is accompanied by a decrease in volume ($\Delta V < 0$) an increase in pressure on the system will increase the equilibrium conversion i.e $P\uparrow x_{Ae}\uparrow$.

For multiple reactions similar conclusions can be drawn. If the activation energy for the reaction of desired product formation is the highest, the optimum yields will be obtained at the highest temperature. If the activation energy for formation of the desired product is the lowest, the lowest temperature will maximize the yield but the rates may be too slow then, so a compromise has to be found. If the activation energy for desired product formation is in between then an optimum temperature profile exists which would maximize the yield.

RATE FORMS OTHER THAN N-TH ORDER

So far we have mainly discussed the n-th order type rate forms. For an irreversible reaction at constant temperature these rate expressions can be plotted against concentration of the limiting reactant or conversion. All these rates have one thing in common, i.e they never decrease with increasing concentration of the limiting reactant (or other reactants), or mathematically,

$$\frac{d(-R_A)}{dC_A} \geq 0 \text{ for all } 0 < C_A \leq C_{A_o}$$

These are "well behaved" rate expressions.

We should watch for possible peculiar behaviour when we encounter a rate which can decrease at least over some reactant concentration range with increasing reactant concentration i.e, for rates where

$$\frac{d(-R_A)}{dC_A} < 0 \text{ for } C_{A_o}'' < C_A < C_{A_o}'$$

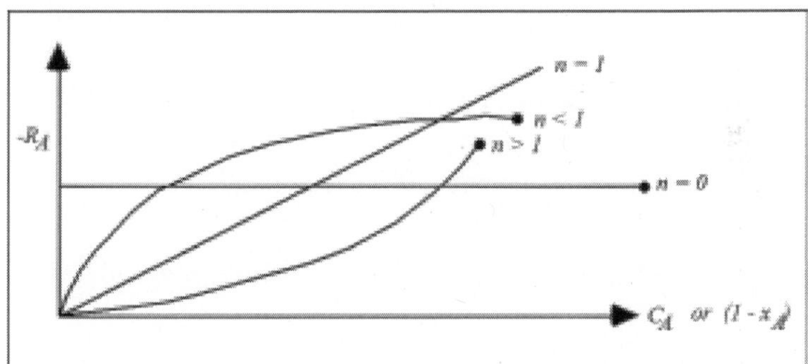

Fig. n-th Order Rate Behaviour

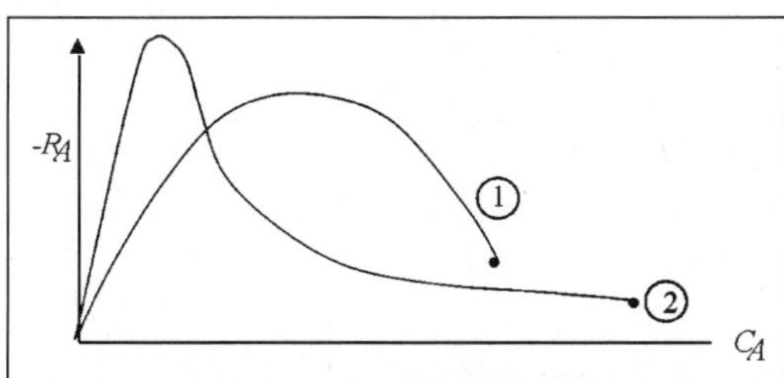

Fig. Non n-th Order Rate Behaviour

The first one (1) results from autocatalytic reactions

$A + P = P + P - R_A = k\, C_A C_P$, the other (2) is observed in a number of catalytic processes with self inhibition $-R_A = \frac{kC_A}{(K+C_A)^2}$. In these particular situations the rate is low at high reactant concentrations, reaches the maximum at some C_A and then decays again as the reactant concentration is further depleted. It is worth considering autocatalytic reactions in a little more detail.

For example,

$$A + P \rightarrow P + P$$

with a rate of $-R_A = k\, C_A\, C_P$

From stoichiometry,

$$C_{A_o} - C_A = C_P - C_{P_o} \quad ; \quad C_P = \overbrace{C_{P_o} + C_{A_o}}^{C_o} - C_A$$

In a batch system then:

$$-R_A = -\frac{dC_A}{dt} = kC_A(C_o - C_A)$$

The rate reaches a maximum value when,

$$\frac{d(-R_A)}{dC_A} = 0 = C_o - 2C_A \Rightarrow C_A = \frac{C_o}{2} \text{ or}$$

$$C_{A_o}(1 - x_A) = \frac{C_o}{2} \Rightarrow x_A = 1 - \frac{C_o}{2C_{A_o}}$$

Then the rate takes its maximum value of,

$$-R_{A_{max}} = k\frac{C_o}{2}\left(C_o - \frac{C_o}{2}\right) = k\frac{C_o^2}{4}$$

If $C_{Ao} = f\, C_o$ where $0 < f < 1$ the ratio of the maximum rate and initial rate is,

$$\frac{-R_{A_{max}}}{-R_{A_{initial}}} = \frac{k\frac{C_o^2}{4}}{kfC_o(C_o - fC_o)} = \frac{1}{4f(1-f)}$$

Thus, if the initial solution contains 99per cent A and 1per cent P (which is reasonable since P serves only to start the reaction) $f = 0.99$ and the maximum rate is over 25 times larger than the initial rate.

The integrated form of the rate expression is:

$$\ell n \frac{C_{Ao}(C - C_A)}{C_A(C_o - C_{Ao})} = kC_o t$$

The maximum rate at $C_A = \frac{C_o}{2}$ would be reached after time of,

$$t_{\max} = \frac{1}{kC_o} \ell n \left(\frac{C_{A_0}}{C_o - C_{A_0}} \right) = \frac{1}{kC_o} \ell n \frac{f}{1-f} = \frac{1}{kC_o} \ell n \left(\frac{f}{1-f} \right)$$

The larger the f is the larger the time t_{max} necessary to reach maximum rates.

This implies that if the product is present as an impurity at 0.1per cent level ($f = 0.999$) the rate of reaction may go unnoticed initially but would peak to 250 times larger value after $k\ C_0 t = 6.9$.

Autocatalytic reactions should be viewed in a broader sense than presented here. Specifically, autocatalytic reactions become especially dangerous when they proceed in the gas phase with an increase in the total number of moles of the reaction mixture (say A + P = 3P) and are exothermic. In that case a slow initial reaction is accelerated on one hand due to the autocatalytic effect of the product and on the other hand due to the increased temperature caused by the heat of reaction.

The rapidly rising pressure of the system due to the expanding reaction mixture volume may, and occasionally does, cause explosions. Even for reactions in the liquid phase rapidly accelerating reaction due to a combined autocatalysis-temperature (caused by exothermicity) effect may lead to rapid build-up of the vapor phase and to explosions. Thus, autocatalytic reactions and all reactions that exhibit rate forms which allow for increases in rates with diminishing reactant concentrations should be watched for and dealt with cautiously.

Autocatalytic reactions can be viewed even in a broader sense than that. For example heat released by a slow reaction causing the same or a similar reaction to accelerate rapidly is also viewed by some as an autocatalytic effect in that reaction system.

As an example of autocatalysis let us consider the hydrolysis of an ester *RCOOR'* in dilute water solution:

$$RCOOR' + H_2O = RCOOH + R'OH$$
$$E + W = A + B$$
$$(ester) + (water) = (acid) + (booze)$$

The reaction proceeds without the catalyst at a very slow rate. However, the acid, product of reaction, catalyzes the same reaction at much faster rates.

$$RCOOR' + H_2O \xrightarrow{RCOOH} RCOOH + R'HO$$
$$E + W \xrightarrow{A} A + B$$

The total rate of disappearance of the ester can be given as the sum of the rate of the spontaneous ("residual") reaction and of the catalytic reaction,

$$-\frac{dC_E}{dt} - k_s C_E + k_c C_E + k_c C_E C_A$$

From stoichiometry,

$$C_{E_o} - C_E = C_A - C_{A_o}$$

also,

$$C_E = C_{E_o}\left(1 - x_E\right) \qquad C_A - C_{A_o} = C_{E_o} x_E$$

If we started initially with pure reactants $C_{A_o} = 0$ in a batch system, the governing equations,

$$-\frac{dC_E}{dt} = C_{E_o}\frac{dx_E}{dt} = k_s C_{E_o}\left(1 - x_E\right) + k_c C_{E_o}^2\left(1 - x_E\right)x_E$$

$$at\ t = 0\ \ C_E = C_{E_o}\ or\ x_E = 0$$

Note: The reaction rates for both the residual and catalytic reaction were presumed independent of water concentration (zeroth order with respect to water) since water is present in large excess.

$$\frac{dx_E}{dt} = k_s\left(1 - x_E\right) + k_c C_{E_o} x_E\left(1 - x_E\right)$$

$$at\ t = 0\ \ x_E = 0$$

Upon integration we get,

$$x_E = \rho\frac{e^{(1+\rho)k_c C_{Eo} t} - 1}{1 + \rho\, e^{(1+\rho)k_c C_{E_o} t}} = \rho\frac{1 - e^{-(1+\rho)k_c C_{E_o} t}}{\rho + e^{-(1+\rho)k_c C_{E_o} t}}$$

where,

$$\rho = \frac{k_s}{k_c C_{E_o}} = \frac{\dfrac{1}{k_c C_{E_o}}}{\dfrac{1}{k_s}} << 1\ \ \frac{1}{k_s}$$

has units of (time) and is a measure of the characteristic reaction time for the residual reaction, i.e gives the time scale over which that reaction takes place. $\dfrac{1}{k_c C_{E_o}}$

has units of time and is a measure of the characteristic reaction time for the catalytic reaction.

Clearly the time scale for the much faster catalytic reaction is much smaller than the time scale for the residual reaction and thus $\rho << 1$.

$\tau = (1+\rho)k_c C_{Eo} t$ becomes a proper dimensionless time for the process overall.

$$x_E = \rho\frac{1 - e^{-\tau}}{\rho + e^{-\tau}}$$

The maximum rate is achieved when,

$$x_E = \frac{1}{2} \; ; \qquad \tau = -\ell n\, \rho \; ; \; t = \frac{1}{(1+\rho) k_c C_{E_o}} \ell n \left(\frac{1}{\rho} \right)$$

$$-R_{E_{\max}} = \frac{C_{E_o}}{2} \left(k_s + k_c \frac{C_{E_o}}{2} \right) \approx k_c \frac{C_{E_o}^2}{4}$$

The ratio of the maximum rate to the initial rate of the residual reaction is:

$$\frac{-R_{E_{\max}}}{-R_{E_{\max}}} = \frac{\frac{C_{E_o}}{2}\left(k_s + \frac{k_c C_{E_o}}{2} \right)}{k_s C_{E_o}} = \frac{1}{2}\left(1 + \frac{1}{2\rho} \right) \approx \frac{1}{4\rho} >> 1$$

VARIABLE VOLUME BATCH SYSTEM

We have derived some time ago how the volume of the reaction system varies with reaction extent for an ideal mixture of ideal gases:

$$V = V_o \left[1 + \frac{\sum_{j=1}^{s} \upsilon_j}{N_{T_o}} X \right]$$

Molar extent can be expressed in terms of fractional conversion since,

$$X = \frac{N_A - N_{A_o}}{\upsilon_A} = \frac{N_P - N_{P_o}}{\upsilon_P} = etc$$

$$x_A = \frac{N_{A_o} - N_A}{N_{A_o}}$$

$$X = \frac{N_{A_o}}{-\upsilon_A} x_A = \frac{N_{A_o}}{|\upsilon_A|} x_A$$

Then $V = V_o\,[1+\varepsilon_A x_A]$ where the coefficient of expression ε_A is defined by:

$$\varepsilon_A = \frac{\left(\sum_{j=1}^{s} \upsilon_j \right) y_{A_o}}{|\upsilon_A|} = \frac{V\big|_{x_A=1} - V\big|_{x_A=o}}{V\big|_{x_A=o}}$$

y_{Ao} – mole fraction of A in the starting mixture.

Let us suppose that we are considering the following irreversible n-th order reaction,

$$aA \rightarrow pP \quad p \neq a$$

in a constant pressure batch reactor starting with pure A.

The material balance on A yields,

$$\frac{dN_A}{dt} = -k\, C_A^n\, V$$

We know that,

$N_A = C_A V$ but the volume now varies with reaction.

$$\frac{d(C_A V)}{dt} = V\frac{dC_A}{dt} + C_A\frac{dV}{dt} = -k\, C_V^n\, V$$

In order to integrate the above expression we need to know how V and $\frac{dV}{dt}$ vary either with concentration $C\,A$ or with time t. We could start from:

$$C_A = \frac{N_A}{V}$$

In any system (constant P, or constant V) by definition of convesion moles of A at time t are given by the initial moles of A times fraction unconverted, i.e: $N_A = N_{Ao}\,(1–x_A)$ and we have seen above:

$$V = V_o\left(1 + \varepsilon_A x_A\right)$$

$$C_A = C_{A_o}\frac{1 - x_A}{1 - \varepsilon_A x_A} \; or \; x_A = \frac{C_{A_o} - C_A}{C_{A_o} + \varepsilon_A C_A}$$

We could substitute x_A in terms of C_{Ao}, C_A in the formula for the volume:

$$V = V_o\frac{(1 + \varepsilon_A)C_{A_o}}{C_{A_o} + \varepsilon_A C_A}$$

$$\frac{dV}{dt} = -V_o\frac{\varepsilon_A(1 + \varepsilon_A)C_{A_o}}{\left(C_{A_o} + \varepsilon_A C_A\right)^2}\frac{dC_A}{dt}$$

After substitution in the first equation for $\frac{d(C_A V)}{dt}$ and some algebra we get,

$$\frac{C_{A_o}}{C_{A_o} + \varepsilon_A C_A}\frac{dC_A}{dt} = -k\, C_A^n \;\; ; \; t = 0 \;\; C_A = C_{A_o}$$

For our example,

$$\varepsilon_A = \frac{(p - a)x1}{a} = \frac{p - a}{a}$$

It is of interest to note that if we did not remember the variation of reaction volume with extent or with conversion, and starting from,

$$\frac{d(C_A V)}{dt} = V\frac{dC_A}{dt} + C_A\frac{dV}{dt} = -k\, C_A^n V$$

we would have had to derive an expression for $\frac{dV}{dt}$. It is instructive to go through that process.

The change in volume of the reaction mixture per unit time $\left(\frac{dV}{dt}\right)\left(\frac{L}{\min}\right)$ is proportional clearly to the total rate of disappearance of $A, -R_A V\left(\frac{moles\ A}{\min}\right)$ multiplied by the change in volume of the reaction mixture per each mole of A reacted $\left(\frac{L}{mole\ A}\right)$.

$$\frac{dV}{dt} = \frac{\varepsilon_A}{C_{A_o}} k\, C_A^n V = (in\ our\ example) \frac{p-a}{aC_{A_o}} k\, C_A^n\, V$$

Now we would have two coupled differential equations to solve simultaneously.

Since differential equations for concentrations lead to relatively complex forms it is advisable to use conversion whenever possible in variable volume systems.

$$\frac{d(C_A V)}{dt} = -k\, C_A^n\, V$$

but,

$$C_A = C_{A_o} \frac{1-x_A}{1+\varepsilon_A x_A} \qquad V = V_o\left(1+\varepsilon_A x_A\right)$$

$$C_A V = C_{A_o} V_o \left(1-x_A\right)$$

$$\frac{d(C_A V)}{dt} = -C_{A_o} \frac{dx_A}{dt} = -k\, C_{A_o}^n \frac{\left(1-x_A\right)^n}{\left(1+\varepsilon_A x_A\right)^n} V_o\left(1+\varepsilon_A x_A\right)$$

$$\frac{dx_A}{dt} = k\, C_{A_o}^{n-1} \frac{\left(1-x_A\right)^n}{\left(1+\varepsilon_A x_A\right)^{n-1}}$$

$$t=0 \qquad x_A = 0$$

The equation for conversion can be more readily integrated for most orders and some of the integrated forms were given earlier.

For example for first order reactions $n = 1$

$$\frac{dx_A}{dt} = k\left(1-x_A\right)$$

$$x_A = 1-e^{-kt}$$

This expression is exactly the same as in a constant volume system, since in the case of first order reactions it does not make any difference to conversion whether the process is performed at constant volume or constant pressure.

The differential equation for concentration is for $n = 1$:

$$\frac{C_{A_o}}{C_{A_o} + \varepsilon_A C_A} \frac{dC_A}{dt} = -k\, C_A$$

$$-\int_{C_{A_o}}^{C_A} \frac{C_{A_o}\, dC_A}{\left(C_{A_o} + \varepsilon_A C_A\right) C_A} = k \int_o^t dt$$

$$\ell n \left[\frac{\left(C_{A_o} + \varepsilon_A C_A\right) C_{A_o}}{\left(1 + \varepsilon_A\right) C_{A_o} C_A} \right] = kt$$

Solved for concentration this yields:

$$C_A = \frac{C_{A_o} e^{-kt}}{1 + \varepsilon_A \left(1 - e^{-kt}\right)} in\ V \neq const$$

We remember that in a constant volume system for first order reaction:

$$C_A = C_{Ao}\, e^{-kt} \quad V = const$$

Thus, although at given reaction time t fractional conversion for a first order reaction is the same in a V = const and $P\,T$ = const system the reactant (or product) concentration is not. This is to be expected since in a V = const system the change in concentration is due exclusively to reaction. In $P\,T$ = const system the change in concentration reflects the combined effect of depletion or formation of the number of moles by reaction and of volume expansion or contraction due to reaction. Naturally for $\varepsilon_A = 0$ (no expansion or contraction) the two expressions are identical.

SURFACE AREA ON REACTION RATES

The more finely divided the solid is, the faster the reaction happens. A powdered solid will *normally* produce a faster reaction than if the same mass is present as a single lump. The powdered solid has a greater surface area than the single lump.

CALCIUM CARBONATE AND HYDROCHLORIC ACID

In the lab, powdered calcium carbonate reacts much faster with dilute hydrochloric acid than if the same mass was present as lumps of marble or limestone.

$$CaCo_{3(s)} + 2HCl_{(aq)} \rightarrow CaCl_{2(aq)} + H_2O_{(i)} + CO_{2(g)}$$

The Catalytic Decomposition of Hydrogen Peroxide

This is another familiar lab reaction. Solid manganese(IV) oxide is often used as the catalyst. Oxygen is given off much faster if the catalyst is present as a powder than as the same mass of granules.

$$2H_2O_{2(aq)} \xrightarrow{MnO_{2(s)}} 2H_2O_{(l)} + O_{2(g)}$$

Catalytic Converters

Catalytic converters use metals like platinum, palladium and rhodium to convert poisonous compounds in vehicle exhausts into less harmful things. For example, a reaction which removes both carbon monoxide and an oxide of nitrogen is:

$$2CO_{(g)} + 2NO_{(g)} \rightarrow 2CO_{2(g)} + N_{2(g)}$$

Because the exhaust gases are only in contact with the catalyst for a very short time, the reactions have to be very fast. The extremely expensive metals used as the catalyst are coated as a very thin layer onto a ceramic honeycomb structure to maximise the surface area.

The Explanation

A reaction if the particles in the gas or liquid collide with the particles in the solid. Increasing the surface area of the solid increases the chances of collision taking place.A reaction between magnesium metal and a dilute acid like hydrochloric acid. The reaction involves collision between magnesium atoms and hydrogen ions.

$$Mg_{(s)} + 2H^+_{(aq)} \rightarrow Mg^{2+}_{(aq)} + H_{2(g)}$$

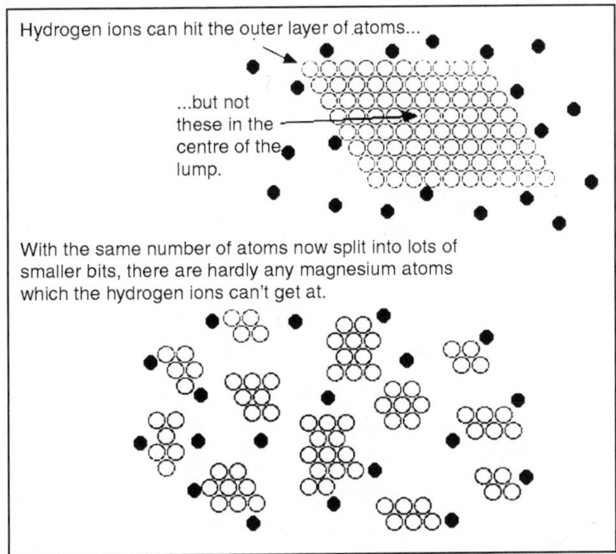

Increasing the number of collisions per second increases the rate of reaction.

THE EFFECT OF PRESSURE ON REACTION RATES

Increasing the pressure on a reaction involving reacting gases increases the rate of reaction. Changing the pressure on a reaction which involves only solids or liquids has no effect on the rate.

An Example

In the manufacture of ammonia by the Haber Process, the rate of reaction between the hydrogen and the nitrogen is increased by the use of very high pressures.

$$N_{2(g)} 3H_{2(g)} \rightleftharpoons 2NH_{3(g)}$$

In fact, the main reason for using high pressures is to improve the percentage of ammonia in the equilibrium mixture, but there is a useful effect on rate of reaction as well.

The Relationship between Pressure and Concentration

Increasing the pressure of a gas is exactly the same as increasing its concentration. If you have a given mass of gas, the way you increase its pressure is to squeeze it into a smaller volume. If you have the same mass in a smaller volume, then its concentration is higher.

The relationship mathematically if you have come across the *ideal gas equation:*

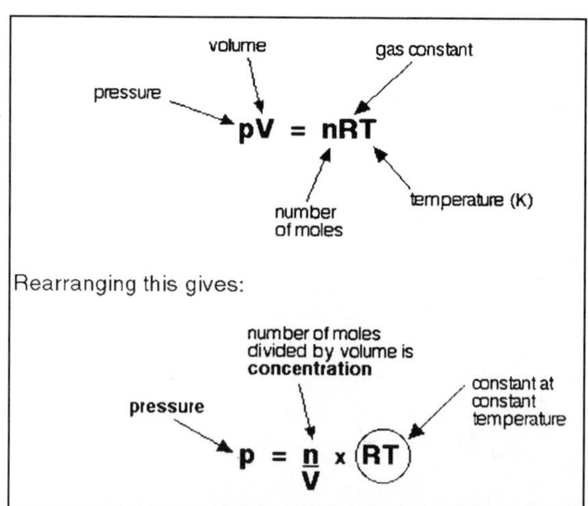

Because "RT" is constant as long as the temperature is constant, this shows that the pressure is directly proportional to the concentration. If you double one, you will also double the other.

THE EFFECT OF INCREASING THE PRESSURE ON THE RATE OF REACTION

Collisions Involving Two Particles

The same argument applies whether the reaction involves collision between two different particles or two of the same particle.

In order for any reaction to happen, those particles must first collide. This is true whether both particles are in the gas state, or whether one is a gas and the other a solid. If the pressure is higher, the chances of collision are greater.

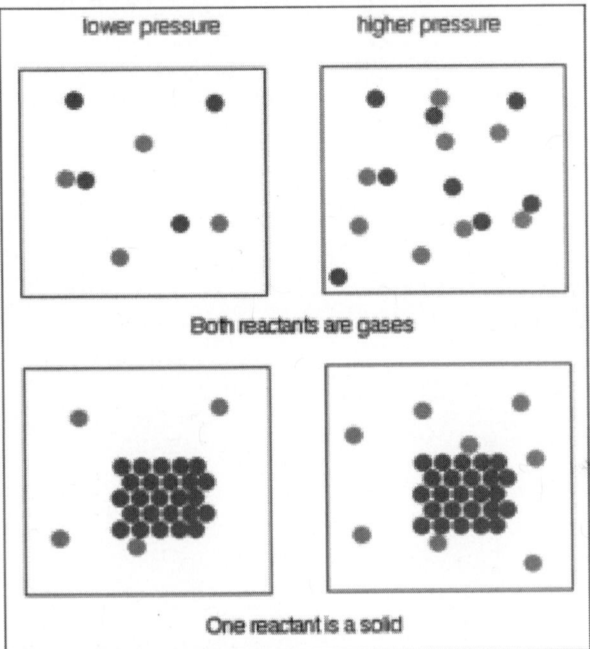

Reactions Involving only One Particle

If a reaction only involves a single particle splitting up in some way, then the number of collisions is irrelevant. What matters now is how many of the particles have enough energy to react at any one time.

Suppose that at any one time 1 in a million particles have enough energy to equal or exceed the activation energy. If you had 100 million particles, 100 of them would react. If you had 200 million particles in the same volume, 200 of them would now react. The rate of reaction has doubled by doubling the pressure.

THE EFFECT OF TEMPERATURE ON REACTION RATES

The temperature the rate of reaction increases. As a rough approximation, for many reactions happening at around room temperature, the rate of reaction doubles for every 10°C rise in temperature. It doesn't apply to all reactions. Even where it is approximately true, it may be that the rate doubles every 9°C or 11°C or whatever. The number of degrees needed to double the rate will also change gradually as the temperature increases.

INCREASING THE COLLISION FREQUENCY

Particles can only react when they collide. If you heat a substance, the particles move faster and so collide more frequently. That will speed up the rate of reaction.

That seems a fairly straightforward explanation until you look at the numbers!

It turns out that the frequency of two-particle collisions in gases is proportional to the square root of the kelvin temperature. If you increase the temperature from 293 K to 303 K (20°C to 30°C), you will increase the collision frequency by a factor of:

$$\sqrt{\frac{303}{293}} = 1.017$$

That's an increase of 1.7per cent for a 10° rise. The rate of reaction will probably have doubled for that increase in temperature - in other words, an increase of about 100per cent. The effect of increasing collision frequency on the rate of the reaction is *very* minor. The important effect is quite different...

The Key Importance of Activation Energy

Collisions only result in a reaction if the particles collide with enough energy to get the reaction started. This minimum energy required is called the activation energy for the reaction.

3

Chemical Reactions of Rate Laws

The rate of a chemical reaction is, perhaps, its most important property because it dictates whether a reaction can occur during a lifetime. Knowing the rate law, an expression relating the rate to the concentrations of reactants, can help a chemist adjust the reaction conditions to get a more suitable rate. If there are two competing reactions for a single reagent, one can, knowing the rate law, favor the exclusive formation of a single product.

To obtain this kind of knowledge about reactions, we will first define what rate means. We will then derive the rate law expression. Using the method of initial rates, we will discuss how to determine the form and order of the rate law. Next, we will probe rate laws in depth and introduce the integrated rate law as an alternative form of the simple rate law that allows us another, more simple, experimental method to determine the order of the rate law. The integrated rate law will also allow us to determine the half-lives of chemical reactions.

CHEMICAL REACTIONS AND RATES

According to the conventions of biochemistry, the rate of a chemical reaction is described by the number of molecules of reactant(s) that are converted into product(s) in a specified time period. Reaction rate is always dependent on the concentration of the chemicals involved in the process and on rate constants that are characteristic of the reaction. For example, the reaction in which A is converted to B is written as follows:

$$A \rightarrow B$$

The rate of this reaction is expressed algebraically as either a decrease in the concentration of reactant A:

$$-[A] = k[B]$$

or an increase in the concentration of product B:

$$[B] = k[A]$$

In the second equation (of the 3 above) the negative sign signifies a decrease in concentration of A as the reaction progresses, brackets define concentration in molarity and the k is known as a rate constant. Rate constants are simply proportionality constants that provide a quantitative connection between chemical concentrations and reaction rates. Each chemical reaction has characteristic values for its rate constants; these in turn directly relate to the equilibrium constant for that reaction. Thus, reaction can be rewritten as an equilibrium expression in order to show the relationship between reaction rates, rate constants and the equilibrium constant for this simple case. The rate constant for the forward reaction is defined as k_{+1} and the reverse as k_{-1}.

At equilibrium the rate (v) of the forward reaction ($A \rightarrow B$) is, by definition, equal to that of the reverse or back reaction ($B \rightarrow A$), a relationship which is algebraically symbolized as:

$$v_{forward} = v_{reverse}$$

where, for the forward reaction:

$$v_{forward} = k_{+1}[A]$$

and for the reverse reaction:

$$v_{reverse} = k_{-1}[B]$$

In the above equations, k_{+1} and k_{-1} represent rate constants for the forward and reverse reactions, respectively. The negative subscript refers only to a reverse reaction, not to an actual negative value for the constant. To put the relationships of the two equations into words, we state that the rate of the forward reaction [$v_{forward}$] is equal to the product of the forward rate constant k_{+1} and the molar concentration of A. The rate of the reverse reaction is equal to the product of the reverse rate constant k_{-1} and the molar concentration of B.

At equilibrium, the rate of the forward reaction is equal to the rate of the reverse reaction leading to theequilibrium constant of the reaction and is expressed by:

$$K_{eq} = \frac{K_{+1}}{K_{-1}} = \frac{[B]}{[A]}$$

This equation demonstrates that the equilibrium constant for a chemical reaction is not only equal to the equilibrium ratio of product and reactant concentrations, but is also equal to the ratio of the characteristic rate constants of the reaction.

EXAMPLES OF COUPLED REACTIONS IN BIOLOGY

Let's look at the hypothetical reaction: $A \rightarrow B$. If this is a thermodynamically unfavorable reaction the $\Delta G^{o'}$ value will be positive. Let's assume it is +4.0 kcal/mol. In order to drive this reaction in the direction written it can be coupled to the hydrolysis of ATP. The free energy of ATP hydrolysis to ADP is shown:

$$ATP + H_2O \rightarrow ADP + P_i: \Delta G^{o'} = -7.3 \text{ kcal/mol}$$

Coupling the two reactions together gives the equation:

$$A + ATP + H_2O \rightarrow B + ADP + P_i + H^+$$

The $\Delta G^{o'}$ for this coupled reaction is the sum of the $\Delta G^{o'}$ values of the two separate reactions, i.e. (-7.3kcal/mol) + (+4.0kcal/mol) = -3.3kcal/mol. This indicates that coupling ATP hydrolysis provides the energy necessary to make the conversion of A to B thermodynamically favorable. Another useful example is to examine one of the reactions of glycolysis. In this case we will look at the oxidation of phosphoenolpyruvate to pyruvate catalyzed by the enzyme pyruvate kinase (PK).

phosphoenolpyruvate $\rightarrow$ pyruvate: $\Delta G^{o'} = -14.7$ kcal/mol

This reaction releases sufficient energy to drive the synthesis of ATP from ADP which would normally be thermodynamically unfavorable with a ÄG$^{o'}$ of +7.3kcal/mol. Note that this value is the reciprocal of the hydrolysis of ATP. This points out another fact that the ÄG$^{o'}$ for a reaction in one direction is equal but mathematically opposite for the reciprocal direction. Coupling the two reactions together yields:

phosphoenolpyruvate + ADP + H^+ $\rightarrow$ pyruvate + ATP: $\Delta G^{o'} = -7.4$ kcal/mol

TEMPERATURE AFFECT A CHEMICAL REACTION

HAVING THE RIGHT CHEMISTRY

In a chemical reaction, two or more substances, called reactants, need to be mixed together to yield a product. A chemical reaction needs many things to work, the main thing being that the reactants come in close contact with each other to mix. Temperature is one factor that allows a chemical reaction to take place under the best conditions.

Heating Things Up

During a chemical reaction, if the reactants stay far away from each other, the reaction will never happen. Temperature itself is a measure of heat or energy. Needed for virtually everything, energy is crucial for a reaction because the more of it there is, the more likely the reaction will work.

Bringing It All Together

The reactants have to come into contact before the reaction takes place. In this instance, the energy needed for the reaction can be determined by the temperature. The higher the temperature, the more energy there is for the two reactants to contact each other for a successful chemical reaction.

EFFECTS OF TEMPERATURE

Kinetic theory says that molecules are in constant motion.The velocity of the molecules is directly proportional to the Kelvin temperature. The kinetic energy and molecule velocity increase with temperature.

KE = [1/2][mv2]

Reactions usually require collisions between reactant molecules or atoms. The formation of bonds requires atoms to come close to one another. New bonds can form only if the atoms are close enough together to share electrons. Some collisions are not successful. These are called ineffective collisions. The particles simply hit and then rebound. This animation illustrates what happens in an ineffective collision.

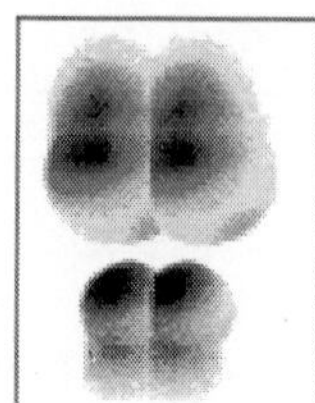

Collisions that lead to products are called effective collisions. An effective collision must happen with a great enough speed, energy and force to break bonds in the colliding molecules. The animation illustrates an effective collision between two diatomic molecules. The two product molecules formed fly outwards.

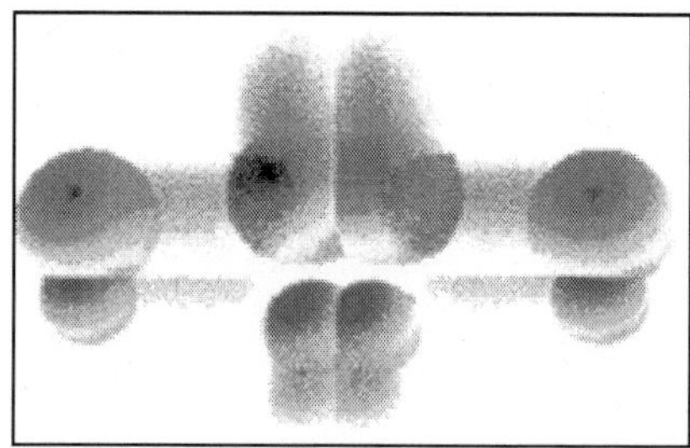

Collisions between molecules will be more violent at higher temperatures. The higher temperatures mean higher velocities. This means there will be less time between collisions. The frequency of collisions will increase. The increased number of collisions and the greater violence of collisions results in more effective collisions. The rate for the reaction increases. Reaction rates are roughly doubled when the temperature increases by 10 degrees Kelvin. This means the rate can be quadrupled if the temperature is raised by 20 degrees Kelvin.

It should be clear that if you can increase reaction rates by increasing temperature you can decrease reaction rates by lowering the temperature. You do this every time you put something in the refrigerator. If you want to see the effect of elevated temperatures increased reaction rates you can leave some dairy product out of the refrigerator for a few days and compare its condition with the same age dairy product that was kept cold.

This illustration shows what happens to an exothermic reaction when the temperature is changed. The dotted blue curve shows the energy for a reaction mixture that is heated. The reactants are"part way" up the energy barrier because they are"hot". The dotted magenta curve shows what cooling does to the reactant energy. The energy goes down and the reaction happens with more difficulty.

Note: The energies of reactants and products have changed. They both have different energies because they were either heated or cooled. The heat of reaction is the slightly different. The relative amounts of reactants and products are slightly different because of the temperature changes.

Concentration

Reaction rates can be increased if the concentration of reactants is raised. An increase in concentration produces more collisions. The chances of an effective collision goes up with the increase in concentration. The exact relationship between reaction rate and concentration depends on the reaction"mechanism". This is the process involving elementary reaction steps. The slowest step controls the rate. The nature of the slow step is not obvious from the balanced equation. Only experimental observation reveals the link between concentration and reaction rates.

Catalysts and Inhibitors

The reaction energy path controls the speed of the reaction. The molecules follow the path of least resistance, but this path may still require a lot of energy. The activation energy for the path may be high and then the reaction will be slow A reaction pathway can be altered by adding nonreacting compounds to the reaction mixture. These molecules can sometimes alter the pathway so the energy needed for reaction is lowered. When this happens the reaction rates are faster. A material that lowers the activation energy is called a catalyst. The four pictures show the effect of a catalyst on hydrogenation of ethylene. The CH_2CH_2 and H_2 molecules are attracted to the metal catalyst. (Graphic by permission of Prentice Hall)

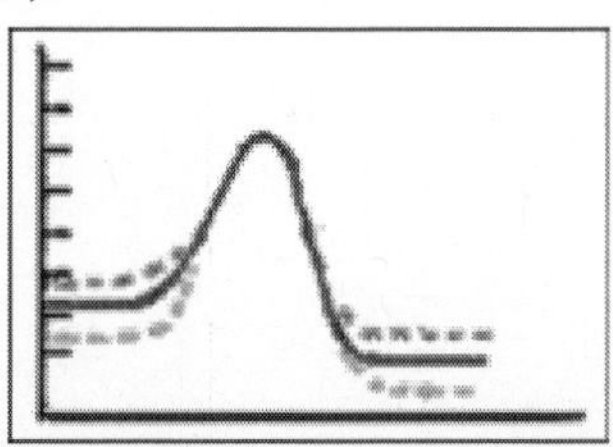

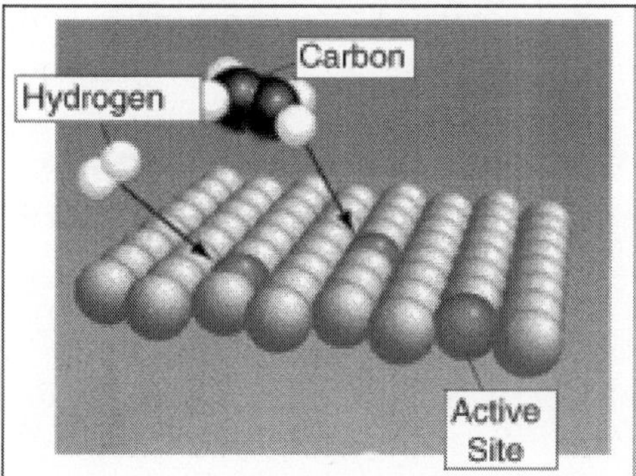

The covalent bonds are weakened because the metal atoms attract electrons away form the"bonded" atoms. The bonds break.

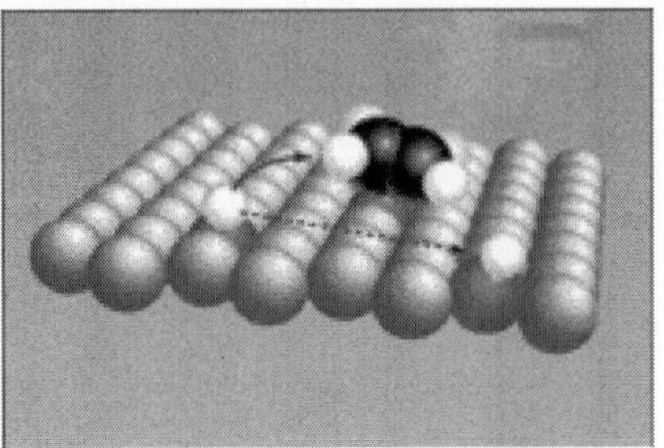

The free radicals from the original molecules move along the surface of the metal until they collide and form products.

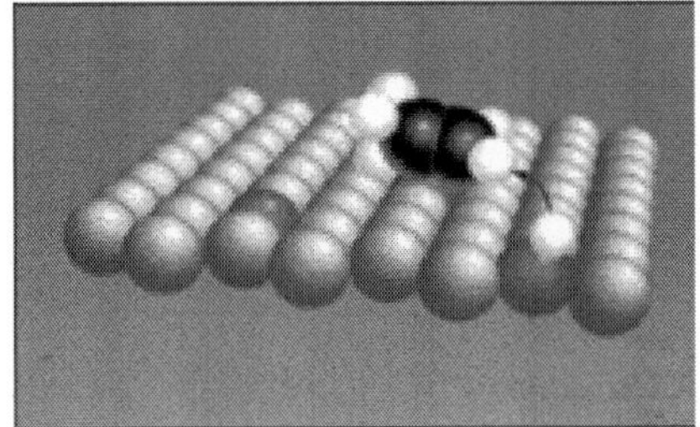

The ethane, CH_3CH_3 is not as"electron rich" as the ethylene, so it is not attracted strongly to the metal. The ethane breaks away from the catalyst.

The metal catalyst is not consumed in the reaction, but the mechanical action of flowing gases wears or erodes the metal away and the catalyst must be replenished.

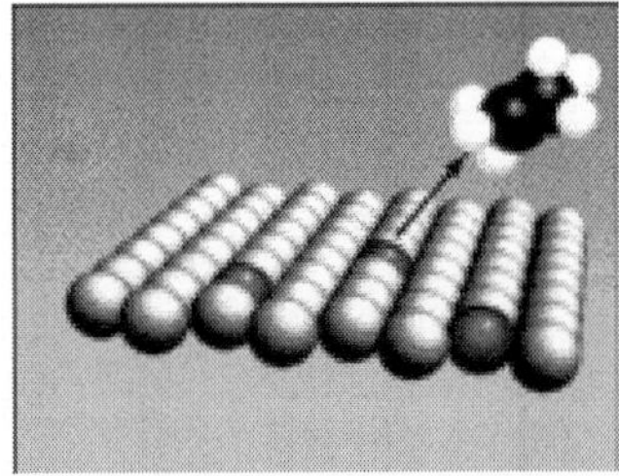

The catalyst is not consumed in the reaction. It merely speeds up the reaction by providing a lower energy path between reactants and products. Catalysts associate with reactant molecules (in biochemistry the reactant molecules are called substrates) and cause a redistribution of electron densities in the reacting molecules (substrates). The bonds that a need to be broken in the reaction are weakened by the association with the catalyst.

This makes the reaction occur faster because the weakened bonds are easier to break. When the substrate reacts the catalyst molecule is released and able to repeat the process with another reactant substrate molecule. Enzymes are biological catalysts.

Enzymes have a big molecular size. The enzyme is usually very large compared to the reacting substrate. Enzymes have folds and creases and the reactant molecules fit into a definite location in these folds and creases. The substrate molecule ties onto the enzyme molecule at a these"active sites". The term"lock and key" is applied to this interaction. The enzyme is the"lock" and the substrate is the"key". Problems occur in biological systems when the catalyst is poisoned. This can happen when the active site is"picked" or tied up by a molecule. This kind of thing sometimes happens when heavy metals react with enzymes. One reason why heavy metals are toxic is that they bind to important metabolic enzymes. This keeps the enzyme from carrying out its normal function and the organism suffers.

Inhibitors are the exact opposite of catalysts. They increase the energy of the reaction pathway. In fact they often block the normal pathway. This is desirable when you want to preserve molecules and slow decomposition reactions.

RATE LAWS

A rate law is an equation that tells us how fast the reaction proceeds and how the reaction rate depends on the concentrations of the chemical species involved. A rate law is an equation of the form,

$$v = f([A],[B],...[E]).$$

Equation is gives us a first order differential equation in t because the reaction velocity is related to a time-derivative of one of the concentrations.

The rate law may contain substances which are not in the balanced reaction and may not contain some things that are in the balanced equation (even on the reactant side).

Usually rate laws take the form,

$$v = k\,[A]^x[B]^y[E]^z...,$$

where x, y, z, are small whole numbers or simple fractions and k is called the "rate constant." The sum of $x + y + z + ...$ is called the "order" of the reaction.

FUNDAMENTALS OF RATE LAWS

DEFINING RATES

In order to introduce the idea of rate, let's suppose we wish to know how fast the following reaction is going:

$$H_2(g) + I_2(g) \rightarrow 2\ HI(g)$$

One way to do this is to define rate as the change in concentration of some species with respect to time, and then measure the concentrations of all species at multiple times to determine the rate. The results of such a hypothetical experiment is given in the for the reaction of hydrogen and iodine. The initial concentrations of H_2 and I_2 are equal at all times and the initial concentration of product is zero:

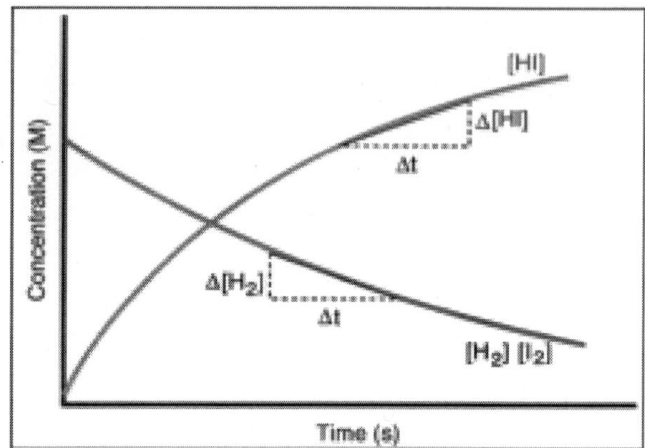

Fig. Graph of Concentration Versus Time for the Reaction of Hydrogen and Iodine

As you can see, the rate of formation of HI is twice the rate of disappearance of H_2 or I_2 at any given time. Also, note that the rate slows in time due to decreasing concentrations of the reactants. Stated mathematically, the relationship between the formation of products and the disappearance of reactants for this reaction is:

$$rate = \frac{\Delta[HI]}{\Delta t} = -\frac{1}{2}\frac{\Delta[H_2]}{\Delta t} = -\frac{1}{2}\frac{\Delta[I_2]}{\Delta t}$$

In general, for the reaction below:

$$aA + bB \rightarrow cC + dD$$

The rate is expressed as follows:

$$rate = -\frac{1}{a}\frac{\Delta[A]}{\Delta t} = -\frac{1}{b}\frac{\Delta[B]}{\Delta t} = \frac{1}{c}\frac{\Delta[C]}{\Delta t} = \frac{1}{d}\frac{\Delta[D]}{\Delta t}$$

Another expression for a rate is called the differential rate law, or simply, the rate law. It expresses the rate of a reaction in terms of the concentrations of the reactants raised to an experimentally determined power. The exponent on each concentration term is called the order of the reaction in that particular reactant. The sum of the exponents in the rate law is called the order of the reaction. *The powers on the concentration terms in a rate law are NOT the stoichiometric coefficients from the balanced equation*! For example, the rate law for the will have the following form:

$$\text{rate} = k\,[A]^p\,[B]^q$$

Note that the exponents are not a and b but some experimentally determined powers p and q which may or may not equal a and b. The order of the is, therefore, p + q. We will discuss in Determining Rate Laws how those exponents can be determined. Also notice in the the presence of the rate constant k. Students often have trouble distinguishing between the rate of a reaction and its rate constant. The rate of a reaction is the total rate of a reaction and is the "rate" in the rate law. The units of rate are always M/ s. The rate constant, k, is an experimentally determined proportionality constant that gives some measure of the intrinsic "reactivity" of the reaction. The units on the rate constant depend on the order of the reaction, n, such that k has units M^{1-n}/ s. In that way, the units of the rate constant, k, are chosen to make the units of rate always M/ s regardless of order. For example, if the order of the reaction is 3, the rate law could be rate = k $[A]^2[B]$. The units of rate are M/ s and the rate constant must, therefore, have units of M^{-2}/ s.

The attentive reader may have noticed that we have only considered the rate of the forward reaction, neglecting any sort of reverse reaction in the rate law. To make our math easier, we have intentionally ignored the reverse reaction and we will continue to do so. This is a justified practice for reactions with negligible reverse rates, such as those with equilibrium constants, K, much greater than 1. For reactions with K's around 1, this is also a valid approximation because they are usually carried out such that products are removed from the reaction mixture as they are formed, keeping concentrations of products low.

Rate laws can also be expressed to relate the concentration of reactants to the time of the reaction. Such an expression is called an integrated rate law because it is the integral of the differential rate law. For those of you who have had some calculus, we present a derivation of an integrated rate law from the differential rate law below for a first order reaction. If you have not had calculus, don't worry. Just pay attention to the end result of the derivation.

$$rate = k\left[A\right] = -\frac{d\left[A\right]}{dt}$$

Rearranging this equation gives the following:

$$\int_{[A]_o}^{[A]} \frac{d\left[A\right]}{[A} = -\int_o^t k\ dt$$

$$In\left[A\right] = -kt + In\left[A\right]_o$$

Using a similar technique, one can derive the integrated rate law for any differential rate law. As we will see in Determining Rate Laws, the fact that the 1st order integrated rate law is linear allows us to test if a reaction is first order by plotting ln [A] versus time.

DETERMINING THE RATE LAW

Kinetics Experiments

The goal of a kinetics experiment is to measure the concentration of a species at a particular time during a reaction so that a rate law can be determined. However, it is exceedingly difficult to get an accurate measurement of a concentration at a known time because the techniques used to measure concentrations don't work instantaneously, but rather take time to perform.

One of the earliest methods used to measure concentrations at specified times is to quench the reaction either by flash freezing it or by adding a substance that severely inhibits the reaction. Both of these techniques are problematic because one can't be sure that the reaction has completely stopped. The reaction may still be going on during the analysis. Additionally, the reaction mixture is destroyed for the purposes of kinetic experiments, so the chemist must make multiple trial runs and waste a large amount of reagents to observe the concentrations at multiple points in time.

A more modern technique to measure concentration is absorbance spectroscopy. This experiment may be used when a product or reactant has an absorbance frequency unique to those of other components of the reaction mixture.

By measuring the absorbance of a particular product or reactant at a variety of known concentrations, you can construct a plot of absorbance versus concentration called a Beer's Law plot. This calibration chart allows you to calculate the unknown concentration given the reaction solution's absorbance. The advantage of this method is that a large number of data points with well known times can be quickly collected using only one reaction mixture.

Method of Initial Rates

When looking at the expression for the, you should notice that the variables in the equation are the concentration terms and the powers p and q:

$$\text{rate} = k\,[A]^p\,[B]^q$$

Because we can measure the concentrations in the rate law using the techniques described above, the unknowns we wish to measure are k, p, and q. One method of directly measuring k, p, and q is called the method of initial rates. By measuring the initial rate (the rate near reaction time zero) for a series of reactions with varying concentrations, we can deduce to what power the rate depends on the concentration of each reagent. For example, let's use the method of initial rates to determine the rate law for the following reaction:

$$C_3H_6O + Br_2 \rightarrow C_3H_5OBr + HBr$$

whose rate law has the form:

$$\text{rate} = k\,[C_3H_6O]^p\,[Br_2]^q$$

Using the following initial rates data, it is possible to calculate the order of the reaction for both bromine and acetone:

Table. Initial Rates Data for the Bromination of Acetone

Experiment	$[Br_2]_o$	$[C_3H_6O]_o$	Initial Rate (M = s^{-1})
1	0.1 M	0.1 M	1.64×10^{-5}
2	0.2 M	0.1 M	1.65×10^{-5}
3	0.1 M	0.2 M	3.29×10^{-5}

To calculate the order of the reaction for bromine, notice that experiments 1 and 2 hold the concentration of acetone constant while doubling the concentration of bromine. The initial rate of the reaction is unaffected by the increase in bromine concentration, so the reaction is zero order in bromine. We can prove this mathematically by taking the ratio of the rates from experiments 1 and two:

$$\frac{rate_2}{rate_1} = \frac{k[Br_2]_2^{\,p}[C_3H_6O]_2^{\,q}}{k[Br_2]_1^{\,p}[C_3H_6O]_1^{\,q}} = \frac{(0.2M)^P (0.1M)^q}{(0.1M)^P (0.1M)^q} = \frac{1.64\times10^{-5}}{1.64\times10^{-5}} = 1$$

As you can see in the above equations, by holding the concentrations of all but one species constant between two experiments, you can calculate the order of the reaction in a single reactant at a time.

By similar reasoning, we can conclude that because the rate of reaction doubled when the concentration of acetone was doubled the reaction must be first order in acetone.

However, had the rate quadrupled or octupled with a doubling of the acetone concentration, the reaction would have been second or third order in acetone, respectively. In practice, you will likely never see a reaction with an order higher than 3. If you calculate an order higher than 3 for a reaction, double check your math because that is highly unusual. If you compute a fractional power for a reactant's order, do not be discouraged; they are quite common (especially half-order reactions).

To calculate the value of k, the rate constant, you simply plug into the rate law the values of the concentrations, the orders, and the rate of the reaction from any one of the three experiments. All three experiments should give a value of $1.64 \times 10^{-4}\ s^{-1}$. You should prove this to yourself.

Integrated Rate Law to Determine the Rate Law

One major disadvantage to using the method of initial rates is the need to perform multiple experiments. Another disadvantage is that it works only for relatively slow reactions.

If your reaction proceeds too fast, the rate that you measure will have a large range of uncertainty. To combat these problems, chemists have developed a method that uses data on concentration versus time during a reaction to infer the order of a reaction.

The strategy works by comparing concentration versus time information to the mathematical predictions made by integrated rate laws. This is usually accomplished by assuming that a reaction has a certain order and making a plot out of the data that should be linear if the assumption about the rate law is correct. If we guess correctly, the graph is linear. If we are wrong, then the graph is curved and we need to choose another order against which to plot our data.

To practice this method, we will first need to know the forms of the integrated rate laws for some common reaction orders. Luckily, only orders zero through two are common so we need to only consider reactions with these three orders. Due to the mathematical complexities of the problem, we will only consider rate laws of the form rate = $[A]^n$. However, with some experimental tricks I will explain below, that treatment does allow us to use the method of integrated rates for a reaction with any arbitrary rate law.

By integrating the differential rate law (the simple rate law from Heading), we derive the integrated rate law for a reaction. Without proof, we present below the forms of the integrated rate laws for reaction orders zero through two:

Table. Integrated Rate Laws for Reaction Orders Zero Through Two.

Order	Rate Law	Integrated Rate Law	Straight Line Plot
1	rate = k	$[A] = -kt + [A]_o$	[A] versus t
2	rate = k [A]	$\text{In }[A] = -kt + \text{In }[A]_o$	In [A] versus t
3	rate = k $[A]^2$	$1/[A] = 2kt + 1/[A]_o$	1/[A] versus t

As you can see, each order of reaction has a unique input and output variable that produces a straight line. For example, if we graph the following rate data for the decomposition of H_2O_2 assuming that it could be zero, first, or second order, we find that only the graph for a 2nd order reaction (1/[A] versus t) gives a straight line. Therefore, the reaction has the rate law rate = k $[H_2O_2]^2$.

Table per cent. Rate Data for the Decomposition of Hydrogen Peroxide.

$[H_2O_2]$	t(min)
0.300 M	0
0.221 M	1
0.107 M	5
0.065 M	10
0.047 M	15
0.037 M	20

Assuming that we did not know that the decomposition is second order, we will make a series of three graphs to determine the order of the reaction:

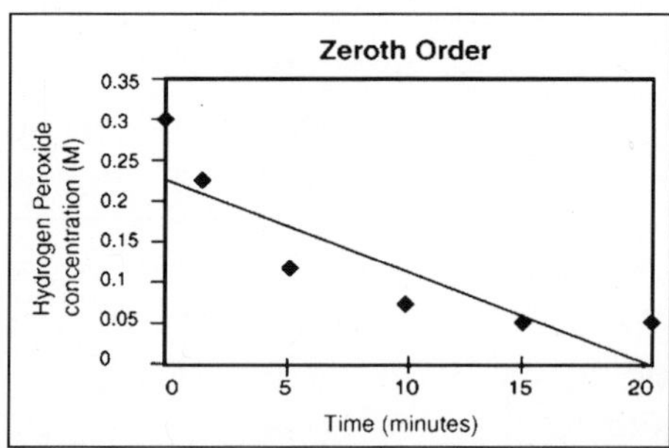

Fig. Plot of Hydrogen Peroxide Concentration versus Time

If the graph was linear, we would conclude that the rate law is zero Order, but it is not. Therefore, let's see if it is first order by plotting ln $[H_2O_2]$ versus time in the following graph:

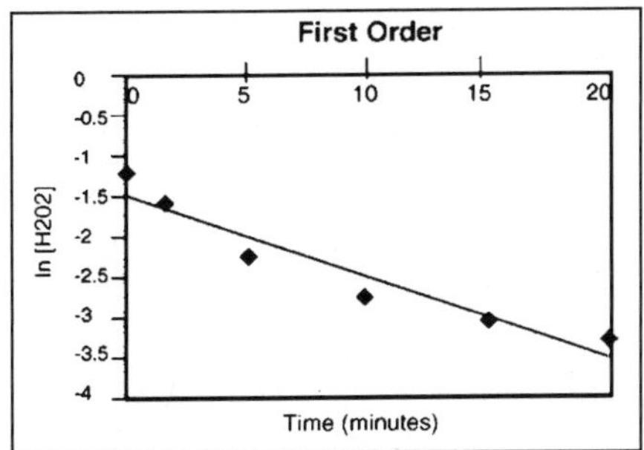

Fig. Plot of the Natural Log of Hydrogen Peroxide Concentration versus Time

Seeing that the reaction is not first order due to the non-linearity of the above graph, we move to plotting $1/[H_2O_2]$ versus time to test whether the decomposition reaction is second order.

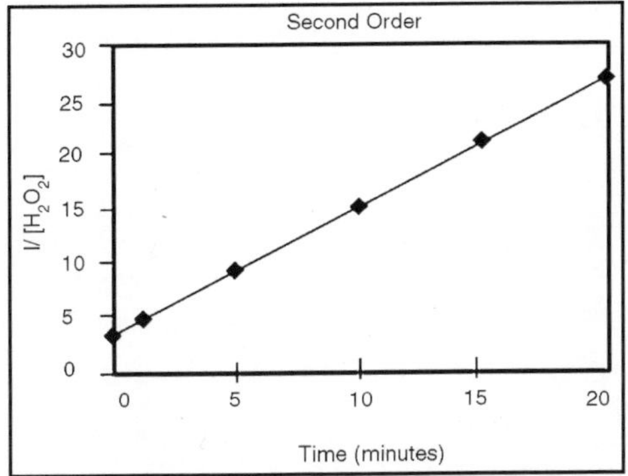

Fig. Plot of Inverse Hydrogen Peroxide Concentration versus Time

Because the above graph is linear, we know that the reaction is second order. The slope of the line is twice the rate constant, k from the rate law.

What you should learn from the above discussion is that you can use integrated rate laws to determine both the rate constant for a reaction and the form of the rate law.

It may seem more complicated to use integrated rate laws rather than the method of initial rates to determine the rate law, but it really does take much longer to perform and analyze the several reactions needed for the method of initial rates than it does to produce the necessary graphs for the integrated rate law method—especially with a good graphing programmes.

You may have noticed that we only list three integrated rate laws, ignoring rate laws like rate = k [A] [B]. That doesn't mean we can't use integrated rate laws to determine the rate law for those types of reactions. We simply must be more clever about how we do it. For a two- component second order reaction with a rate law rate = k [A] [B], we can make the concentration of B so large as compared to A that the concentration of B is almost constant. Assuming that the concentration of B is constant, the reaction becomes pseudo-first order—that is, the reaction will behave as if it was first order. The kinetics data for this reaction will give a graph of ln [A] versus time that is linear, telling us that the reaction is first order in A.

If the reaction happened to be second order in A, in the present example, then a graph of 1/[A] versus time would be linear. We can, similarly, determine the order of B by making the concentration of A large. For multiple component rate laws, you can simply make the concentration of all reagents but the one of interest large to iteratively determine the order of the reaction in each component. (Chemists are fond of using the word 'iterative' to say 'repetitive'.)

Half-Lives

An earlier and less common way to measure rate is by the half-life of a reaction. A half-life is the time it takes for one half of the starting material to be transformed into its products. Often you will hear half-life associated with radioactive decay phenomena (which follow first order kinetics), but the term can be applied to any reaction.

The half-life of a reaction not only depends on the rate constant of the reaction (those with larger k's have shorter half-lives) but also on the integrated rate law for the reaction. To derive the form of the half-life expression for a first order reaction, we start with its integrated rate law, and then substitute the value 0.5 for the ratio of [A] to $[A]_o$:

$$In\left[A\right] = -kt + In\left[A\right]_o \quad or \quad In\left(\frac{\left[A\right]}{\left[A\right]_o}\right) = -kt$$

$$In 0.5 = -kt_{1/2} \quad so \quad t_{1/2} - \frac{In 2}{k}$$

Using the same techniques with the different integrated rate laws, you can derive the half-life expression for a reaction of any order. Summarized below are the half-lives for reactions of orders zero through two.

Table. Half-lives of Reactions with Orders Zero Through Two

Order	Half-Life
0	$[A]_o$ / 2k
1	In 2/k
2	1/2k $[A]_o$

COMMON TYPES OF RATE LAWS

First Order Reactions

In a first order reaction the rate is proportional to the concentration of one of the reactants. That is,

$$v = \text{rate} = k[\text{B}],$$

where B is a reactant. If we have a reaction which is known to be first order in B, such as

B + other reactants → products,

we would write the rate law as,

$$-\frac{d[B]}{dt} = k[B]$$

The constant, k, in this rate equation is the first order rate constant.

Second Order Reactions

In a second order reaction the rate is proportional to concentration squared. For example, possible second order rate laws might be written as

$$\text{Rate} = k[\text{B}]^2$$

or as,

$$\text{Rate} = k[\text{A}][\text{B}].$$

That is, the rate might be proportional to the square of the concentration of one of the reactants, or it might be proportional to the product of two different concentrations.

Third Order Reactions

There are several different ways to write a rate law for a third order reaction. One might have cases where,

$$\text{Rate} = k[\text{A}]^3,$$

or,

$$\text{Rate} = k[\text{A}]^2[\text{B}],$$

or,

$$\text{Rate} = k[\text{A}][\text{B}][\text{C}],$$

and so on.

INTEGRATED FORMS OF RATE LAWS

In order to understand how the concentrations of the species in a chemical reaction change with time it is necessary to integrate the rate law (which is given as the time-derivative of one of the concentrations) to find out how the concentrations change over time.

First Order Reactions

Suppose we have a first order reaction of the form,

B +.... → products.

Then we can write the rate law and integrate it as follows (recall that the derivative is negative because the concentration of the reactant, B, is decreasing):

$$-\frac{d[B]}{dt} = -k[B]$$

$$\frac{d[B]}{[B]} = -k\,dt$$

$$In\frac{[B]}{[B]_o} = -kt$$

$$[B] = [B]_o e^{-kt}$$

The first order rate law is a very important rate law, radioactive decay and many chemical reactions follow this rate law and some of the language of kinetics comes from this law. This form appears in many places in nature. One of its consequences is that it gives rise to a concept called "half-life."

Half-life

The half-life, usually symbolized by $t_{1/2}$, is the time required for [B] to drop from its initial value $[B]_o$ to $[B]_o/2$.

Using the integrated form of the first order rate law we find that,

$$\frac{[B]_o}{2} = [B]_o e^{-kt_{1/2}},$$

or,

$$\frac{1}{2} = e^{-kt_{1/2}}$$

Taking the logarithm of both sides gives,

$$In\frac{1}{2} = -kt_{1/2}$$

$$In\,2 = kt_{1/2},$$

or,

$$t_{1/2} = \frac{In\,2}{k}$$

(You can also write,

$$[B] = [B]_o 2^{-\frac{1}{t_{1/2}}} = \frac{[B]_o}{2^{\frac{1}{t_{1/2}}}},$$

which may actually give a little more insight into what is meant by half-life. This equation demonstrates clearly that the concentration drops by a factor of two for every $t_{1/2}$ increment in time.) For first order processes it is common to define a "relaxation time." τ, by,

$$\tau = \frac{1}{k},$$

so that one can write the integrated form of the rate law as,

$$[B] = [B]_o e^{-t/\tau}$$

τ is the time required for [B] to drop from $[B]_0$ to $[B]_0/e$. Sometimes ô is called the "one over *e*" time. Although the half-life is almost always used to describe the decay rate of radioactive elements, it is common for chemists to talk about the rate of first order processes in chemistry in terms of the relaxation time.

RATE LAWS AND REACTION ORDER

The relation between the rate of a reaction and the concentrations of reactants is expressed by its *rate law*. For example, the rate of the gas-phase decomposition of dinitrogen pentoxide,

$$2N_2O_5 \rightarrow 4NO_2 + O_2$$

has been found to be directly proportional to the concentration of N_2O_5:

$$rate = k\,[N_2O_5]$$

More generally, for a reaction of the form,

$$n_A\,A + n_B\,B + ... \rightarrow products$$

the rate law will be in which the exponents *a* and *b* are usually (but not always) integers and, we must emphasize once again, bear no relation to the coefficients n_A, n_B.

Since the rate of a reaction has the dimensions of (concentration/time), the dimensions of the rate constant *k* will depend on the exponents of the concentration terms in the rate law. To make this work out properly, if we let *p* be the sum of the exponents of the concentration terms in the rate law,

$$p = a + b + ...$$

then *k* will have the dimensions (concentration^{1-p}/time).

REACTION ORDER

The *order* of a rate law is the sum of the exponents in its concentration terms. For the N_2O_5 decomposition with the rate law $k[N_2O_5]$, this exponent is 1 (and thus is not explicitly shown); this reaction is therefore a first order reaction. We can also say that the reaction is "first order in N_2O_5".

For more complicated rate laws, we can speak of the overall reaction order and also the orders with respect to each component. As an example, consider a reaction,

$$A + 3B + 2C \rightarrow \text{products}$$

whose experimental rate law is,

$$rate = k[A]\,[B]^2$$

We would describe this reaction as third-order overall, first-order in A, second-order in B, and zero-order in C.

Observing Rate-vs.-concentration Proportionality

In order to determine the value of the exponent in a rate equation term, we need to see how the rate varies with the concentration of the substance. For a single-reactant decomposition reaction of the form,

$$A \rightarrow \text{products}$$

in which the rate is $-d[A]/dt$, we simply plot [A] as function of time, draw tangents at various intervals, and see how the slopes of these tangents (the instantaneous rates) depend on [A].

- If doubling the concentration of A doubles the rate, then the reaction is first-order in A.
- If doubling the concentration results in a four-fold rate increase, the reaction is second-order in A.

Initial Rate Method

When there is more than one reactant, the method described above is rarely practical, since the concentrations of the different reactants will generally fall at different rates, depending on the stoichiometry. Instead, we measure only the rate near the beginning of the reaction, before the concentrations have had time to change significantly. The experiment is then repeated with a different starting concentration of the reactant in question, but keeping the concentrations of any others the same. After the order with respect to one component is found, another series of trials is conducted in which the order of another component is found.

Dealing with Multiple Reactants: The Isolation Method

It is not always practical to determine orders of two or more reactants by the method illustrated in the preceding example. Fortunately, there is another way to accomplish the same task: we can use excess concentrations of all the reactants except the one we wish to investigate. For example, suppose the reaction is,

$$A + B + C \rightarrow \text{products}$$

and we need to find the order with respect to [B] in the rate law. If we set $[B]_0$ to 0.020 M and let $[A]_0 = [C]_0 = 2.00$M, then if the reaction goes to completion, the change in [A] and [C] will also be 0.020 M which is only 1 percent of their original values. This will often be smaller than the experimental error in determining the rates, so it can be neglected. By "flooding" the reaction mixture with one or more reactants, we are effectively *isolating* the one in which we are interested.

THE RATES OF CHEMICAL REACTIONS

Chemical reactions vary greatly in the speed at which they occur. Some are essentially instantaneous, while others may take years to reach equilibrium.

DEFINITION OF REACTION RATE

The speed of a chemical reaction may be defined as the change in concentration of a substance divided by the time interval during which this change is observed: .

$$rate = \frac{\Delta(concentration)}{\Delta(time)}$$

For a reaction of the form A + B → C, the rate can be expressed in terms of the change in concentration of any of its components,

$$rate = -\frac{\Delta[A]}{\Delta t}$$

$$rate = -\frac{\Delta[B]}{\Delta t}$$

$$rate = \frac{\Delta[C]}{\Delta t}$$

in which Δ[A] is the difference between the concentration of A over the time interval $t_2 - t_1$:

$$\Delta[A] = [A]_2 - [A]_1$$

Notice the minus signs in the first two examples above. The concentration of a reactant always decreases with time, so Δ[A] and Δ[B] are both negative. Since negative rates don't make much sense, rates expressed in terms of a reactant concentration are always preceded by a minus sign in order to make the rate come out positive.

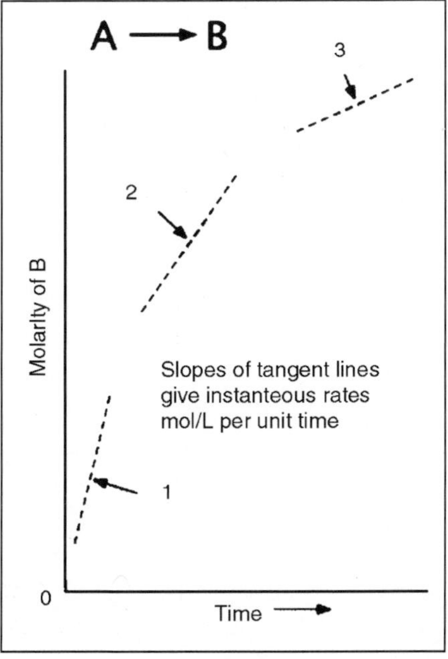

Consider now a reaction in which the coefficients are different:

$$A + 3B \rightarrow 2D$$

It is clear that [B] decreases three times as rapidly as [A], so in order to avoid ambiguity when expressing the rate in terms of different components, it is customary to divide each change in concentration by the appropriate coefficient:

$$rate = \frac{-[A]}{\Delta t} = \frac{-[B]}{3\Delta t} = \frac{[D]}{2\Delta t}$$

Instantaneous Rates

Most reactions slow down as the reactants are consumed. Consequently, the rates given by the expressions shown above tend to lose their meaning when measured over longer time intervals Δt.

Note: Instantaneous rates are also known as differential rates.

Thus for the reaction whose progress is plotted here, the actual rate (as measured by the increasing concentration of product) varies continuously, being greatest at time zero. The *instantaneous rate* of a reaction is given by the slope of a tangent to the concentration-vs.-time curve. Three such rates have been identified in this plot.

An instantaneous rate taken near the beginning of the reaction ($t = 0$) is known as an *initial rate* (label (1)here). As we shall soon see, initial rates play an important role in the study of reaction kinetics.

RATE LAW AND EQUATION

For nearly all forward, irreversible reactions, the rate is proportional to the product of the concentrations of the reactants, each raised to some power. For the general reaction:

$$aA + bB \rightarrow cC + dD$$

The rate is proportional to $[A]^m[B]^n$ that is:

$$\text{rate} = k[A]^m[B]^n$$

This expression is the rate law for the general reaction above, where k is the rate constant. Multiplying the units of k by the concentration factors raised to the appropriate powers give the rate in units of concentration/time.

The dependence of the rate of reaction on the concentrations can often be expressed as a direct proportionality in which the concentrations may appear to be the zero, first, or second power. The power to which the concentration of a substance appears in the rate law is the order of the reaction with respect to that substance. In the reaction above the order of reaction is:

$$m + n$$

The order of the chemical equation can only be determined by experiment. In other words, one cannot determine what m and n are by just looking at a balanced chemical equation; m and n must be determined by the use of data.

The overall order of a reaction is the sum of the orders with respect to the sum of the exponents. Furthermore, the order of a reaction is stated with respect to a named substance in the reaction. The exponents in the rate law are not equal to the stoichiometric coefficients unless the reaction actually occurs via a single step mechanism. However, the exponents are equal to the stoichiometric coefficients of the rate-determining step. In general, the rate law can calculate the rate of reaction from known concentrations for reactants and derive an equation that expresses a reactant as a function of time.

The proportionality factor, k, called the rate constant is a constant at a fixed temperature. Nonetheless, the rate constant varies with temperature. There are dimensions to k and can be determined with simple dimensional analysis of the particular rate law. The units should be expressed when the k-values are tabulated. The higher the k value, the faster the reaction proceeds.

EXPERIMENTAL DETERMINATION OF RATE LAW

The values of k, x, and y in the rate law equation ($r = [A]^m[B]^n$) must be determined experimentally for a given reaction at a given temperature. The rate is usually measured as a function of the initial concentrations of the reactants, A and B.

Order of Reactions

Chemical reactions are often classified on the basis of kinetics as zero-order, first-order, second-order, mixed order, or higher-order reactions. The general reaction $aA + bB \rightarrow cC + dD$ will be used in the discussion next.

First lets note what each of these orders means in terms of initial rate of reaction effect:

- Zero-order in the reactant—there is no effect on the initial rate of reaction
- First-order in the reactant—the initial rate of reaction doubles
- Second order in the reactant—the initial rate of the reaction quadruples
- Third order in the reactant—the initial rate of reaction increases eightfold

Zero-Order Reactions

A zero-order reaction has a constant rate, which is independent of the reactant's concentrations. Thus the rate law is:

$$\text{rate} = k = \text{constant}$$

where k has the units of $M(\text{sec}^{-1})$. In other words, a zero-order reaction has a rate law in which the sum of the exponents is equal to zero. An increase in temperature or a decrease in in temperature is the only factor that can change the rate of a zero-order reaction. In addition, a reaction is zero order if concentration data are plotted versus time and the result is a straight line. The slope of this resulting line is the negative of the zero order rate constant k. At times, chemists and researchers are also concerned with the relationship between the concentration of a reactant and time. Such expression is called the integrated rate law in which the equation expresses the concentration of a reactant as a function of time (remember, each order of reaction has its own unique integrated rate law). The integrated rate law of a zero-order reaction is:

$[A_t] = -kt + [A_0]$ (See page on zero-order reactions to see how this is derived)

Notice, however, that this model cannot be entirely accurate since this equation predicts negative concentrations at sufficiently large times. In other words, if one were to graph the concentration of A as a function of time, at some point, the line will cross below 0. This is of course, physically impossible since concentrations cannot be negative. Nevertheless, this model is a sufficient model for ranges of time where concentration is predicted as greater than zero.

The half life ($t_{1/2}$) of a reaction is the time needed for the concentration of the radioactive substance to decrease to one-half of its original value. The half-life of a zero-order reaction can be derived as follows:

Given a reaction involving reactant A and from the definition of a half-life, we know that $t_{1/2}$ is the time it takes for half of the initial concentration of reactant A to react. So we can now substitute new conditions into the integrated rate law form to obtain:

$$\left[\frac{A_0}{2}\right] = -kt + \left[A_0\right]$$

We now solve for $t_{1/2}$ to obtain the following:

$$t_{1/2} = \frac{\left[A\right]_0}{2k}$$

First-Order Reactions

A first-order reaction has a rate proportional to the concentration of one reactant.

$$\text{rate} = k[A] \text{ or rate} = k[B]$$

First-order rate constants have units of sec^{-1}. In other words, a first-order reaction has a rate law in which the sum of the exponents is equal to 1.

The integrated rate law of a first-order reactions is:

$$\ln[A]_t = -kt + \ln[A]_0$$

or,

$$\ln([A]_t/[A]_0) = -kt$$

or,

$$\left[A\right] = \left[A\right]_0 e^{-kt}$$

Moreover, a first-order reaction can be determined by plotting a graph of ln[*A*] vs. time *t* and a straight line is produced with a negative slope of k.

The classic example of a first-order reaction is the process of radioactive decay. The concentration of radioactive substance A at any time t an be expressed mathematically as:

$$[A_t] = [A_0]e^{-kt}$$

where $[A_0]$ = initial concentration of A

$[A_t]$ = concentration of A at time t

k = rate constant

t = elapsed time

The half-life of a first order reaction can be calculated in a similar fashion as with the half-life of the zero order reaction and one would obtain the following:

$$t_{1/2} = \frac{In(2)}{k}$$

where k is the first order rate constant. Notice that the half-life associated with the first-order reaction is the only case where half-life is independent of concentration of a reactant or product. In other words, [A] does not appear in the half-life formula above.

Second-Order Reactions

A second-order reaction has a rate proportional to the product of the concentration of two reactants, or to the square of the concentration of a single reactant. For example:

$$\text{rate} = k[A]^2$$
$$\text{rate} = k[B]^2$$
$$\text{rate} = k[A][B]$$

are all second-order reactions. Therefore, a second-order reaction has rate law in which the sum of the exponents are equal to 2.

The integrated rate law of a second-order reaction is as follows:

$$\frac{1}{[A]} = \frac{1}{[A]_0} + kt$$

(See page on second-order reactions to see how this is derived)

<meta charset=”utf-8"/>

The half-life of a second-order reaction is:

$$t_{\frac{1}{2}} = \frac{1}{k[A]_0}$$

Determining Reaction Rate

In the laboratory, one may collect a sample of data consisting of measured concentrations of a certain reactant A at different times. This sample data may look like the following (Sample data obtain from ChemElements Post-Laboratory Exercises):

Time (min)	Concentration (M)
0	0.906
0.9184	0.8739
9.0875	0.5622
11.2485	0.5156
17.5255	0.3718
23.9993	0.2702
27.7949	0.2238
31.9783	0.1761
35.2118	0.1495
42.973	0.1029
46.6555	0.086
50.3922	0.0697
55.4747	0.0546
61.827	0.0393
65.6603	0.0324
70.0939	0.026

One can then plot [A] versus time, ln[A] versus time, and 1/[A] versus time to see which plot yields a straight line. The reaction order will then be the order associated with the plot that gives a straight line. While it may seem that doing this seems tedious and difficult, the process becomes quite simple with the use of Excel, or any other similar programmes.

By utilizing the formula capabilities of Excel, we can obtain two more data tables of ln[A] vs. time and 1/[A] vs. time very easily.

Time (min)	ln([A])
0	
0.9184	-0.085122253
9.0875	2.206899843
11.2485	2.420234786
17.5255	2.863656963
23.9993	3.178024663
27.7949	3.324852551
31.9783	3.465057548
35.2118	3.561381254
42.973	3.760572012
46.6555	3.84279082
50.3922	3.919836401
55.4747	4.015927061
61.827	4.124340162
65.6603	4.184494481
70.0939	4.249835772

And

Time (min)	1/[A]
0	
0.9184	1.088850174
9.0875	0.110041265
11.2485	0.088900742
17.5255	0.057059713
23.9993	0.041667882
27.7949	0.035977823
31.9783	0.031271206

35.2118	0.028399571
42.973	0.023270426
46.6555	0.0214337
50.3922	0.019844341
55.4747	0.018026235
61.827	0.016174163
65.6603	0.015229903
70.0939	0.014266577

We now plot the three data sets to get

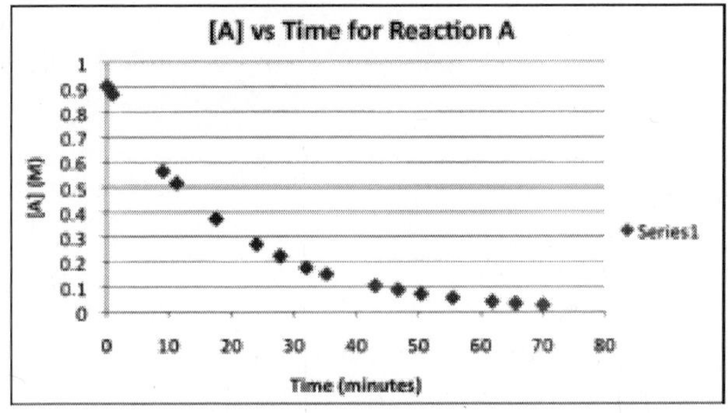

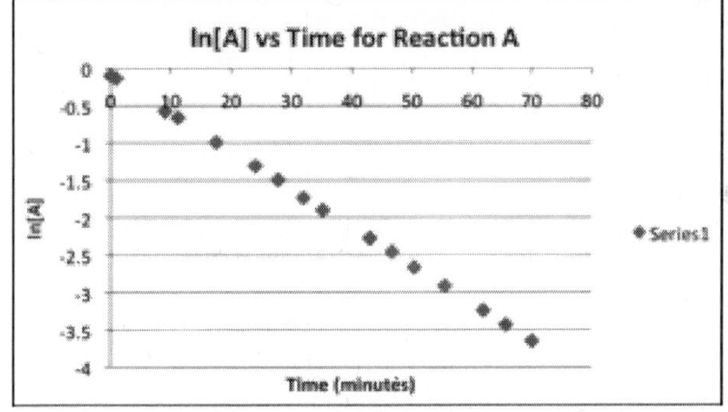

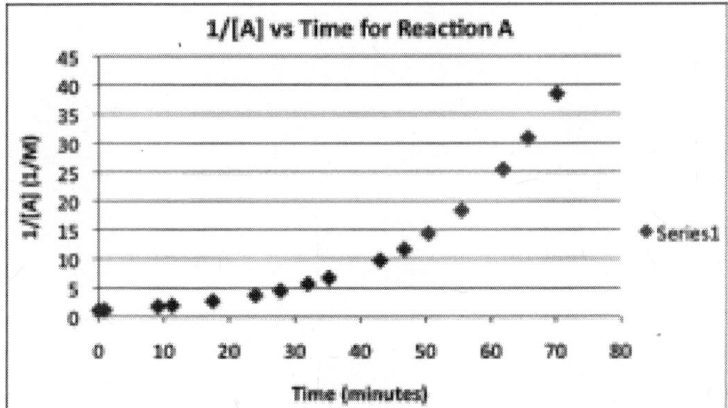

We can see clearly that the graph of ln[A] vs time is a straight line. Therefore the reaction associated with the given data is a first order reaction.

ORDERS OF REACTION AND RATE EQUATIONS

RATE EQUATIONS

There are several simple ways of measuring a reaction rate. For example, if a gas was being given off during a reaction, you could take some measurements and work out the volume being given off per second at any particular time during the reaction.

A rate of 2 $cm^3 s^{-1}$ is obviously twice as fast as one of 1 $cm^3 s^{-1}$.

However, for this more formal and mathematical look at rates of reaction, the rate is usually measured by looking at how fast the*concentration of one of the reactants* is falling at any one time.

For example, suppose you had a reaction between two substances A and B. Assume that at least one of them is in a form where it is sensible to measure its concentration - for example, in solution or as a gas.

$$A + B \rightarrow \text{products}$$

For this reaction you could measure the rate of the reaction by finding out how fast the concentration of, say, A was falling per second.

You might, for example, find that at the beginning of the reaction, its concentration was falling at a rate of 0.0040 mol $dm^{-3} s^{-1}$. This means that every second the concentration of A was falling by 0.0040 moles per cubic decimetre. This rate will decrease during the reaction as A gets used up.

Orders of Reaction

Orders of reaction are always found by doing experiments. You can't deduce anything about the order of a reaction just by looking at the equation for the reaction. So let's suppose that you have done some experiments to find out what happens to the rate of a reaction as the concentration of one of the reactants, A, changes. Some of the simple things that you might find are:

One Possibility: The Rate of Reaction is Proportional to the Concentration of A

That means that if you double the concentration of A, the rate doubles as well. If you increase the concentration of A by a factor of 4, the rate goes up 4 times as well.

You can express this using symbols as:

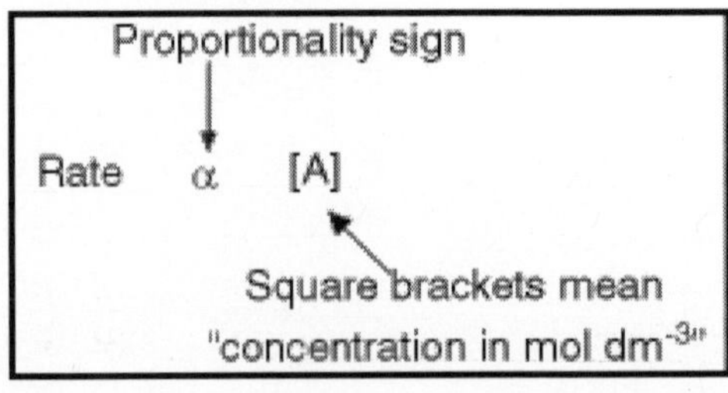

Writing a formula in square brackets is a standard way of showing a concentration measured in moles per cubic decimetre (litre).

You can also write this by getting rid of the proportionality sign and introducing a constant, k.

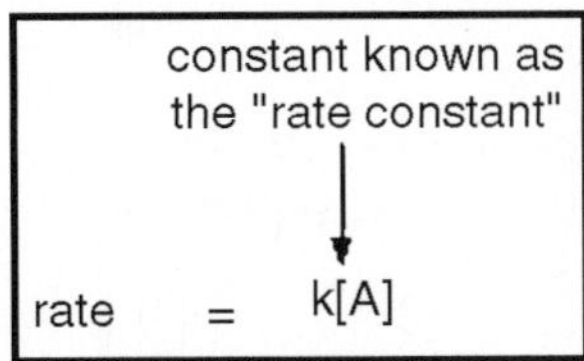

Another Possibility: The Rate of Reaction is Proportional to the Square of the Concentration of A

This means that if you doubled the concentration of A, the rate would go up 4 times (2^2). If you tripled the concentration of A, the rate would increase 9 times (3^2). In symbol terms:

$$\text{Rate} \propto [A]^2$$
$$\text{Rate} = k[A]^2$$

Generalising This

By doing experiments involving a reaction between A and B, you would find that the rate of the reaction was related to the concentrations of A and B in this way:

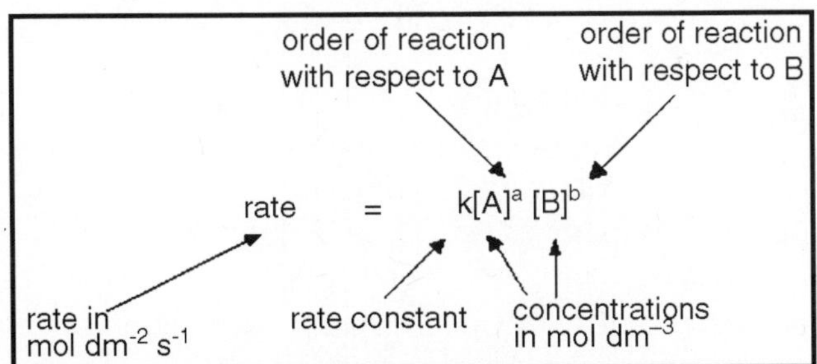

This is called the *rate equation* for the reaction.

The concentrations of A and B have to be raised to some power to show how they affect the rate of the reaction. These powers are called the *orders of reaction* with respect to A and B.

For UK A' level purposes, the orders of reaction you are likely to meet will be 0, 1 or 2. But other values are possible including fractional ones like 1.53, for example. If the order of reaction with respect to A is 0 (zero), this means that the concentration of A doesn't affect the rate of reaction. Mathematically, any number raised to the power of zero (x^0) is equal to 1. That means that that particular term disappears from the rate equation.

The *overall order of the reaction* is found by adding up the individual orders. For example, if the reaction is first order with respect to both A and B (a = 1 and b = 1), the overall order is 2. We call this an overall second order reaction.

The Rate Constant

Surprisingly, the rate constant isn't actually a true constant! It varies, for example, if you change the temperature of the reaction, add a catalyst, or change the catalyst.

RELATION TO RATE LAW

For the reaction:

$$\alpha A + \beta B \rightarrow P$$

The rate law reads:

$$\text{rate}=k[A]^x[B]^y$$

where [A] is the concentration of species A, x is the order with respect to species A. [B] is the concentration of species B, y is the order with respect to species B, and k is the rate constant.

n, is the reaction order for the whole chemical reaction. This can be found by adding the reaction orders with respect to the reactants. In this case, n = x + y.

Simple Rules

The order of a reaction does NOT need to be an integer. The following orders are possible:

- *Zero:* A zero order tells you that the concentration of that species does NOT affect the rate of a reaction
- *Negative integer:* A negative order tells you that the concentration of that species Inversely affects the rate of a reaction
- *Positive integer:* A positive order tells you that the concentration of that species Directly affects the rate of a reaction
- *Non-Integer:* Non-integer orders, both positive and negative, represent more intricate relationships between concentrations and rate in more complex reactions.

Methods to Determining Reaction Order

For chemical reactions that require only one elementary step, the values of x and y are equal to the stoichiometric coefficients in front of the reactant. For other chemical reactions that require more than one elementary state, this may not always be the case. In saying this, there are many simple ways of determining the order of a reaction. One very popular method is known as the differential method.

The Differential Method

The Differential Method, also known as the Initial Rates Method, uses an experimental data table to determine the order of a reaction with respect to the reactions that were used. Below is an example of a table that can be looked at from the following chemical reaction:

$$A + B \rightarrow P$$

Experiment	[A] M	[B] M	Rate M Min^{-1}
1	0.100	0.100	1.0×10^{-3}
2	0.200	0.100	$1.0 \text{ X } 10^{-3}$
3	0.100	0.200	2.0×10^{-3}

When looking at the experiments in the table above, what needs to be noticed are factors that are changing in each experiment. So, in order to determine the reaction order with respect to A, you need to find where A is changing. That is between experiments 1 and 2. Write a rate law equation based on the chemical reaction above (can be seen below)

$$\text{Rate} = k[A]^x[B]^y$$

Once this is done, you must divide the rate law equation from experiment 2 by the rate law equation for experiment 1. You will notice that the $[B]^y$ will cancel out and you are left with "x" being the unknown variable. By doing simple algebra, you can determine through the steps that $x = 0$.

The same steps must be taken for determining the reaction order with respect to B. Instead, however, you will use experiment 1 and 3 for this process. After working through the problem and canceling out $[A]^x$ from the equation, you should find that $y = 1$.

Finding the reaction order for the whole process is the easy addition of x and y, $n = 0 + 1$. Therefore $n = 1$

After finding your reaction order, several pieces of information can be solved for, such as half-life for example.

4

Reaction Rates and Collision Theory

It should be clear now that the rate of a reaction varies depending on a number of factors. But how can we *explain* why reactions take place at different speeds under different conditions? Why, for example, does an increase in the surface area of the reactants also increase the rate of the reaction? One way to explain this is to use collision theory.

For a reaction to occur, the particles that are reacting must collide with one another. Only a fraction of all the collisions that take place actually cause a chemical change. These are called 'successful' collisions. When there is an increase in the *concentration* of reactants, the chance that reactant particles will collide with each other also increases because there are more particles in that space. In other words, the *collision frequency* of the reactants increases. The number of *successful* collisions will therefore also increase, and so will the rate of the reaction. In the same way, if the *surface area* of the reactants increases, there is also a greater chance that successful collisions will occur.

When the *temperature* of the reaction increases, the average kinetic energy of the reactant particles increases and they will move around much more actively. They are therefore more likely to collide with one another. Increasing the temperature also increases the number of particles whose energy will be greater than the activation energy for the reaction.

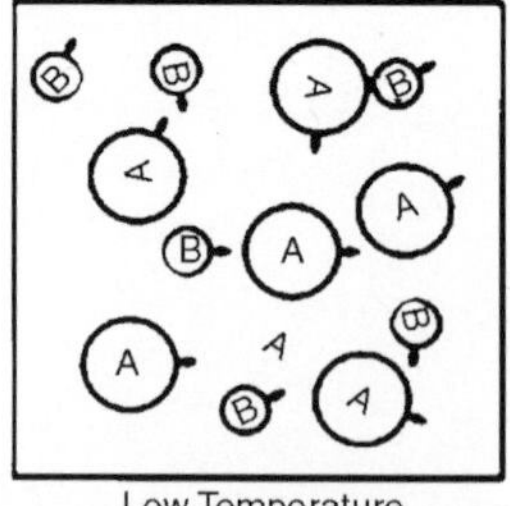
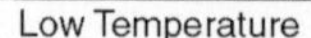

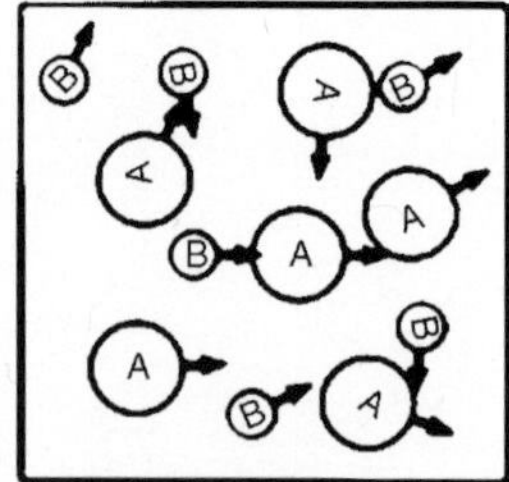
High Temperature

Fig. An Increase in the Temperature of a Reaction Increases the Chances that the Reactant Particles (A and B) will Collide because the Particles have more Energy and Move Around More.

MEASURING RATES OF REACTION

How the rate of a reaction is measured will depend on what the reaction is, and what product forms. Look back to the reactions that have been discussed so far. In each case, how was the rate of the reaction measured? The following examples will give you some ideas about other ways to measure the rate of a reaction:

- *Reactions that produce hydrogen gas:*

When a metal dissolves in an acid, hydrogen gas is produced. A lit splint can be used to test for hydrogen. The 'pop' sound shows that hydrogen is present. For example, magnesium reacts with sulphuric acid to produce magnesium sulphate and hydrogen.

$$Mg(s) + H_2SO_4 \rightarrow MgSO_4 + H_2$$

- *Reactions that produce carbon dioxide:*

When a carbonate dissolves in an acid, carbon dioxide gas is produced. When carbon dioxide is passes through limewater, it turns the limewater milky. This is a simple test for the presence of carbon dioxide. For example, calcium carbonate reacts with hydrochloric acid to produce calcium chloride, water and carbon dioxide.

$$CaCO_3(s) + 2HCl(aq) \rightarrow CaCl_2(aq) + H_2O(I) + CO_2(g)$$

- *Reactions that produce gases such as oxygen or carbon dioxide:*

Hydrogen peroxide decomposes to produce oxygen. The volume of oxygen produced can be measured using the gas syringe method. The gas collects in the syringe, pushing out against the plunger. The volume of gas that has been produced can be read from the markings on the syringe. For example, hydrogen peroxide decomposes in the presence of a manganese(IV) oxide catalyst to produce oxygen and water.

$$2H_2O_2(aq) \rightarrow 2H_2O(I) + O_2(g)$$

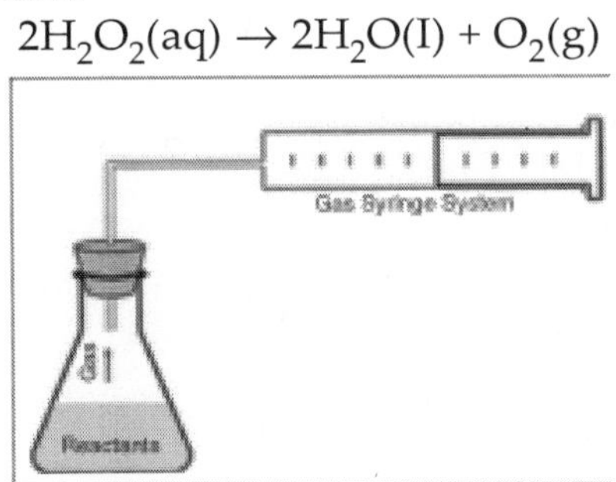

Fig. Gas Syringe Method

- *Precipitate reactions:*

In reactions where a *precipitate* is formed, the amount of precipitate formed in a period of time can be used as a measure of the reaction rate. For example, when sodium thiosulphate reacts with an acid, a yellow precipitate of sulphur is formed. The reaction is as follows:

$$Na_2S_2O_3(aq) + 2HCl(aq) \rightarrow 2NaCl(aq) + SO_2(aq) + H_2O(I) + S(s)$$

One way to estimate the rate of this reaction is to carry out the investigation in a conical flask and to place a piece of paper with a black cross underneath the bottom of the flask. At the beginning of the reaction, the cross will be clearly visible when you look into the flask. However, as the reaction progresses and more precipitate is formed, the cross will gradually become less clear and will eventually disappear altogether. Noting the time that it takes for this to happen will give an idea of the reaction rate. Note that it is not possible to collect the SO_2 gas that is produced in the reaction, because it is very soluble in water.

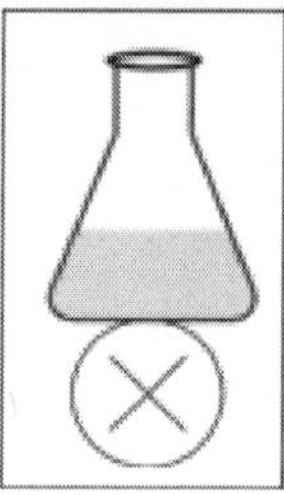

Fig. At the Beginning of the Reaction between Sodium Thiosulphate and Hydrochloric Acid, when no Precipitate has been Formed, the Cross at the Bottom of the conical Flask can be Clearly Seen.

- *Changes in mass:*

The rate of a reaction that produces a gas can also be measured by calculating the mass loss as the gas is formed and escapes from the reaction flask. This method can be used for reactions that produce carbon dioxide or oxygen, but are not very accurate for reactions that give off hydrogen because the mass is too low for accuracy. Measuring changes in mass may also be suitable for other types of reactions.

THINGS TO MEASURE

The measurement itself depends on the nature of the reactant or product:

- The mass of a substance - solid, liquid or gas - is measured with a balance
- The volume of a gas is usually measured with a gas syringe, or sometimes an upside-down measuring cylinder or burette

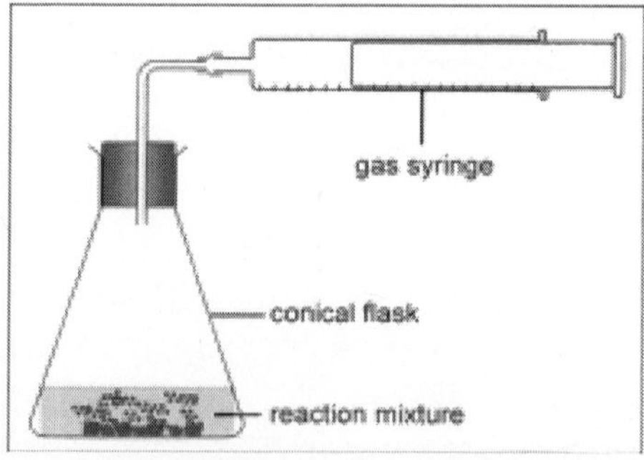

Fig. Measuring the Production of a Gas Using a Gas Syringe

Effect of Temperature and Concentration

The rate of a chemical reaction can be increased by raising the temperature. It can also be increased by increasing the concentration of a reactant in solution, or the pressure of a reactant gas.

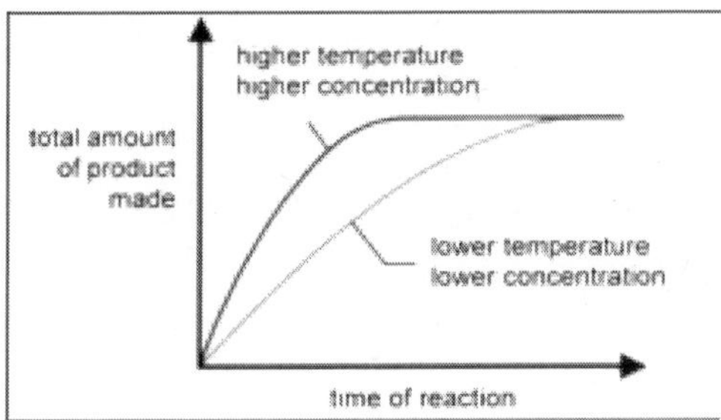

Changing the Temperature

If the temperature is increased:

- The reactant particles move more quickly
- They have more energy
- The particles collide more often, and more of the collisions result in a reaction
- The rate of reaction increases

Changing the Concentration or Pressure

If the concentration of a dissolved reactant is increased, or the pressure of a reacting gas is increased:

- The reactant particles become more crowded
- There is a greater chance of the particles colliding
- The rate of reaction increases

Effect of Surface Area

The rate of a chemical reaction can be raised by increasing the surface area of a solid reactant. This is done by cutting the substance into small pieces, or grinding it into a powder.

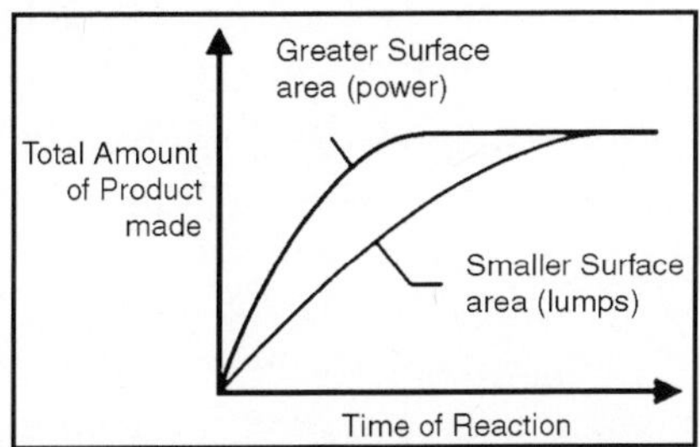

Fig. Lumps react more slowly than powders

Changing the Surface Area

If a solid reactant is broken into small pieces or ground into a powder:

- Its surface area increases
- More particles are exposed to the other reactant
- There are more collisions
- The rate of reaction increases

Explosions

Explosions are very fast reactions in which a lot of gaseous product is released very quickly. For example:

- Burning hydrogen
- Explosions from TNT or dynamite

Fine powders can also cause explosions - for example, custard powder, flour or sulfur. This is a particular hazard in factories. Great care is needed to avoid naked flames or other sources of ignition.

Effect of Catalysts

A catalyst is a substance that can increase the rate of a reaction. The catalyst itself remains unchanged at the end of the reaction it catalyses. Only a very small amount of catalyst is needed to increase the rate of reaction between large amounts of reactants.

Different catalysts catalyse different reactions. The table summarises some common catalysts used in industry and the reactions they catalyse:

Table. Some Common Catalysts Used in Industry and the Reactions they Catalyse

Catalyst	Rreaction Catalysed
iron	making ammonia from nitrogen and hydrogen
platinum	making nitric acid from ammonia
vanadium(V) oxide	making sulphuric acid

MOLECULAR COLLISIONS

In analyzing the gas so far, we've ignored collisions between molecules, and in fact for air at ordinary temperatures the relationship between pressure, volume and temperature came out correctly. Furthermore, Maxwell's speed distribution can be used to find what fraction of the molecules in a planet's atmosphere are moving at above escape velocity, so we can predict what gases will remain surrounding a planet, given the gravitational force near the surface, and the temperature. But there are other phenomena for which an understanding of collisions is all-important.

For example, if two different gases, say oxygen and nitrogen, in a container at the same pressure and temperature are separated by a partition, how quickly will they mix once the partition is removed? Assuming room temperature, the molecules will be moving at hundreds of meters per second, so one might imagine the mixing will be over in hundredths of a second. But that is not the case at all—observationally, it might take an hour, for a box holding a few liters. This surprisingly slow penetration of one gas by another is called diffusion. The reason it takes the gases so long to mix becomes evident on tracking one molecule as it enters the other gas. Think of an oxygen molecule moving into nitrogen. We'll take 02 and N2 to be little spheres of diameter d. So now visualize the little 02 sphere shooting into this space where all these other spheres are moving around. Temporarily, for ease of visualization, let's imagine all the other spheres to be at rest.

How far can we expect the O_2 to get before it hits an N_2? The average distance before a collision is called the mean free path. Let's try to picture how much room there is to fly between these fixed N_2 spheres. We do know that if it were liquid nitrogen, there would be very little room: liquids are just about incompressible, so the molecules must be touching. Roughly speaking, a molecule of diameter d will occupy a cubical volume of about d3 (there has to be some space left over—we can pack cubes to fill space, but not spheres).

We also know that liquid nitrogen weighs about 800 kg per cubic meter, whereas N_2 gas at room temperature (and pressure) weighs about 1.2 kg per cubic meter, a ratio of 670. This means that on average each molecule in the gas has 670 times more room—that is, it has a space 670 times the volume d3 we gave it in the liquid. So in the gas, the average center-to-center separation of the molecules will be the cube root of 670, which is about 8.75d. So the picture is a gas of spheres of diameter d, placed at random, but separated on average by distances of order 10d. It's clear that shooting an oxygen molecule into this it will get quite a way.

We now try to estimate just how far an 02 will get, on average, as it shoots into this forest of spheres. Picture the motion of the center of the oxygen molecule. Before any collision, it will be moving on a straight-line path. Just how close does the O_2 center have to get to an N2 center for a hit? Taking both O_2, N_2 to be spheres of diameter d, if an N2 center lies within d of the O_2 center's path, there will be a hit. So we can think of the O_2 as sweeping out a volume, a cylinder of radius d centered on its path, hitting and deflecting if it encounters an N_2 centered within that cylinder. So how far will it get, on average, before a hit? In traveling a distance x, it sweeps out a volume $\pi d^2 x$. Now picture it going through the gas for some considerable length of time, so there are many collisions.

The volume swept out will look like a stovepipe, long straight cylindrical sections connected by elbows at the collisions. The total volume of this stovepipe (ignoring tiny corrections from the elbows) will be just $\pi d^2 L$, L being the total length, that is, the total distance the molecule traveled. If the density of the nitrogen is n molecules per cubic meter, the number of N2's in this stovepipe volume will be $\pi d^2 Ln$, in other words, this will be the number of collisions. Therefore, the average distance between collisions, the mean free path l, is given by:

$$\text{mean free path } l = \frac{\text{total distance traveled}}{\text{number of collisions}} = \frac{L}{\pi d^2 Ln} = \frac{L}{\pi d^2 n}.$$

So what is n? We estimated above that each molecule has space 670d 3 to itself, so n is just how many of those volumes there are in one cubic meter, that is, $n = 1/670d^3$.

Therefore, the mean free path is given by

$$l = \frac{1}{\pi d^2 n} = \frac{670d^3}{\pi d^3} = 200d.$$

Notice that this derivation of the mean free path in terms of the molecular diameter depends only on knowing the ratio of the gas density to the liquid density—it does not depend on the actual size of the molecules! But it does mean that if we can somehow measure the mean free path, by measuring how fast one gas diffuses into another, for example, we can deduce the size of the molecules, and historically this was one of the first ways the size of molecules was determined, and so Avogadro's number was found.

Let us now put in some numbers to find this mean free path: for O2, N2, d = 0.3 nm, so the mean free path l = 60nm, or 6×10^{-8}m. The speed of the molecules at room temperature v is approximately 500 meters per sec., so the molecule has of order 1010 collisions per second!

Actually, there is one further correction we should make. We took the N2 molecules to be at rest, whereas in fact they're moving as fast as the oxygen molecule, approximately.

This means that even if the O2 is temporarily at rest, it can undergo a collision as an N2 comes towards it. Clearly, what really counts in the collision rate is the relative velocity of the molecules. Defining the average velocity as the root mean square velocity, if the O2 has velocity $\vec{v}_1$ and the N2 $\vec{v}_2$, then the square of the relative velocity $\overline{(\vec{v}_1 - \vec{v}_2)^2} = \overline{\vec{v}_1^2} - 2\overline{\vec{v}_1 - \vec{v}_2}^2 + \overline{\vec{v}_2^2} = \overline{\vec{v}_1^2} + \overline{\vec{v}_2^2}$, since $\overline{\vec{v}_1 . \vec{v}_2}$ must average to zero, the relative directions being random. So the average square of the relative velocity is twice the average square of the velocity, and therefore the average root-mean-square velocity is up by a factor "2, and the collision rate is increased by this factor. Consequently, the mean free path is decreased by a factor of "2 when we take into account that all the molecules are moving.

Our final result, then, is that the mean free path

$$1 = \frac{1}{\sqrt{2}\pi d^2 n}.$$

Finding the mean free path is—literally—the first step in figuring out how rapidly the oxygen atoms will diffuse into the nitrogen gas, and of course vice versa.

What we really want to know is just how much we can expect the gases to have intermingled after a given period of time. We'll just follow the one molecule, and estimate how far it gets. To begin with, let's assume for simplicity that it tales steps all of the same length l, but after each collision it bounces off in a random direction. So after N steps, it will have moved to a point

$$\vec{L} = \vec{l}_1 + \vec{l}_2 + \vec{l}_3 + \cdots + \vec{l}_N,$$

where each vector $\vec{l}_j$ has length l, but the vectors all point in random different directions.

If we now imagine many of the oxygen molecules following random paths like this, how far on average can we expect them to have drifted after N steps? (note that they could with equal likelihood be going backwards!) The appropriate measure is the root-mean-square distance,

$$\overline{\vec{L}^2} = \overline{\left(\vec{l}_1 + \vec{l}_2 + \vec{l}_3 ... \vec{l}_N\right)^2} = Nl^2 + \sum_{i.j} \overline{\vec{l}_i . \vec{l}_j}$$

Since the direction after each collision is completely random, $\overline{\vec{l}_i . \vec{l}_j} = 0$, and the root-mean-square distance $\sqrt{\overline{\vec{L}}^2} = \sqrt{N} l.$

If we allow steps of different lengths, the same argument works, but now l is the root-mean-square path length.

The important factor here is the $\sqrt{N}$. Recall from above that l = 60 nm, or 6×10^{-8} m., and there are of order 1010 collisions per second. This means that the average distance diffused in one second is $\sqrt{10^{10}} l = 10^5 l$, say half a centimeter. The average distance in one hour would be only 60 times this, or 30 cm., one foot, and in a day about five feet—the average distance traveled is only increasing as the square root of the time elapsed!

This is a very general result. For example, suppose we have a gas in which the mean free path is l and the average speed of the molecules is v. Then the average time between collisions $\tau = l/v$. The number of collisions in time t will be t/τ, so the average distance a molecule moves in time t will be $r = l\sqrt{t/\tau}$.

A famous mystery cleared up by arguments like this was that Newton predicted the speed of sound would be given by $c^2 = B/\rho$, as we discussed earlier in the course, with B the bulk modulus. But when B was measured carefully by slowly compressing air, the result was in error by about 30per cent! The speed of sound predicted a higher (stiffer) bulk modulus.

The explanation turned out to be that in a slow measurement of the bulk modulus, the gas stays at the same temperature—the heating caused by slow compression leaks away. But if the compression is rapid, the gas heats up and so the pressure goes up more than if it had stayed at the same temperature. So the question is whether the compression and decompression as a sound wave passes through is so rapid that the heated-up gas doesn't have time to spread to the cooled regions. For sound at say 1000Hz, the wavelength is 34 cm. If compression heats gas locally, the hot molecules will diffuse away in a similar manner to that discussed above. They will be slightly faster than the average molecules. In 1/1000 th of a second, they will have 107 collisions, so will travel about $\sqrt{10^7}\, l = 30C\ 0l = 0.2$mm. This tiny distance compared with the wavelength of the sound wave means that during the compression/ decompression cycles as the wave passes through, the heat has no chance to dissipate—so, effectively, it's like compressing a gas in an insulated container, it's harder to compress than it would be if the heat generated could flow away, and the bulk modulus is higher by an amount (around 30per cent) we shall work out in a forthcoming lecture.

The simple kinetic picture of a gas leads to the Gas Law, $PV = nRT$, with the macroscopic variable T now seen as proportional to the average energy of a molecule. Putting heat into a gas simply means increasing the molecules' average kinetic energy (plus rotational energy, *etc.*, for more complicated molecules) and the work done by an expanding gas is just a transfer of energy from this internal energy to whatever the gas is pushing.

The First Law of Thermodynamics is nothing but conservation of total energy, now that we know heat is really microscopic kinetic energy for an ideal gas (and rotational energy, *etc.*). By applying these findings to the Carnot Cycle, we discovered that this reversible ideal engine had efficiency $1 - T_C/T_H$. Furthermore, we found that no heat engine could be more efficient, at least if we granted the truth of the Second Law of Thermodynamics, which states that it is impossible to construct an engine that just takes heat energy out of the air, say, and converts it into work, without any available "cold reservoir" to dump waste heat into.

The incredible thing is that Carnot, in the 1820's, who believed heat was a conserved caloric fluid, and therefore didn't know the First Law (that heat is just microscopic energy, which is not conserved by itself, but can be transformed into other forms of energy) correctly found the efficiency of his cycle to be $1 - T_C/T_H$, and then argued—also correctly—that this set the absolute limit on heat engine efficiency! How did he do that? It's worth a brief examination of his ideas, because they give a clue about something we haven't mentioned so far—entropy.

Remember, Carnot saw his cycle as a water wheel, the caloric fluid being the water, the temperature difference being the height difference from the top to the bottom of the waterfall. But he did know one difference from a waterfall: he knew there was an absolute zero, a level below which nothing could "fall". (This was from the well known results on the contraction of a gas with cooling. He took the absolute zero to be -267 Celsius, only a few degrees off the correct value). This meant that one could assign to the caloric fluid an absolute value of "potential energy" in this imaginary water wheel scenario: at temperature

T, an amount of fluid F would have "potential energy" FT. This is the total possible amount of heat energy in the fluid, and to extract it all you would have to "drop" the fluid all the way down the temperature scale to the absolute zero. But in our real (if idealized) heat engine, it can only drop as far as the lowest temperature available, that of the cold reservoir TC. Hence the efficiency factor $1-T_C/T_H$: it's just the ratio of how far the caloric fluid is able to fall in our engine to how far it would have to fall (down from TH to absolute zero) to give up all its heat energy.

In this picture, the heat energy delivered from the hot reservoir $Q_H = FT_H$, so the amount of actual "caloric fluid" must be $F = Q_H/T_H$. But the fluid is conserved, in Carnot's picture, so the waste heat is $Q_C = FT_C$, with the same F, and therefore $Q_H/T_H = Q_C/T_C$. This equation is the important one: even though the argument he used is incorrect—there is no conserved caloric fluid—the equation is right, and relates directly to the efficiency of his engine: the heat utilized is $Q_H - Q_C = Q_H(1- T_C/T_H)$. But how did he know no engine could do better?

Ironically, although he didn't know the First Law of Thermodynamics, he had figured out the Second Law, and he presented an argument exactly like that given in the last lecture to demonstrate that no engine could outperform his reversible one.

Let's go back to the Carnot cycle and consider a bit more how the gas gains heat from a reservoir as it follows a path in the (P, V) plane. We begin with the first half of the cycle, from a to c:

ISOTHERM

This does not mean that we can say the gas at (P_C, V_C) has QH more heat than the gas at (P_a, V_a). Why not? Because we could equally well have gone from a to c by a route which is the second half of the Carnot cycle traveled backwards:

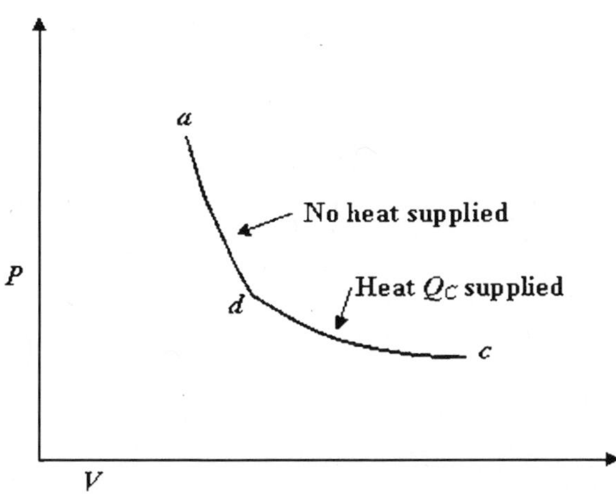

This is a perfectly well defined reversible route, ending at the same place, but with quite a different amount of heat supplied. So we cannot say that a gas at a given (P, V) contains a definite amount of heat. It does of course have a definite internal energy, but that energy can be increased by adding a mix of external work and supplied heat, and the two different routes from a to c have the same total energy supplied to the gas, but with more heat and less work along the top route.

However, notice that one thing is the same over the two routes in the diagrams above: the ratio of the heat supplied to the temperature at which is was delivered: $Q_H/T_H = Q_C/T_C$.

Of course, we've chosen two particular reversible routes from a to c, but it turns out that for any reversible route from a to c, the integral of increments of heat supplied divided by the temperature of delivery, $\int_a^c dQ/T$, is the same! To see how this can be, consider cutting a corner in the previous route:

If we follow the path aefgc instead of adc, where ef is an isotherm and fg an adiabat, how does that affect $\int_a^c dQ/T$? The answer is it doesn't: look at the little Carnot cycle efgd. We've just changed from the bottom route to the top route around this cycle from e to g, so from the previous argument about the original big Carnot cycle, $\int_e^g dQ/T$ is the same.

But we can now cut corners on the corners: any zigzag route from a to c, with the zigs isotherms and the zags adiabats, in other words, any reversible route, can be constructed by adding little Carnot cycles to the original route. In fact, any path you can draw in the plane from a to c can be approximated arbitrarily well by a reversible route made up of little bits of isotherms and adiabats.

The value of $\int_a^c dQ/T$ is evidently the same along any reversible path from a to c: in contrast to $\int_a^c dQ$, the heat change, which was different for different reversible routes.

We can define a new state variable, S, such that the difference

$$S(P_2, V_2) = S(P_1, V_1) + \int_{(P_1,V_1)}^{(P_2,V_2)} \frac{dQ}{T}$$

where the integral is understood to be along a reversible path. Given the value of S at a single point (P_1,V_1), we can find its value everywhere.

KINETIC MOLECULAR THEORY

The rates of chemical reactions increase with increasing temperature, often dramatically. Coal reacts with oxygen at a negligible rate at room temperature, but burns spontaneously with the liberation of heat and light at 500°C.

A sirloin steak "cooks" (oxidizes) very slowly at room temperature, but requires only 20 minutes at 180°C. Reactions are accelerated by temperature increase because the elementary processes by which they occur take place more rapidly.

We begin this section with two questions:

1. Why are some elementary processes fast and others slow at a particular temperature?
2. Why are reaction rates (rate constants) so dramatically temperature dependent?

We can arrive at answers to these questions by looking in more detail at the collision process.

Requirements for Collision to Lead to Reaction

As discussed earlier, rate is proportional to the number of molecular collisions occuring per unit volume per unit time. For the decomposition of hydrogen iodide to hydrogen and iodine, assuming that [HI] = 0.001 M and a temperature of 500 °C, the rate would

$$2HI(g) \rightarrow H_2(g) + I_2(g)$$

be 5.8×10^4 M/s if all collisions led to formation of products. The observed rate, however, is only 1.2×10^{-3} M/s, 5×10^7 times slower than it would be if all collisions between HI molecules were effective.

Our conclusion must be that only a small fraction of molecular collisions lead to formation of products; *i.e.*, are effective. In general, for the elementary process,

$$A + B \rightarrow C + D$$

Rate = Number of effective collisions/volume-time, which should be proportional to the total number of collisions per unit volume per unit time. We call this total number of collisions Z. We can relate Z to the concentrations of A and B via equation:

$$Z = Z_0[A][B]$$

where Z_0 is the number of collisions per volume per time when A and B have 1 molar concentrations. The number of effective collisions is less than the total number of collisions for two reasons. First, two colliding molecules must be properly oriented if they are to react.

Reaction requires that bonds within reactants be broken and that new bonds form between atoms in the colliding molecules. A bond cannot form between atoms in separate molecules unless they are near each other during collision. Two of the many possible collision orientations of CO and NO_2, which are reactants.

$$CO + NO_2 \rightarrow CO_2 + NO$$

This bimolecular elementary process involves transfer of an oxygen atom from NO_2 to CO. For this to occur, it is necessary that an oxygen atom of NO_2 approach the carbon atom of CO during collision, so that a bond may form between them. The relative orientation brings the nitrogen atom of NO_2 and the oxygen atom of CO together, so cannot be effective in advancing the process. Similarly, many other possible orientations of these two molecules are unsuitable for transfer of an oxygen atom from N to C, so will not be effective. We symbolize the fraction of collisions having the proper orientation of molecules as p, and recognize that p is a small number.

Second, for reaction to occur, it is necessary that colliding molecules have some minimum total kinetic energy. This is called the activation energy, symbolize E_a. In almost all cases, successful reaction requires that existing bonds be broken and new bonds made. For the bond to break, an input of energy at least equal to the depth of the well is required. This energy is drawn from the total kinetic energy possessed by the colliding species. If the kinetic energy is insufficient to overcome the bonding interactions, the reactant molecules separate without change.

Even if the total kinetic energy is sufficient for reaction, the collision orientation may still be inappropriate for reaction. Both factors—sufficient kinetic energy and correct orientation—are necessary for reaction to occur. Assuming that the minimum energy is available and the orientation correct, bonds will be ruptured, new bonds will form with the release of energy, and new molecules will separate from the collision center. Between reactants and products, both of which are relatively low in potential energy, lies a state of maximum potential energy called the transition state.

The arrangement of atoms that exists in this state is called the activated complex. We may think of it as a "molecule" in which chemical bonds are in flux, some being broken and others being made. We can depict this situation graphically in terms of a reaction coordinate diagram, which is a plot of potential energy versus the progress of reaction, which is usually called the "reaction coordinate." A reaction coordinate diagram for the bimolecular elementary process. Potential energy is plotted on the vertical axis, with the reaction coordinate on the horizontal axis.

The reactants, A and B are located to the left on the reaction coordinate. As reaction proceeds, potential energy rises to the maximum at the transition state, then falls again as bond formation in the product species is completed and the product molecules separate. The activation energy, E_a, for conversion of A and B to C and D (the forward reaction) is the difference in potential energy between the transition state and the reactants. This is labelled in the diagram. In drawing this diagram for the generic process, it has been assumed that products have lower potential energy than reactants; that is, that the process is exothermic. Thus DE for the process is negative.

The activation energy for the reverse process, also indicated in the diagram, is the difference in PE between the transition state and the potential energy of C + D. It is clear from the figure that DE for the reaction is the difference between the activation energies for the forward and reverse processes.

This is generally true:

E_a(forward) = PE(activated complex) – PE(A+B)

E_a(reverse) = PE(activated complex) – PE(C+D)

E_a(forward) – E_a(reverse) = DE for elementary process

Every elementary process can be depicted as a single-humped potential energy curve, with an activated complex at the maximum. Letting the fraction of collisions with sufficient kinetic energy to react be f, then

Rate = Number of effective collisions/volume-time = pfZ

We are now in a position to answer the first of the two questions posed at the beginning of this section. Some elementary processes are slower than others either because they have very restrictive orientation requirements, corresponding to a very small value for p; or they have high potential energy barriers (E_a), corresponding to a small value for f. Next we consider the effect of an increase in temperature. The distribution of kinetic energies in a collection of molecules is governed by the Maxwell-Boltzmann distribution. The M-B distribution curve for temperature T_1. The value of the activation energy for elementary process has been indicated on the graph. The fraction of molecules with kinetic energy greater than E_a is given by the area under the M-B curve to the right of E_a.

As temperature is increased to some new value T_2, the M-B curve shifts to the right and flattens, indicating the increase in both the average molecular kinetic energy and the breadth of the distribution. Notice that the area under the curve to the right of E_a is much larger for the high-temperature curve, leading to a dramatic increase in the number of effective molecular collisions. Quantitatively, the fraction of effective collision can be expressed as:

$$f = \text{area under curve right of } E_a \; \alpha \; e^{-Ea/RT}$$

Thus:

$$\text{Rate} = \text{pfZ} = pe^{-Ea/RT}Z_o[A][B]$$

Equation has the form of a rate law if we identify the terms preceding the concentrations of A and B as the rate constant, k:

$$k = pZ_oe^{-Ea/RT} = Ae^{-Ea/RT}$$

Where A contains information about the orientation requirements. Taking the logarithm of both sides, called the Arrhenius equation.

$$\ln k = \ln A - E_a/RT$$

The value of the activation energy for a process is obtained by measuring the rate constant at each of several temperatures, and constructing a plot of lnk versus 1/T. The plot should be linear, with slope = $-E_a/R$ and intercept = lnA.

We are now able to offer an answer to the second of the two questions posed at the outset: reaction rates are temperature dependent because of the requirement of a minimum kinetic energy for effective collisions.

Example:. Aqueous hydrogen peroxide and iodide ion react according to the following equation.

$$H_2O_2(aq) + 3I^-(aq) + 2H^+(aq) \rightarrow I_3^-(aq) + 2H_2O$$

The reaction is first order in peroxide and first order in iodide, second order overall. The reaction was studied at several temperatures, with the following results. What is the activation energy for the reaction?

T, °C	k_{obs}, $M^{-1}s^{-1}$
2	3.03×10^{-3}
17	1.0×10^{-2}
50	7.6×10^{-2}

Solution: The activation energy can be obtained from the slope of a plot of ln k_{obs} versus reciprocal Kelvin temperature. The activation energy obtained from the slope is 53.6 kJ/mole.

Overall (Multistep) Reactions

Thus far we have discussed the temperature dependence of a single elementary process. Similar considerations apply to overall (net) reactions consisting of several steps.

We make several general statements about reaction coordinate diagrams for net reactions before looking at a specific example.

- The number of maxima (activated complexes) in a reaction coordinate diagram is equal to the number of steps in the mechanism.
- The slowest (rate-determining) step usually has the highest E_a.
- The rate law for the overall reaction indicates the atoms present in the activated complex for the slowest elementary step.
- Intermediates occur at minima in the reaction coordinate diagram.
- The E_a measured experimentally for the multistep reaction is the difference between the potential energy at the highest point in the diagram and the potential energy of the reactants. (This is analogous to the process of rolling a ball over a succession of hills of varying heights; to make it, the ball must have kinetic energy sufficient to get over the highest hill.)

Example: For the reaction

$$2NO_2 + F_2 \rightarrow 2NO_2F$$

the rate law is $-d[F_2]/dt = k_{obs}[NO_2][F_2]$. Propose a mechanism for the reaction, and draw an appropriate reaction coordinate diagram.

Solution: The rate law tells us that the activated complex of the slow step contains two oxygen atoms, two fluorine atoms, and one nitrogen atom. This suggests the following mechanism:

$$NO_2 + F_2 \rightarrow NO_2F + F \; [k_1 \text{ slow}]$$

$$F + NO_2 \rightarrow NO_2F \; [k_2 \text{ fast}]$$

Because reaction involves the breaking of the F-F bond and formation of a bond between N and F, the structure is a reasonable arrangement of atoms in the activated complex. Finally, There are two "humps" (maxima), one corresponding to each elementary step in the mechanism; the intermediate, F, is located at the minimum between maxima; the activation energy for the overall reaction is indicated; and the overall reaction is exothermic. The activation energy for the first step is substantially larger than that for the second step, consistent with the first step being rate-determining. Thus the reaction coordinate diagram reveals by its profile which is the slow step in the overall reaction.

Example: The reaction coordinate diagram for an unspecified reaction. From the details of the diagram, say as much as possible about the net reaction to which it corresponds.

Solution: Because there are three maxima in the diagram, the mechanism must consist of three steps. Step 2 must be the rds, because it has the largest E_a. There is one activated complex for each step, for a total of three. There are two minima, indicating two intermediates. The experimental rate law for the reaction should indicate the composition of the activated complex of the second step.

The overall reaction is exothermic, because products are lower on the energy axis than reactants. Finally, the measured E_a is the energy difference between the highest maximum in the diagram and the reactants. It is shown in the diagram. The temperature dependence of the observed rate law for an overall reaction consisting of several steps is usually qualitatively similar to that for a single elementary process. However, there are complex mechanisms for which this is not the case. Indeed, examples are known of reactions for which the observed rate constant is either independent of temperature or decreases with increasing temperature.

KINETICS AND EQUILIBRIUM

The connection between the equilibrium constant for a reaction and the rate constants of the elementary steps by which it occurs is an important one. We begin by stating the relationship. We will then demonstrate, for a specific case, that it is true. The equilibrium constant for a net reaction is the ratio of the product of the forward rate constants for all steps in the mechanism to the product of the reverse rate constants for all steps in the mechanism:

$$K_{eq} = k_1k_2k_3\ldots/k_{-1}k_{-2}k_{-3}\ldots$$

We now show for a specific net reaction and a specific (proposed) mechanism that this statement is true. Consider again the reaction of NO_2 and F_2 to produce NO_2F, reaction:

$$2NO_2 + F_2 \Leftrightarrow 2NO_2F$$

We have written the reaction with the double arrow because we are now interested in the forward and reverse rates at equilibrium. At equilibrium, the overall forward rate and overall reverse rate must be equal:

$$rate_f = rate_r$$

This must be true regardless of the detailed pathway by which the reaction occurs. All available experimental evidence indicates that the mechanism is the two-step process:

$$NO_2 + F_2 \Leftrightarrow NO_2F + F \; [k_1, k_{-1}]$$
$$NO_2 + F \rightarrow NO_2F \; [k_2, k_{-2}]$$

When the overall reaction is in equilibrium, so must each elementary step be. Consequently, we have used double arrows in the elementary steps as well. The equality of the forward and reverse rates of the individual steps is expressed mathematically as follows.

$$k_1[NO_2]_e[F_2]_e = k_{-1}[NO_2F]_e[F]_e$$
$$k_2[NO_2]_e[F]_e = k_{-2}[NO_2F]_e$$

Eliminating the concentration of the intermediate, F.

$$[NO_2F]_e^2/[NO_2]_e[F_2]_e = k_1k_2/k_{-2}k_{-2} = K_{eq}$$

The statement made at the outset is thus seen to be true for this specific case: the equilibrium constant is the ratio of the product of forward rate constants to the product of reverse rate constants. For the general overall reaction occuring by an n-step mechanism,

$$K_{eq} = k_1k_2...k_n/k_{-1}k_{-2}...k_{-n}$$

Objection is frequently made to the method of derivation that we have just used, because the mechanism for is not known with certainty; it is and will remain a hypothesis, however well based in experiment. If it is found that the two step mechanism is indeed NOT correct, then the relationship is invalid. This is certainly true. However, even if the mechanism above is incorrect, Reaction must occur by some mechanism. At equilibrium each step in the mechanism must be balanced in rate, enabling us to eliminate concentrations of any and all intermediates just as we did above, to arrive at a modified version. The validity of the general relationship is independent of the details of mechanism.

Thus the intuitively-expected connection between k and K does indeed exist. In fact, there are many examples in which the equilibrium constant for a reaction has been obtained from kinetics studies. This process can never be reversed however; it is not possible to obtain k by performing equilibrium studies. Before leaving this matter, we make one more very important point.

There is a tendency to believe that reactions with large equilibrium constants are fast, whereas those with small equilibrium constants are slow. However, neither of these beliefs is true. A large value for K_{eq} means that the product of forward rate constants is much larger than the product of reverse rate constants. It does not follow, however, that the forward rate constants are LARGE. It means only that they are larger than the reverse rate constants.

$k_1k_2...k_n >> k_{-1}k_{-2}...k_{-n}$ does not mean that $k_1, k_2, ..., k_n$ are large in the absolute sense.

A large K_{eq} can result from the ratio of a slow forward rate to an even slower reverse rate.

CATALYSIS

A number of chemical reactions that ordinarily occur slowly can be induced to occur more rapidly by the addition of a suitable substance called a catalyst. We will define a catalyst as a substance that speeds up a reaction without itself being consumed or chemically changed in the overall reaction process. This definition is in practice rather restrictive, because eventually, all real catalysts become deactivated by "poisoning" or via irreversible structural changes; thus they cannot function indefinitely.

For our purposes, here, however, this definition will serve. The reactant molecule that is affected by the catalyst is called the substrate. The process by which a catalyst effects its action is called catalysis. Catalysis is one of the most intensely studied areas in science because it is of tremendous biological and industrial importance. We begin our exploration of catalysts by discussing how they work.

The Mode of Catalyst Function

A catalyst functions either by entering into a slow step of the uncatalyzed reaction mechanism, or by creating an entirely new mechanism for the reaction. In either case, it is thought that the catalyzed pathway has a lower activation energy than the uncatalyzed path. We can illustrate catalytic action generically in terms of the following reaction scheme, where A, B, D, and F are reactants and products in the overall reaction, and C is a catalyst for the reaction.

$$A + B \rightarrow D + F \text{ [overall]}$$

Uncatalyzed mechanism:

$$A + B \rightarrow AB \text{ [slow]}$$
$$AB \rightarrow D + F$$

Catalyzed mechanism:

$$A + C \rightarrow CA \text{ [fast]}$$
$$CA + B \rightarrow CAB \text{ [less slow than first step of uncatalyzed mechanism]}$$
$$CAB \rightarrow D + CF \text{ [fast]}$$
$$CF \rightarrow C + F \text{ [fast]}$$

The effect of the catalyst on the activation energy of the slow step of the reaction is shown schematically. The catalyst affects only E_a; it does not affect the energy of either the reactants or products of the reaction. The intermediate, CAB, in which both reactants are bound to the catalyst, is of particular interest. A possible (and fairly common) structural motif for this intermediate.

Notice that, although A and B are bound to different sites on the catalyst, C, they are in proximity and can effectively interact. Binding of A and B to C causes shifts in electron density that may facilitate bond breaking within A and/

or B and bond formation between A and B or fragments of them. Thus the catalyst provides an organizing center for A and B, facilitating their interaction. In this manner the catalyst can overcome the orientation factor discussed earlier.

As soon as the molecule of catalyst is regenerated in the last step of the 4-step mechanism, it may bind another molecule of A and proceed once again through the sequence, converting A and B to D and F. This sequence is repeated a large number of times, so that generally only a small amount of catalyst is required to convert a substantial quantity of reactant to product.

Chemists call the process by which the catalyst cycles through the reaction over and over again a catalytic loop. The unbound form of the catalyst, C, is placed at the 12 oclock position of the loop, and the various forms in which the catalyst is found are placed more or less evenly around the remainder of the loop. Arrows show the direction of reaction around the loop, with reactants brought in from outside the loop, and products ejected out of the loop.

This is a very effective visual presentation of catalyst action. The number of times that a molecule of catalyst cycles through the loop per unit time is called the turnover number of the catalyst. An effective measure of turnover number is moles product produced per mole catalyst present per time. Let's look at a specific example of the effect of a catalyst on a simple reaction, from which we can draw some general conclusions.

Example: The reaction of ethanol (ethyl alcohol) with bromide ion to produce ethyl bromide and hydroxide ion.

$$C_2H_5OH + Br^- \rightarrow C_2H_5Br + OH^-$$

The reaction is quite slow in neutral or basic solution, but occurs quite rapidly in acidic solution because it is catalyzed by the hydronium ion, H_3O^+. Discuss the mechanisms for the uncatalyzed and catalyzed processes.

Solution: The uncatalyzed reaction is thought to occur in a single elementary step, in which bromide ion attacks the hydroxyl carbon atom of ethanol while the hydroxide group simultaneously departs:

$$Br^- + C_2H_5OH \rightarrow \text{activated complex} \rightarrow C_2H_5Br + OH^-$$

$$\text{Rate} = k_{obs}[C_2H_5OH][Br^-]$$

This mechanism is proposed based on the experimental rate law, which is overall second order. The observed rate law for the catalyzed process is,

$$\text{Rate} = k_{obs}[Br^-][C_2H_5OH][H_3O^+]$$

The following three-step mechanism is consistent with this rate law if k_{obs} is identified with k_2K_1:

$$C_2H_5OH + H_3O^+ \rightarrow C_2H_5OH_2^+ + H_2O \text{ [fast, } K_1]$$

$$C_2H_5OH_2^+ + Br^- \rightarrow C_2H_5Br + H_2O \text{ [slow } k_2]$$

$$2H_2O \rightarrow H_3O^+ + OH^-$$

The reaction coordinate diagrams for the uncatalyzed and catalyzed pathways. Several general statements about catalysis are evident from this example.

- Because the catalyst does not appear in the overall equation for the reaction, it does not affect the equilibrium; it affects only the kinetics. The catalyst speeds up not only the forward reaction, but also the reverse, by the same factor. The position of equilibrium remains the same.
- the activation barrier is lower in the catalyzed mechanism. Even though the catalyzed path involves more steps, the overall reaction occurs more rapidly.
- The concentration of the catalyst appears in the rate law, even though the catalyst does not appear as a reactant or product in the overall equation.
- Although a catalyst and an intermediate share the property of not appearing in the overall equation, they are different types of things. First, the catalyst appears in the rate law; an intermediate cannot. Second, the catalyst appears first as a reactant, then is generated later as a product. An intermediate appears first as a product, then later as a reactant.

Examples of Catalysis

We will now briefly discuss several examples of catalysis that are actually used on a mammoth scale in the chemical industry. Each of these processes, in its own way, has a major impact on the quality of our lives.

The Polymerization of Ethylene

Ethylene is a very simple molecule with formula C_2H_4. It is a gas, obtained as a byproduct during the catalytic cracking of petroleum. Under certain conditions, ethylene molecules can be made to join together end-to-end to form very long chain-like molecules of polyethylene, so called because it consists of many (poly) ethylenes. The process is represented.

$$2n\ CH_2 = CH_2 \xrightarrow{TiCl_3} -(CH_2CH_2CH_2CH_2)_n-$$

Even at high temperature and pressure of ethylene, this process occurs negligibly slowly. In the presence of a small amount of a modified form of $TiCl_3$ (called tickle-3 in the plastics industry), however, it occurs rapidly at only moderately high temperature and pressure. The process is referred to as Ziegler-Natta catalysis after its two coinventors, who jointly received the Nobel Prize in Chemistry in 1963. Since the discovery of this process in the 1950's, the entire plastics industry has developed and grown to huge proportions.

To this day, the mechanism of Ziegler-Natta catalysis is not fully understood. Mechanistic studies are difficult for several reasons, one of which is that the process is heterogeneous; that is, the catalyst ($TiCl_3$) and substrate (ethylene) are in different phases. The catalytic process takes place on the surface of crystals of $TiCl_3$, which rapidly become covered with and blocked from view by the resulting polyethylene. Much effort is ongoing in the US chemical industry to develop more efficient and easily handled Ziegler-Natta catalysts.

The Production of Sulfuric Acid

Year after year, sulfuric acid ranks first on the list of the top ten chemical substances produced in the United States: billions of pounds are produced annually. Sulfuric acid is synthesized by the so-called Contact Process, which involves the four sequential steps below:

$$S + O_2 \rightarrow SO_2 \text{ [fast]}$$

$SO_2 + O_2 \rightarrow SO_3$ [slow, because it occurs by a termolecular elementary process catalyzed by V_2O_5]

$$SO_3 + H_2SO_4 \rightarrow H_2S_2O_7 \text{ [fast]}$$

$$H_2S_2O_7 + H_2O \rightarrow 2H_2SO_4 \text{ [fast]}$$

The reaction of SO_3 with water is very exothermic and causes extensive spattering and production of a fine mist of highly acidic water. For this reason, direct reaction of SO_3 with water in the third step is impractical. Instead, SO_3 is bubbled into pure sulfuric acid, with which it reacts smoothly to give fuming sulfuric acid, $H_2S_2O_7$. This can then be treated with the stoichiometrically correct amount of water to give sulfuric acid.

The contact process has been so perfected that sulfuric acid is very inexpensive to produce. Consequently, it is used in any industrial process requiring acid. It finds it major uses in the production of phosphate fertilizers; in paper manufacture; in the petroleum industry; in steel production; and in the production of detergents. The importance of these products in our lives is obvious.

Catalytic Converters

For some years now, catalysts have been placed within the exhaust systems of automobiles to reduce the amount of poisonous or otherwise harmful emissions. These noble metal catalysts (based on platinum and palladium) carry out a dual function. First, they facilitate oxidation of carbon monoxide, resulting from incomplete hydrocarbon combustion, to carbon dioxide:

Pt

$$CO(g) + 1/2\ O_2(g) \rightarrow CO_2(g)$$

Second, they catalyze decomposition of nitric oxide, produced during engine operation and oxidized rapidly to toxic NO_2 by atmospheric oxygen, to N_2 and O_2. Unfortunately, catalytic converters also facilitate oxidation of SO_2 to SO_3, which is the precursor of acid rain. Low-sulfur petroleum distillates are therefore essential. The Haber Process. Vast quantities of ammonia are synthesized each year for use as fertilizer. Currently the most efficient process for ammonia synthesis is the Haber-Bosch Process, developed during the ten-year period preceding 1913, in which nitrogen and hydrogen react directly at high temperature and pressure and in the presence of an activated iron catalyst to form ammonia.

$$N_2(g) + 3H_2(g) \xrightarrow{Fe} 2NH_3(g)$$

A catalyst and high temperature are necessary to cause the reaction to go at a reasonable rate. Unfortunately, high temperature makes the exothermic reaction less favoured, so very high pressure is used to favour products. Even with conditions optimized, reaction is incomplete, and unreacted hydrogen gas is recycled for maximum efficiency of ammonia production. Nitrogen fertilizers produced from ammonia are largely responsible for the incredible growth in agricultural production over the decades since the First World War, when the process was first put on line in Germany. Ironically, the original motivation for development of the process was the requirement of explosives for the war effort.

The processes discussed above have at least three features in common. First, catalysis is heterogeneous; the catalyst is in all cases a solid, interacting with the substrate in the gas phase. Second, the mechanisms for these processes are incompletely understood. Thus catalysts are used successfully on a huge scale, even though we do not understand how they work. Third, in all cases the catalyst involves a transition metal, from the d block of the periodic table.

Transition metals are often versatile catalysts because they are flexible in coordination number (that is, the number of atoms, ions, or molecules to which they may bind in a Lewis acid-base interaction); in stereochemistry (that is, in the shapes that their adducts assume); and in oxidation state (that is, in the charge that they carry). Nature has chosen transition metals to serve as the centerpieces for many of its catalysts, the enzymes, most probably for these same reasons.

DEVELOPED COLLISION THEORY

This theory was developed during the course of World War I which was in itself a 'Collision' of imperialist nations vying for power. It was invented by two people independently on the opposite sides of the war, one a brilliant German chemist named Max Trautz in 1916 and the other a British professor of chemistry named William Lewis in 1918.

BASICS

This theory models a chemical reaction as an inelastic and chaotic collision of millions of reactant molecules simultaneously. Some collisions just occur at the right energies and at right angles to break certain old chemical bonds and make new ones. Breaking an old bond requires energy from a collision. Whereas, making a new bond, releases energy in the form of heat.

The exact energy required for a particular old bond to be broken and a new one to be formed is called the 'Activation Energy'. This is the threshold energy at or beyond which, if the collisions occur, new bonds form. However the energies of molecules are not same and very few possess the energy to create a product. Only the right activation energy is not sufficient for the process to occur. There are many more factors that contribute and determine the ultimate fate of a chemical reaction. A systematic study of all the multifarious factors that affect the rate and outcome of a chemical reaction is collision theory.

Explaining the Factors of Collision Theory

The Maxwell-Boltzmann distribution, named after two great physicists who discovered it, is a graph of the energy of reactant molecules against the number of molecules possessing that amount of energy. You cannot actually measure the energy and count every molecule in a solution. It is mathematically proven and observed fact that any random distribution over large numbers looks like a Gaussian distribution. So the graph of energy distribution of molecules is a Gaussian distribution. If you picture the graph, it is bell shaped, with the bell having a steep slope on the left which reaches a round peak and then a very slow declining slope.

In this graph, if you mark the activation energy on the x-axis, you will observe that a very marginal slice of the area under the graph has energy higher than the activation energy. It is only the last small triangular area at the high energy end of the graph. So majority of the reactant molecules do not have the required energy to make a reaction possible.

Collision theory has developed mathematical models that say that the rate of reaction is directly proportional to the temperature and concentration roughly. Let us see how one can increase the probability of reaction of reactant molecules which are below the threshold of activation energy. This can be done by studying every individual factor and its effect on the rate of reaction

Effect of Activation Energy

As already mentioned above, the activation energy is an energy hill that the reactants must scale in order to react and create reactants. The activation energy largely determines the rate of a reaction and the quantity of products created. This threshold energy is determined by the energy of the chemical bonds that need to be broken in order to form new ones.

Temperature Effects

One way of increasing the fraction of molecules that undergo reaction, by crossing of the energy barrier is, heating and raising the temperature of the reaction solution. When the reactants are heated, they vibrate and collide more vigorously with other reactant molecules. This increases the rate of reaction and gives a higher yield invariably.

Effect of Concentration

Concentration is defined as the number of molecules of reactants per unit volume. The more the concentration of reactant molecules, more is the probability of collision due to their sheer number. Increasing the concentration of reactant molecules may or may not increase the rate of chemical reaction. It depends on the particular chemical reaction. Chemical reactions happen in specific ratios of chemical reactants. Excess concentration may have no effect if one of the reactants is used up and the reaction goes into saturation.

Catalyst Effect

There is one ingenious way of getting round the activation energy threshold or energy hill. Instead of going over the hill, one can tunnel through it. This is exactly what catalysts do. Catalysts are chemicals or substances that catalyze or promote a chemical reaction to occur and themselves remaining neutral in the end. They are like parts of an assembling mechanism that help making the final product but then detach themselves from it.

Effect of Pressure

If one is dealing with a chemical reaction involving gaseous reactants, an increase in pressure accelerates the rate of reaction. Compression of the gas due to increase in pressure, automatically increases the concentration of a gas. This causes the probability of gaseous reactant interaction to increase, which ultimately increases the rate of reaction.

Effects of Collision Angles

Not only the energy, but the angle of collision also matters. The reactant molecules or atoms must attack the right side of another molecule to break chemical bonds. The morphology and electron concentration centers of molecules matter. For a bond to break and new one to form, the appropriate parts of two molecules must bump into each other.

All these relationships of various factors to the rate of chemical reaction is formulated in terms of definite mathematical equations. By using the quantitative relations they represent, one can have a rough idea of the rate of reaction by knowing the reactant concentration, temperature and other factors. This is an empirical theory. It cannot predict results accurately when the chemical reactions increase in complexity. However, that does not make it wrong, only inaccurate. Collision theory like many other physical theories is an approximation to truth.

REACTION ENVIRONMENTS AND CONDITIONS

Typically laboratory experiments can be performed in one of the reactor types described below. Each of the reactor types can be operated (more or less successfully) under isothermal, adiabatic, or non-isothermal conditions. Under isothermal experimental conditions the temperature of any point of the reaction mixture in the reactor should be the same and constant (equal to the desired experimental temperature) at all times during the run.

Adiabatic experimental conditions are achieved when the reaction mixture does not exchange any heat with the surroundings (other than sensible heat of the inflow and outflow stream in flow reactors) and no work is done on or by the reaction mixture (other than PV work of the fluid entering and leaving the vessel in flow reactors).

Non-isothermal experimental conditions are achieved:

- Either when a temperature programmes with respect to time or position in the reactor is established
- Or when the reactor establishes its own temperature profile in time and space while exchanging heat with the surroundings.

From the point of view of determination of kinetic rates, isothermal conditions are preferred. Kinetic data for simple reactions with well defined stoichiometry can also be obtained from adiabatic runs, but interpretation of non-isothermal runs is usually extremely tedious and is to be avoided. The typical reactor types in which experiments are performed are:

Batch Reactor

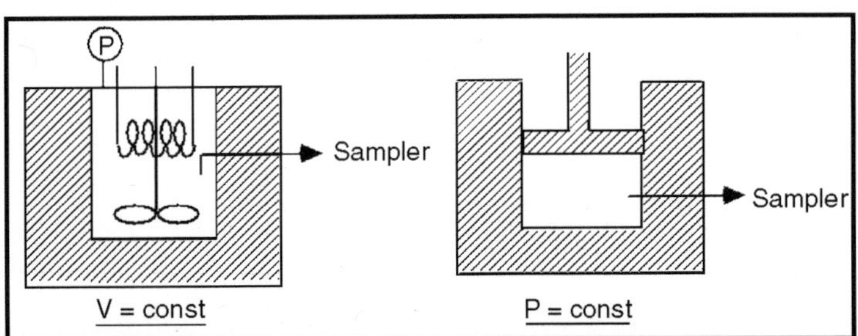

A typical batch reactor (V = *const*) is a vessel of constant volume (i.e., a flask, autoclave, *etc.*) into which the reactants are charged at the beginning of the run. The reactor is equipped either with a cooling/heating coil or jacket, or is well insulated, and can be run isothermally (or close to it) or adiabatically. A mixer provides for vigorous agitation of the reaction system. The progress of reaction can be monitored by taking samples of the reaction mixture in specified time intervals and analyzing their composition, i.e. concentrations of certain components are observed as a function of time. In case of gas phase reactions, which proceed with the change in the number of moles, the change of the overall pressure in the system can be monitored in time and tied to reaction progress in case of single reactions.

Another type of batch reactor, which is infrequently used in practice to generate rate data but which illustrates an important concept to be used later, is the constant pressure batch system where the volume of the reaction mixture may change in time (i.e. in case of gas phase reactions which proceed with the change in the total number of moles).

Semibatch Reactor

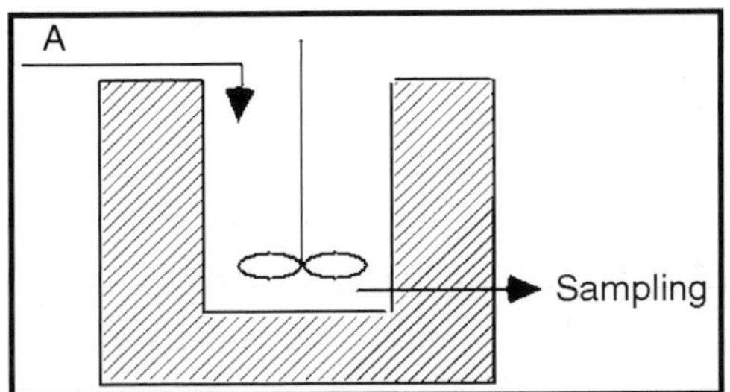

In the case of a semibatch reactor some of the reactants are charged at the beginning of the run while one or more reactants are added continuously throughout the run. Again, the reactor can be run isothermally or adiabatically, and sampling of the reaction mixture is performed in time in order to monitor the progress of reaction. This reactor type may be quite useful when one is trying to determine the reaction order with respect to say reactant B and constantly keeps adding reactant A in large excess. It is also a convenient device in complex reaction systems to study the effect of the order of reactant addition on selectivity and yield *etc.*

Continuous Flow Reactor

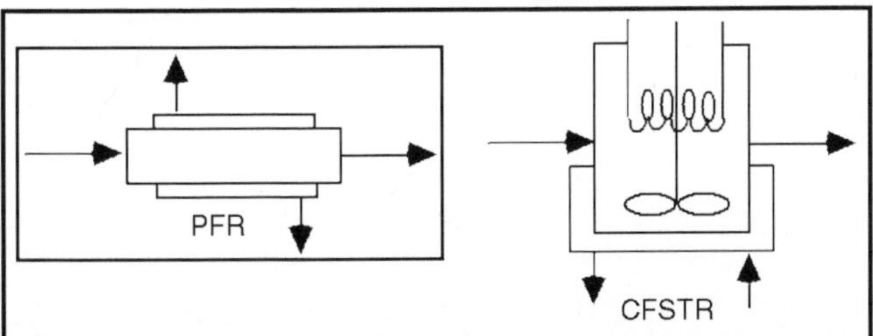

Reaction rates and rate expressions can also be determined in continuous flow systems run at steady state. Two basic types of continuous flow reactors are: the ideal plug flow reactor (PFR) and the ideal continuous flow stirred tank reaction (CFSTR).

Plug Flow Reactor (PFR)

The main assumption of the ideal plug flow reactor is that the fluid is perfectly mixed in the direction perpendicular to main flow and that there is absolutely no mixing in the axial direction, i.e. in the direction of flow. Thus, it is visualized that all fluid molecules move at a uniform velocity u, i.e. the molecules that enter at $t = 0$ form a front (plug) that moves at velocity u all the way to the exit. Therefore, there are no variations in composition and temperature in the direction perpendicular to flow, while concentration changes in the axial or z direction as one proceeds downstream.

The assumptions of the plug flow reactor are frequently met in industrial practice and in the laboratory. High Reynolds number flow in sufficiently long tubes, i.e. tubes of high L/d_t (length/diameter) ratio, will usually approximate well the plug flow reactor. Flow in packed beds can also be treated as plug flow. This reactor can be operated isothermally or adiabatically.

Continuous Flow Stirred Tank Reactor (CFSTR)

The CFSTR is also frequently called the backmixed reactor, ideal stirred tank reactor (ISTR), *etc.* The main assumption is that the reaction mixture in a CFSTR is ideally or perfectly mixed. Thus, when the reactor operates at steady state every point (i.e. every portion) in the reactor has the same composition and temperature. Since there is nothing to distinguish the points in the exit line from the points in the reactor, this implies that the composition and temperature in the exit stream is identical to the composition and temperature of the reaction mixture in the reactor. This ideal reactor in practice is approached by devices that provide a very vigorous mixing of the reaction mixture. It can be operated isothermally or adiabatically.

Other flow type reactors which do not approach a PFR or CFSTR behaviour are not useful for evaluation of rate expressions.

We have mentioned before that it is desirable to perform kinetic runs at constant temperature. Of the above reactor types CFSTR is the easiest to run at isothermal conditions. When operated at steady state, the composition in the reactor is constant and the heat released (or taken) per unit time is constant and can be readily removed or provided. In contrast, in a batch reactor, since the composition changes in time so will the heat released or absorbed; thus, one must have sufficient flexibility to meet varying heat requirements. Similarly, in a PFR fluid composition changes along the reactor, and thus the heat released or taken per unit reactor length changes along the reactor. Cooling or heating must meet this varying requirements.

It also should be pointed out that flow reactors (especially laboratory ones) operate essentially at constant pressure conditions since the pressure drops are usually negligible. In making energy balances, friction losses, or the work done on the fluid by the impeller, can generally be neglected in comparison to heats of reaction. At the same time, since the volume of the reaction mixture is fixed in flow reactors and also $P = const$, this means that in the case of gas phase reactions which proceed with the changes in the total number of moles there would be a change in the volumetric flow rate of the mixture as it moves through the PFR. For such gas phase reactions the exit and entrance volumetric flow rates in a CFSTR are different.

DETERMINATION OF THE RATE

When we run any of the above reactor types we determine the rate of reaction by making a material balance on the reactor. Let us for simplicity consider a single reaction,

$$aA + bB = dD;$$

or,

$$\sum_{j=1}^{s} v_j A_j = 0$$

Since for a single reaction,

$$\frac{N_{Ao} - N_A}{a} = \frac{N_{BO} - N_B}{b} = \frac{N_D - N_{DO}}{d};$$

or,

$$X = \frac{N_j - N_{jo}}{v_j}$$

it is sufficient to monitor only one variable of the system, i.e. one concentration, or any property of the system such as its pressure, viscosity, refactive index, *etc.* that is directly and uniquely related to its composition. The rate of reaction $r\left(T, C_A, C_B, C_D\right) = r\left(T, C_j\right)$ is a function of temperature and composition.

EVALUATIONS OF RATE FORMS USING BATCH

We will concentrate now on evaluation of rate expressions in homogeneous systems using batch reactors. The problem, if we start from a simple single reaction, can be stated as follows:

For a reaction $aA + bB + dD + gG$ find the reaction rate, i.e. find the rate's dependence on concentration and temperature.

The experiments are selected to be performed at isothermal conditions so that the rate's dependence on concentrations is established first. Once the rate form is known, isothermal runs can be repeated at different temperature levels so that the dependence of the specific rate constant(s) on temperature can be established.

Let us assume further that these hypothetical experiments are performed in an enclosed batch vessel so that the volume of the reaction mixture is constant $V = const$.

Since $V = const$ the stoichiometric relationship,

$$\frac{N_{Ao} - N_A}{a} = \frac{N_{Bo} - N_B}{b} = \frac{N_D - N_{Do}}{d} = \frac{N_G - N_{Go}}{g} = X$$

or for a general single reaction,

$$\sum \nu_j A_j = 0$$

$$\frac{N_j - N_{jo}}{\nu_j} = X$$

can be extended by dividing it by V and since,

$$\frac{N_{Ao}}{V} = C_{Ao}, \ \frac{N_A}{V} = C_A$$

etc., while,

$$\xi = \frac{X}{V}$$

is the extent per unit volume we get,

$$\frac{C_{Ao} - C_A}{a} = \frac{C_{Bo} - C_B}{b} = \frac{C_D - C_{Do}}{d} = \frac{C_G - C_{Go}}{g} = \xi$$

or,

$$\frac{C_j - C_{jo}}{\nu_j} = \xi$$

Let us that suppose we have charged the reactor with N_{Ao}, N_{Bo} moles of A and B and no D and G. A balance on A gives,

$$ar\left(C_j\right) V = -\frac{dN_A}{dt}$$

But $N_A = C_A V$ and $V = const$ so that,

$$-r_A = ar\left(C_j\right) = -\frac{dC_A}{dt}$$

The rate is in general a function of concentration of all species. Let us assume that we expect to find the rate of the form,

$$r = k_f\, C_A^{\alpha} C_B^{\beta} - k_b\, C_D^{\gamma} C_G^{\delta}$$

where $k_f = k_{fo}\, e^{-Ef/RT}$ and $k_b = k_{bo}\, e^{-Eb/RT}$. We have assumed the rate constants to obey the Arrhenius form but since we operate at constant temperature (isothermal runs) both k_f, k_b are constants to be determined from the run as well as the powers $\alpha,\beta,\gamma,\delta$.

Using the stoichiometric relationship we could eliminate all other concentrations in terms of C_A.

$$C_B = C_{Bo} - \frac{b}{a}\left(C_{Ao} - C_A\right)$$

$$C_D = C_{Do} + \frac{d}{a}\left(C_{Ao} - C_A\right)$$

$$C_G = C_{Go} + \frac{g}{a}\left(C_{Ao} - C_A\right)$$

In General:

$$C_j = C_{jo} + \frac{\nu_j}{\nu_l}\left(C_l - C_{lo}\right)$$

Thus we would have,

$$a\left\{k_f\, C_A^{\alpha}\left[C_{Bo} - \frac{b}{a}\left(C_{Ao} - C_A\right)\right]^{\beta} - k_b\, \underbrace{\frac{dg}{a^2}\left(C_{Ao} - C_A\right)}_{\substack{\text{this follows nice since} \\ C_{Do} = C_{Go} = 0 \text{ were assumed}}}^{\gamma+\delta}\right\} = -\frac{dC_A}{dt}$$

At $t = 0\ C_A = C_{Ao}$.

Clearly, we could monitor the concentration of A, C_A in time continuously, or as is more often the case discretely. We can try to evaluate the slope of C_A vs. t line, $\frac{dC_A}{dt}$, at various values of C_A.

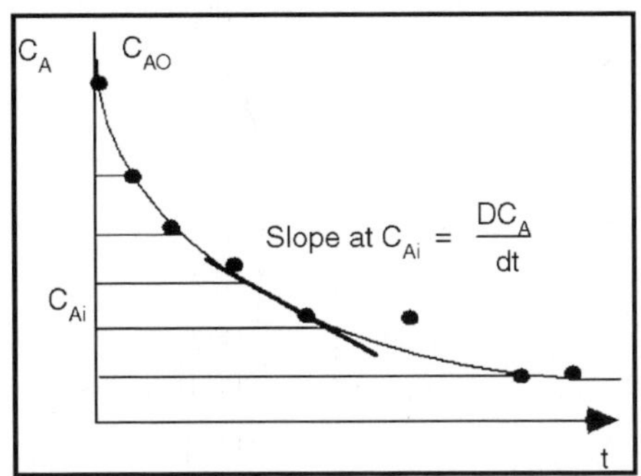

Plugging the C_A and corresponding $\frac{dC_A}{dt}$ values in the above equation we would get a large set of nonlinear equations from which we would have to determine a great number of parameters, namely six all together: α, β, γ, δ, k_f, k_b.

This method of attack due to large errors hidden in experimental data does not seem too fruitful. Instead we can decide to run the experiment only up to low conversions $x_A \doteq \frac{C_{Ao} - C_A}{C_{Ao}}$. In this situation we can neglect the rate of the reverse reaction - we are using the method of initial rates.

$$a\, k_f\, C_A^{\alpha}\left[C_{Bo} - \frac{b}{a}\left(C_{Ao} - C_A\right)\right]^{\beta} = -\frac{dC_A}{dt}$$

Now we would have to determine only three parameters from the experimental run: α, β, and k_f.

In addition, we can decide, if experimentally feasible, to make the job even easier and use reactant B in large excess over the stoichiometrically required amount i.e.,

$$\frac{C_{Bo}}{C_{Ao}} >> \frac{b}{a}.$$

The mass balance can be rewritten in the following form:

$$a\,k_f\,C_A^{\alpha}\,C_{Ao}^{\beta}\left[\frac{C_{Bo}}{C_{Ao}} - \frac{b}{a}x_A\right]^{\beta} = -\frac{dC_A}{dt}$$

but since $x_A << 1$ (low conversion) and $\frac{C_{Bo}}{C_{Ao}} >> \frac{b}{a}$ we can safely ignore $\frac{b}{a}x_A = \frac{b}{a}\frac{C_{Ao} - C_A}{C_{Ao}}$ in comparison to $\frac{C_{Bo}}{C_{Ao}}$

Thus, we get $\left(\text{with } k_f^1 = k_f\,C_{Bo}^{\beta}\right)$:

$$a\left(k_f\,C_{Bo}^{\beta}\right)C_A^{\alpha} = -\frac{dC_A}{dt} = a\,k_f^1\,C_A^{\alpha}$$

and we have used the method of excess in addition to the method of initial rates. Now we have only two parameters to determine: α and k_f^1. This seems a reasonable task. We can afterwards proceed to repeat the runs at different temperatures, still measuring rates at small conversions and using B in excess, in order to determine E_f from the Arrhenius plot. After that, experiments can be performed in excess of A to determine β and k_{fo}, and then runs starting with only C and D can be performed to determine the reverse rate's parameters γ, δ, k_{bo}, E_b.

The previous discussion indicates that if we want to determine a rate expression for a single reaction, and we monitor the concentration of one component (one species) as a function of time (say of limiting reactant A), we can almost always, by proper selection of experimental conditions, reduce the problem to one of the type:

$$-\frac{dC_A}{dt} = k\,C_A^{\alpha};\ t = 0\quad C_A = C_{Ao}$$

where from a set of C_A vs. t, i.e. C_{Ai} vs. t_i, data parameters α and k have to be determined.

For the analysis of this type we can employ two basic methods:

1. Differential method or
2. Integral method.

Let us discuss each of them.

Differential Method

Given a set of data C_A vs. t, i.e. C_{Ai} vs. t_i, determine α and k in the hypothesized rate form, kC_A^{α}.

- Tabulate the data and evaluate the differences $\Delta C_{Ai} = C_{Ai+1} - C_{Ai}$ and $\Delta t_i = t_{i+1} - t_i$.
- Evaluate the approximation to the derivative $\frac{dC_A}{dt}$.

This can be done either directly from the table of differences of the data i.e.

$$\frac{dC_A}{dt}\left(\tilde{t}_i\right) \approx \frac{\Delta C_{Ai}}{\Delta t_i} = \frac{C_{Ai+1} - C_{Ai}}{t_{i+1} - t_i}; t_i < \tilde{t} < t_{i+1}$$

or by numerical differentiation of the data, or by plotting a smooth curve through the C_A vs. t data and evaluating its slope graphically, or by matching the C_A vs. t data by some function and evaluating its derivative. Using the ratio of the differences from the table of data leads to very poor estimates of the derivative when Δt_i are uneven and large and ΔC_{Ai} are large. At the same time the question arises to what value of $\tilde{t}\left(t_i < \tilde{t} < t_{i+1}\right)$, i.e. to what value of $\tilde{C}_A\left(t_i > \tilde{C}_A > C_{Ai+1}\right)$ does the derivative approximation $\frac{\Delta C_{Ai}}{\Delta t}$ correspond. The quality (i.e., poor quality) of error prone kinetic data usually does not warrant the use of more sophisticated numerical differentiation methods.

The best practical procedure is to plot $\left(-\frac{\Delta C_{Ai}}{\Delta t_i}\right)$ verses t as a stepwise curve shown below. Since $\frac{\Delta C_{Ai}}{\Delta t_i}$ was evaluated by using C_{Ai+1}, C_{Ai} and t_{i+1}, it is valid for the interval $t_{i+1} - t_i$

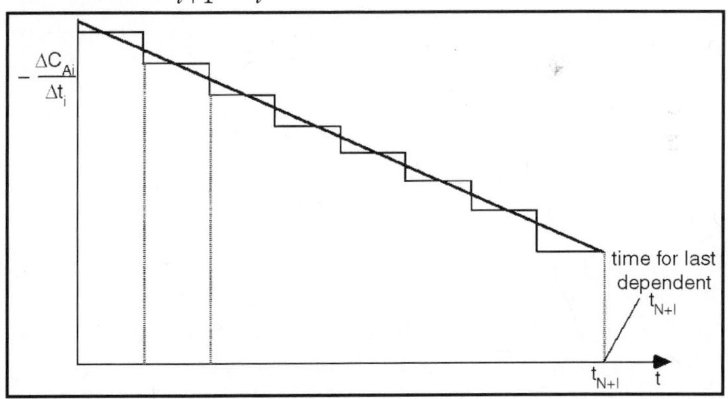

The area under every step is the value of the ordinate $\frac{\Delta C_{Ai}}{\Delta t_i}$ times the value of the corresponding time interval $t_{i+1} - t_i$, i.e., $-\frac{\Delta C_{Ai}}{\Delta t_i}\Delta t_i = C_{A_i} - C_{A_{i+1}}$ which measures the variation in concentration during one time interval. The area under the whole stepwise curve is:

$$\sum_{i=0}^{N}\left(-\frac{\Delta C_{Ai}}{\Delta t_i}\right)\Delta t_i = C_{A_O} - C_{A_{N+1}}$$

and measures the overall concentration change of A during the experiment. Yet we know that the overall concentration change we can obtain by integrating the derivative $\frac{dC_A}{dt}$ with respect to time:

$$C_{A_O} - C_{A_{N+1}} = \int_{t_o}^{t_{n+1}} \left(-\frac{dC_{Ai}}{dt}\right) dt$$

This tells us to plot a smooth curve on our step-wise plot in such a way that the area under the smooth curve is equal to the area under the stepwise curves. Since both areas are $C_{Ao} - C_{AN+1}$ the smooth curve gives us our best estimate of $\left(-\frac{dC_A}{dt}\right)$. Now for every t_i we can read the value of the derivative $\left(-\frac{dC_A}{dt}(t_i)\right)$ by going to the smooth curve at that value of t_i and reading of the corresponding value of the ordinate. Since for every t_i we have a measured C_{Ai} value, we can now pair the values $\underbrace{-\frac{dC_A}{dt}(t_i)}_{\text{means } -\frac{dC_A}{dt} \text{ evaluated at } t_i}$ and C_{Ai}.

- Now plot $-\frac{dC_A}{dt}(t_i)$ versus C_{Ai} on log-log paper. By doing this $\underbrace{\log\left(-\frac{dC_A}{dt}\right)}_{y} = \underbrace{\log k}_{a} + \alpha \underbrace{\log C_A}_{x}$

we should be able to pass a straight line through the data points provided that the rate indeed was of n-th order form and not of a more complex form.

The slope from the log-log plot gives our α. From any point then we can determine $\log k$ and k. It is advisable to determine k from a number of points and to take an average value.

The strength of the differential method is that all that we have to assume is that the rate is of a particular form (n-th order form). Both the reaction rate order α and the kinetic constant k are then extracted by the method. The main weakness of the method is poor accuracy.

Integral Method

Given a set of data C_A vs. t, i.e. C_{Ai} vs. t_i, determine α and k in the hypothesized rate form $k\, C_A^{\alpha}$.

Procedure:

- Tabulate the C_A vs. t_i data.

- Assume a particular order α, i.e. $\alpha = 1$, or $\alpha = \frac{1}{2}$ or $\alpha = 2$, etc.
- Integrate the batch mass balance.

$$\underbrace{-\int_{C_{Ao}}^{C_A} \frac{dC_A}{C_A^{\alpha}}}_{F(C_A,\, C_{Ao})} = k\int_o^t dt \qquad F(C_A, C_{Ao}) = kt$$

The form of $F(C_A, C_{Ao})$ depends on the chosen value of α:

for $\alpha = 1, F(C_A, C_{Ao}) = \ln \frac{C_{Ao}}{C_A}$

for $\alpha = 2$, $F(C_A, C_{Ao}) = \frac{1}{C_A} - \frac{1}{C_{Ao}}$ *etc.*

- Form the pairs of values $F(C_{Ai}, C_{Ao})$ vs t_i, i.e. evaluate the known function $F(C_A, C_{Ao})$ at all measured values C_{Ai} and pair them with corresponding t_i by extending the previous table of data.
- Plot $F(C_{Ai}, C_{Ao})$ on a linear plot $\underbrace{F(C_A, C_{Ao})}_{y} = k\underbrace{t}_{x}$

If the assumed order α was correct one will be able to pass a straight line through the above plotted points, the slope of the lines gives k and the assumed order α has been confirmed.

If the assumed order α was incorrect, a straight line plot cannot be obtained, and one has to go back to step 2 and assume another α and repeat all other steps.

The strength of the integral method is that the integration procedure smoothes the error of experimental data and allows more accurate evaluation of k's. The main weakness is that the method is very tedious to use unless one has a good indication of what α may be. Clearly the knowledge of the mechanism would help.

In general, if one does not have any idea what α may be, it is advisable to get an estimate of α and k using the differential method, and then retest the found α and upgrade the accuracy of the k value by the integral method.

In the above discussion we have assumed that the concentration of a reactant was monitored in time. Naturally, all the above procedures hold if instead a product concentration is monitored, or a physical quantity proportional to the composition of the system, or total pressure of the system (in case of gas phase reactions which proceed with the change in the total number of moles). Let us consider this last case:

$$aA + bB = dD + gG$$

or,

$$\sum_{j=1}^{S} \nu_j A_j = 0$$

$$\Delta n = d + g - a - b$$

$$\Delta n = \sum_{j=1}^{S} \nu_j$$

Let us assume a rate of the form:

$$-r_A = k\, C_A^{\alpha}\, C_B^{\beta}$$

Assuming ideal gases,

$$C_{Ao} = \frac{P_{Ao}}{RT} \quad C_{Bo} = \frac{P_{Bo}}{RT}$$

The total initial pressure, if there were no inerts and products present, is given by:

$$C_{Ao} + C_{Bo} = \frac{P_{Ao} + P_{Bo}}{RT} = C_{To} = \frac{P_{To}}{RT}$$

Concentrations are related to extent per unit volume ξ, as shown below:

$$C_A = \frac{P_A}{RT} = \frac{N_A}{V} = \frac{N_{Ao} + aX}{v} = C_{Ao} - a\xi$$

$$C_B = \frac{P_B}{RT} = C_{Bo} - b\xi$$

$$C_D = \frac{P_D}{RT} = d\xi$$

$$C_G = \frac{P_G}{RT} = g\xi$$

$$C_T = \left(C_A + C_B + C_D + C_G\right) = \frac{\left(P_A + P_B + P_D + P_G\right)}{RT} = \frac{P_T}{RT} = C_{To} + (\Delta n)\xi$$

$$C_T = \frac{P_T}{RT} = \frac{P_{To}}{RT} + (\Delta n)\xi \Rightarrow \xi = \frac{P_T - P_{To}}{(\Delta n)\, RT}$$

or in general,

$$C_T = C_{To} + \sum_{j=1}^{S} \nu_j\, \xi \quad \xi = \frac{P_T - P_{To}}{\sum \nu_j\, RT}$$

Now let us get an equation in terms of total pressure:

$$-\frac{dC_A}{dt} = k\, C_A^{\alpha}\, C_B^{\beta}$$

$$-\frac{1}{RT}\frac{dP_A}{dt} = k\, P_A^{\alpha}\, P_B^{\beta}\, \frac{1}{(RT)^{\alpha+\beta}}$$

$$-\frac{dP_A}{dt} = \underbrace{\left[k\,(RT)\right]^{1-(\alpha+\beta)}}_{kp}\, P_A^{\alpha}\, P_B^{\beta}$$

$$P_A = P_{Ao} - a\xi = P_{Ao} - \frac{a}{\Delta n}\left(P_T - P_{To}\right)$$

$$P_{Ao} = y_{Ao}\, P_{To}$$

↑ Mole fraction of A initially

$$\frac{dP_A}{dt} = -\frac{a}{\Delta n}\frac{dP_T}{dt}$$

$$\frac{a}{\Delta n}\frac{dP_T}{dt} = k_p\left[y_{Ao}P_{To} - \frac{a}{\Delta n}\left(P_T - P_{To}\right)\right]^{\alpha}\left[y_{Bo}P_{To} - \frac{b}{\Delta n}\left(P_T - P_{To}\right)\right]^{\beta}$$

$$\frac{dP_T}{dt} = \left(\frac{k_p\Delta n}{a}\right)\left[y_{Ao}P_{To} - \frac{a}{\Delta n}\left(P_T - P_{To}\right)\right]\alpha\left[y_{Bo}P_{To} - \frac{b}{\Delta n}\left(P_T - P_{To}\right)\right]^{\beta}$$

This equation relates now the change of total pressure P_T with time and from the $P_T(t_i)$ vs. t_i data α, β, k_p, can be determined by the previously outlined methods. Method of excess can be again used in order to further simplify the apparent rate form.

Clearly, both the integral and differential method have one thing in common - they attempt to find a relationship between properly arranged experimental data which is linear. The key is to find and plot a straight line relationship.

Differential method (n-th order reaction)

$$\underbrace{\log\left(-\frac{dC_A}{dt}\right)}_{y} = \log k + n\underbrace{\log C_A}_{x}$$

Integral method (n-th order reaction)

$$F\left(C_{A1}\right) = F\left(C_{Ao}\right) + kt\,f(n)$$

$$\underbrace{\frac{1}{C_A^{\,n+1}}}_{y} = \frac{1}{C_{Ao}^{\,n+1}} + k(n+1)\underbrace{t}_{x}$$

For rate forms other than n-th order straight line relationships are also sought. For example for Michaelis-Menten rate form:

$$-\frac{dC_A}{dt} = -r_A = \frac{k\,C_A}{K + C_A}$$

Differential Analysis

$$\underbrace{\left[\frac{1}{\left(-\frac{dC_A}{dt}\right)}\right]}_{y} = \frac{1}{k} + \frac{K}{k}\underbrace{\left(\frac{1}{C_A}\right)}_{x}$$

or,

$$\underbrace{\left[\frac{C_A}{\left(-\dfrac{dC_A}{dt}\right)}\right]}_{y} = \frac{K}{k} + \frac{1}{k}\underbrace{C_A}_{x}$$

Integral Analysis

$$\underbrace{\left(\frac{\ln \dfrac{C_{Ao}}{C_A}}{C_{Ao} - C_A}\right)}_{y} = \frac{k}{K}\underbrace{\left(\frac{t}{C_{Ao} - C_A}\right)}_{x} - \frac{1}{K}$$

Thus any combination that yields a straight line relationship is sought.

COLLISION THEORY AND RATE OF REACTION

According to collision theory, chemical reactions can occur only if energy is provided to break those bonds; the source of that energy is the kinetic energy of the molecules.

Concepts of Collision Theory

- Chemical systems are composed of atoms, ions and molecules, which randomly move at various speeds.
- The average kinetic energy of these particles is proportional to the temperature of the system.
- A chemical reaction takes place only when particles collide with each other or with the walls of the container.
- A collision will be effective if the particles have enough energy to break the bonds and collide in correct orientation.
- A collision is ineffective if the energy possessed by particles is not sufficient to break the bonds to form new type of molecules.
- The rate of reaction depends on the frequency of collision and the fraction of those collisions that are effective.

Rate =frequency of collisions × fraction of collisions that are effective

Increasing either the collision frequency or the fraction of effective collisions will increase the rate.

Reaction Mechanisms reaction mechanism a series of elementary steps that makes up an overall reaction elementary step a step in a reaction mechanism that only involves one-, two-, or three-particle collisions.

Analogy to a Reaction Mechanism

Imagine a car being assembled in a plant by a series of workers. One assembles the chassis, another adds wheels, another the seats, and so on. The worker who controls the overall rate of production of cars will not necessarily be the first or the last but, rather, the slowest worker.This worker could be called the "rate-determining worker."

If the slowest worker is at the beginning of the line, other workers will wait for cars to arrive at their station; if the slowest worker is at the end, partially assembled cars will stack up waiting to be finished. Adding workers—increasing concentration—at the fast steps will have no effect on car production. However, increasing the "concentration" of workers at the slowest step should increase the rate of production.

Chemical Example of a Reaction Mechanism

$$4HBr_{(g)}+O_{2(g)} \rightarrow 2H_2O_{(g)}+2Br_{2(g)}$$

It is highly unlikely that this reaction would occur in a single step, because a total of five reactant molecules would have to collide simultaneously with the proper alignment and sufficient energy to break and form new bonds. Experimental evidence shows that increasing the concentration of oxygen increases the reaction rate, just as we would expect. Quantitatively, the empirically determined rate equation for this system is

$$r=k(HBr)(O_2)$$

The elementary steps involved in the above chemical reaction can be theorized as follows.

$$HBr_{(g)} + O_{2(g)} \rightarrow HOOBr_{(g)} \quad (slow)$$

$$HOOBr_{(g)} + HBr_{(g)} \rightarrow 2HOBr_{(g)} \quad (fast)$$

$$2\left\{HOBr_{(g)} + HBr_{(g)} \rightarrow H_2O_{(g)} + Br_{2(g)}\right\} \quad (fast)$$

$$4HBr_{(g)} + O_{2(g)} \rightarrow 2H_2O_{(g)} + 2Br_{2(g)}$$

Theoretical interpretations for the above reaction mechanisms:

- The first elementary step is slow because it needs higher activation energy.
- The overall rate of reaction is controlled by the first reaction.
- The slowest reaction step is known as rate determining step.
- HOOBr and HOBr are known a reaction intermediates, which are formed during the reaction but immediately react again and are not present when the reaction is complete.

A potential energy diagram for the above reaction can be constructed as follows.

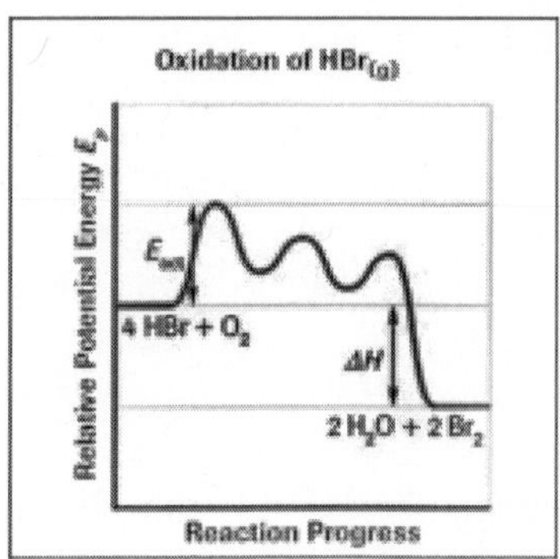

Unstable activated complexes exist at the "peaks" and slightly more stable reaction intermediates exist at the small "valleys" within a mechanism. Reaction mechanisms are only "best guesses" at the behaviour of molecules, but there are three rules that must be followed in proposing a mechanism:

- Each step must be elementary, involving no more than three reactant (and more usually only one or two) molecules;
- The slowest or rate-determining step must be consistent with the rate equation; and
- The elementary steps must add up to the overall equation.

Practice Problem

Consider the overall reaction involving three elements as reactants, and a compound as the product:

$$X + 2Y + 2Z \rightarrow XY_2Z_2$$

When a series of reactions is performed with different initial concentrations of reactants, the results are as follows:

- Doubling the concentration of X has no effect on the overall rate
- Doubling the concentration of Y multiplies the overall rate by 4
- Doubling the concentration of Z doubles the overall rate

5

Reaction Mechanisms and the Rate-Law Expression

INTRODUCTION

Knowledge about involved steps in a reaction, and to determine which step is slowest or rate determining, is called mechanism.

The intelligent guess depending on the observed rates of reactions about the series of steps (known as elementary processes) leading to the formation of products is called the reaction mechanism.

PROPERTIES OF MECHANISMS

Mechanisms describe in a stepwise manner the exact collisions and events that are required for the conversion of reactants into products. Mechanisms achieve that goal by breaking up the overall balanced chemical equation into a series of elementary steps. An elementary step is written to mean a single collision or molecular vibration that results in a chemical reaction. The following picture of an elementary step shows a single collision between water and boron trifluoride:

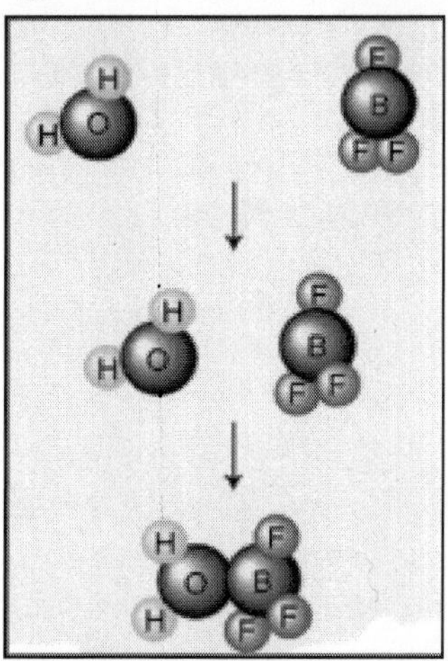

Fig. Schematic representation of an elementary step

The molecularity of an elementary step describes the number of reactive partners in the elementary step. For example, the above elementary step is called bimolecular because two molecules collide. Commonly, elementary steps are mono-, bi-, or termolecular.

The probability of four molecules colliding at exactly the same place and time is so small that we can safely assume that no reaction will ever be tetramolecular. Because take up a large amount of space, we will represent elementary steps in this SparkNote as normal reactions with molecular formula line equations. You will know from the context (*i.e.* talking about the steps of a mechanism) whether the reaction is an elementary step or an overall reaction.

To better understand mechanisms, let's consider the following mechanism for the decomposition of ozone, O_3:

$$O_3 \rightarrow O_2 + O \quad (slow)$$
$$O + O_3 \rightarrow 2O_2$$
$$Net: \quad 2O_3 \rightarrow 3O_2$$

The above mechanism exhibits a property of all mechanisms: it is a series of elementary steps whose sum is the overall balanced reaction. Note the presence of the oxygen atom, O, intermediate in the above equation. It is an intermediate because it is both created and destroyed in the mechanism and does not appear in the net equation.

Another property of mechanisms is that they must predict the experimentally determined rate law. To calculate the rate law from a mechanism you need to first know the rate limiting step. The rate limiting step determines the rate of the reaction because it is the slowest step. You can rationalize that a reaction can only go so fast as its slowest step by thinking about what happens when you encounter an accident on the highway that closes all but one lane. You may have been able to race along at 65 m.p.h. (depending on your state's laws) before you reached the lane closure but the slow passage of cars past the accident limits your rate. You can only go as fast through that one lane as the slowest car in front of you.

In the above, the first reaction is labeled as "slow". This reaction is the rate determining step because it is the slowest step. As we have stated, that means that the rate of the overall reaction is equal to the rate of the rate determining step.

The rate of an elementary step is the rate constant for that step multiplied by the concentrations of the reactants raised to their stoichiometric powers. Note that this rule only applies for elementary steps. The rate of an overall reaction is NOT the product of the concentrations of the reactants raised to their stoichiometric powers. The rate law for the first elementary step in the is rate = k [O_3]. Because this step is the rate determining step, the rate law is also the rate law for the overall reaction. Using similar techniques we can calculate the rate law predicted by any mechanism. We then check the predicted rate law against the experimentally determined rate law to test the validity of the proposed mechanism.

REACTION COORDINATE DIAGRAMS

We can follow the progress of a reaction on its way from reactants to products by graphing the energy of the species versus the reaction coordinate. We will be vague in describing the reaction coordinate because its definition is a mess of other variables composed to best make sense of the progress of the reaction. The value of the reaction coordinate is between zero and one. Understanding the meaning of the reaction coordinate is not important, just know that small values of reaction coordinate (0-0.2) mean little reaction has taken place and large values (0.8-1.0) mean that the reaction is almost over. It is a kind of scale of the progress of a reaction. A typical reaction coordinate diagram for a mechanism with a single step is shown below:

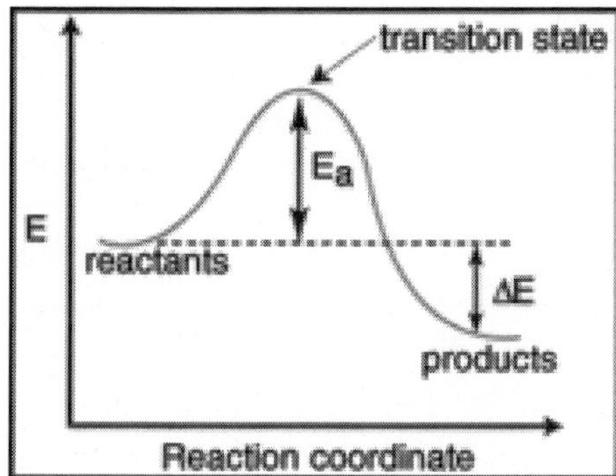

Fig. A Reaction Coordinate Diagram for a Single-step Reaction

Note that the reactants are placed on the left and the products on the right. The choice of the energy levels of the reactants and products is dictated by their energies, those with higher energies are higher on the diagram and those with lower energies are lower on the diagram. The difference is energy between the reactants and the transition state is called the activation energy. The activation energy is the height of the energy barrier of the reaction. The transition state is the point of maximum energy on the diagram which represents a species possessing both reactant-like and product-like properties. Because it is so high in energy, the transition state is very reactive and can never be isolated due to its extremely short lifetime.

The relative energy of the reactants and products, the ΔE on the diagram, determines whether the reaction is exothermic or endothermic. A reaction will be exothermic if the energy of the products is less than the energy of the reactants. A reaction is endothermic when the energy of the products is greater than the energy of the reactants. The is for an exothermic reaction. Below is a reaction coordinate diagram for an endothermic reaction.

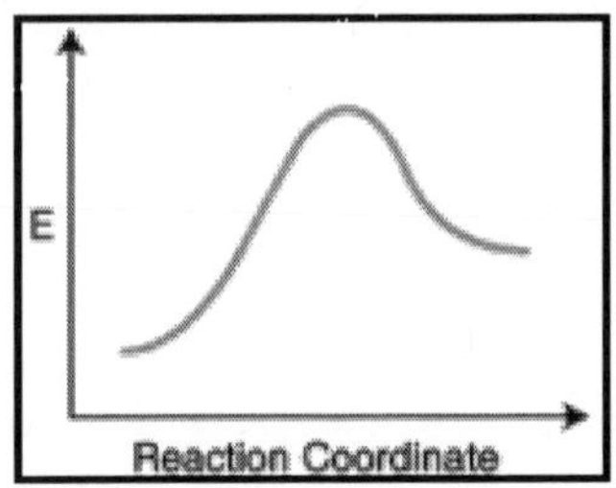

Fig. Reaction Coordinate Diagram for an Endothermic Reaction

If a reaction has *n* elementary steps in its mechanism, there will be *n*–1minima between the products and reactants representing intermediates. There will also be *n* maxima representing the *n* transition states. For example, a reaction with three elementary steps could have the following reaction coordinate diagram.

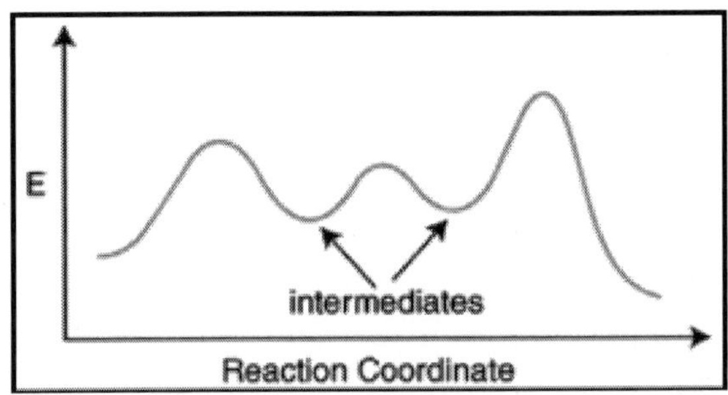

Fig. Reaction Coordinate Diagram for a Three-step Reaction

One confusing point about reaction coordinate diagrams is how to determine what the rate determining step is. Even experienced chemists consistently get this type of problem wrong. The rate determining step is not the one with the highest activation energy for the step. The rate determining step is the step whose transition state has the highest energy.

Activation Energy and the Arrhenius Equation

Intuitively, it makes sense that a reaction with a higher activation barrier will be slower. Think of how much harder you must roll a ball up a large hill than a smaller one. Let's consider chemical reactions more deeply to derive an equation which describes the relationship between the rate constant of a reaction and its activation barrier.

To simplify our derivation, we will assume that the reaction has a one-step mechanism. This elementary step represents a collision as shown in. Therefore, ,the frequency of the collisions,*f*, will be important in our equation. Notice that only a certain orientation of the molecules will lead to a reaction. For example, the following collision will not lead to a reaction. The reagent molecules simply bounce off of one another:

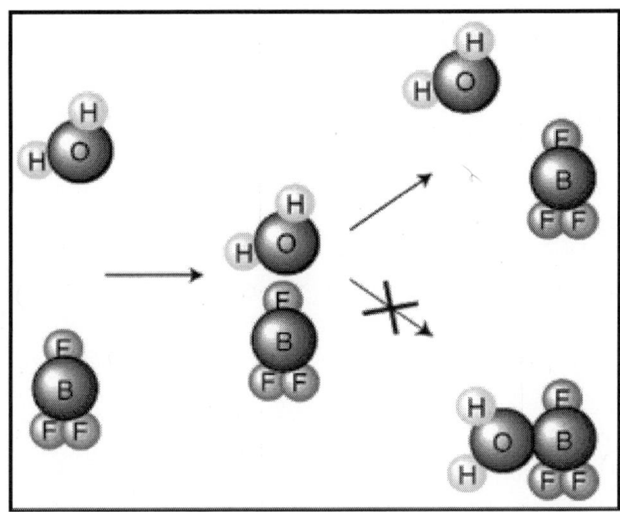

Fig. Only Specific Orientations During a Collision will Lead to a Reaction

Therefore, we will need to include an orientation factor (or steric factor), *p,* that takes into account the fact that only a certain fraction of collisions will lead to reaction due to the orientation of the molecules. Another factor we must consider is that only a certain fraction of the molecules colliding will have enough energy to overcome the activation barrier. The Boltzmann distribution is a thermodynamic equation that tells us what fraction of the molecules have a certain amount of energy. As you know, at higher temperatures the average kinetic energy of the molecules increases. Therefore, at higher temperatures more molecules have an energy greater than the activation energy—as shown in.

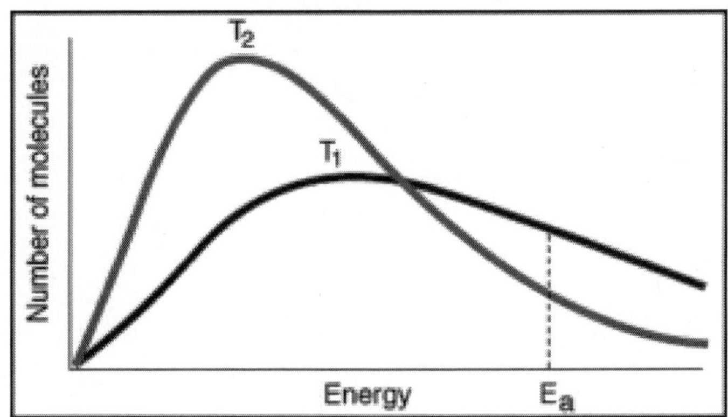

Fig. Boltzmann Distributions for T_1 Greater than T_2

Combining the above considerations, we state the following relationship between the rate constant and the activation energy, called the Arrhenius equation:

$$k = f\,p\,e^{-Ea/RT}$$

The variable k is the rate constant, which is dependent on the frequency of the collisions *f*, orientation factor *p*, activation energy E_a, and temperature T. From the expression for the Arrhenius equation you should note that a small increase in activation energy leads to a large decrease in rate constant. Furthermore, temperature has a similarly exponential effect on the rate constant. An experimental rule of thumb is that a 10°C increase in temperature leads to a doubling of the rate constant.

One application of the Arrhenius equation that is useful is the determination of the activation energy for a reaction. Taking the natural log of the Arrhenius equation gives a linear equation:

$$In\,k = -\frac{E_a}{R}\left(\frac{1}{T}\right) + In\left(f\,p\right)$$

A graph of ln k versus 1/ T should give a straight line whose slope is - E_a/ R. By measuring the rate constant at a range of different temperatures, you can construct a graph to determine the activation energy of a reaction.

Catalysis

A catalyst speeds up a reaction without being explicit in the overall balanced equation. It does this by providing an alternate mechanism for the reaction that has a lower activation barrier than does the uncatalyzed pathway. Compare the catalytic and regular mechanisms for the hydrogenation of ethylene to ethane and their associated reaction coordinate diagrams in:

	Uncatalyzed mechanism:		*Pd catalyzed mechanism:*
	$H_2 \rightleftharpoons 2H$		$H_2 + 2Pd \rightleftharpoons 2Pd - H$
	$C_2H_4 + H \rightarrow C_2H_5$		$C_2H_4 + Pd - H \rightarrow C_2H_5 - Pd$
	$C_2H_5 + H \rightarrow C_2H_6$		$C_2H_5 - Pd + Pd - H \rightarrow C_2H_6 + 2Pd$
Net:	$C_2H_4 + H_2 \rightarrow C_2H_6$	*Net:*	$C_2H_4 + H_2 \rightarrow C_2H_6$

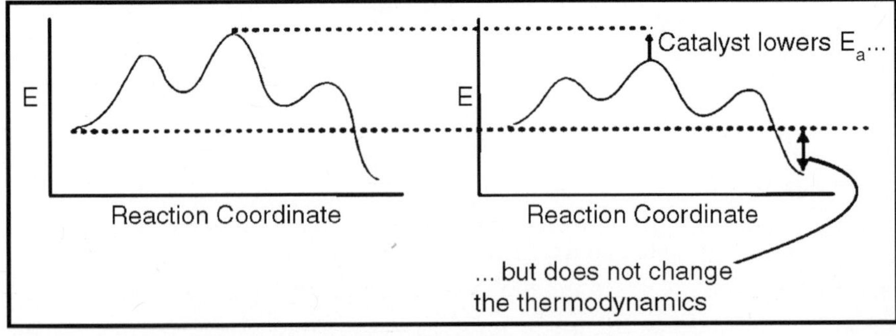

Fig. Mechanisms of Ethylene Hydrogenation

As you can see, the catalyst changes the mechanism of the reaction and lowers the activation energy. The catalyst, because it does not appear in the overall balanced equation has absolutely no effect on the thermodynamics of the reaction.

There are two types of catalysts—heterogeneous catalysts and homogeneous catalysts. There is no fundamental difference in how these catalysts work.

The difference lies in whether the catalyst is in the same phase (solid, liquid, or gas) as the reagents. A homogeneous catalyst is in the same phase as the reactants while a heterogeneous catalyst is not. An enzyme is a biological homogeneous catalyst.

DESCRIPTION OF A REACTION MECHANISM

Because a reaction mechanism is used to describe what occurs at each step of a reaction, it also describes the transition state, or the state when the maximum of potential energy is reached, in a reaction.

A mechanism must show the order that the bonds form or break and the rate of each elementary step. Also accounted for in a mechanism are the reaction intermediates, which are stable molecules that do not appear in the experimentally determined rate law because they are formed in one step and consumed in a subsequent step. Since a reaction cannot proceed faster than the rate of slowest elementary step, the slowest step in a mechanism establishes the rate of the overall reaction. This elementary step is known as the rate-determining step.

A mechanism must satisfy the following two requirements:

1. The elementary steps must add up to give the overall balanced equation for the reaction.
2. The rate law for the rate-determining step must agree with the experimentally determined rate law.

ELEMENTARY PROCESSES

Elementary processes are almost always either unimolecular or bimolecular. A unimolecular elementary process is when a single molecule dissociates. A bimolecular elementary process is when two molecules collide with each other to form products. A third process, called termolecular, is rare because it involves three molecules colliding at the same time. In the bimolecular and termolecular processes, the molecules may be different or the same.

Although a rate law for an overall reaction can only be experimentally determined, the rate law for each elementary step can be deduced from the chemical equation through inspection. A unimolecular elementary step has a first order rate law, whereas a bimolecular elementary step has a second order rate law. The table below summarizes the types of elementary steps and the rate laws that they follow. A, B, and C here represent the reactants or reaction intermediates.

Elementary Steps and Rate Laws

Molecularity	Elementary Step	Rate Law for Elementary Step
Unimolecular	A → products	rate = k[A]
Bimolecular	A + B → products	rate = k[A][B]
	A + A → products	rate = $k[A]^2$
Termolecular	A + A + B → products	rate = $k[A]^2[B]$
	A + A + A → products	rate = $k[A]^3$
	A + B + C → products	rate = k[A][B][C

Elementary processes are also reversible. Some may reach conditions of equilibrium where both the foward and reverse reactions are equal.

BIMOLECULAR ELEMENTARY PROCESSES

Mechanism Example with a Slow Step and a Fast Step

$$H_{2\,(g)} + 2\,ICl_{\,(g)} \rightarrow I_{2\,(g)} + 2\,HCl_{\,(g)}$$

The reaction between the gases iodine-monochloride and hydrogen produce two gases, iodine and hydrogen chloride as products. Through experiment, the rate law is found to be:

rate of reaction = $k\,[H_2]\,[ICl]$

Using a proposed reaction mechanism:

- *Slow reaction:* $H_2 + ICl \rightarrow HI + HCl$ Rate(1): $k_1\,[H_2]\,[ICl]$
- *Fast reaction:* $HI + ICl \rightarrow I_2 + HCl$ Rate(2): $k_2\,[HI]\,[ICl]$

Overall reaction: $H_2 + 2\,ICl \rightarrow I_2 + 2\,HCl$

The mechanism chosen for the reaction above proposes that Step (1) is the rate-determining step, since it occurs slower. According to our mechanism, the rate law agrees that the mechanism works for this reaction. The rate of Step (1) = $K_1\,[H_2]\,[ICl]$ is in agreement with the experimentally determined rate law, so the mechanism matches the stoichiometry of the overall reaction. However, this does not mean in any manner that the mechanism is proven to be correct. It only shows that the chosen mechanism works perfectly for this reaction. This is a reasonable proposal.

When there is enough energy for the slow reaction to react, the reaction moves to a faster step, Step (2). Step (2) occurs rapidly and our mechanism suggests that HI is consumed in the second step as fast as it is being formed. HI is, therefore, the reaction intermediate.

The diagram below shows the reaction progress of the two-step mechanism proposed.

Notice how the reaction intermediate and the transition states do not last very long, but the intermediate can be isolated due to its fully formed bonds.

Diagram of a Two-Step Reaction

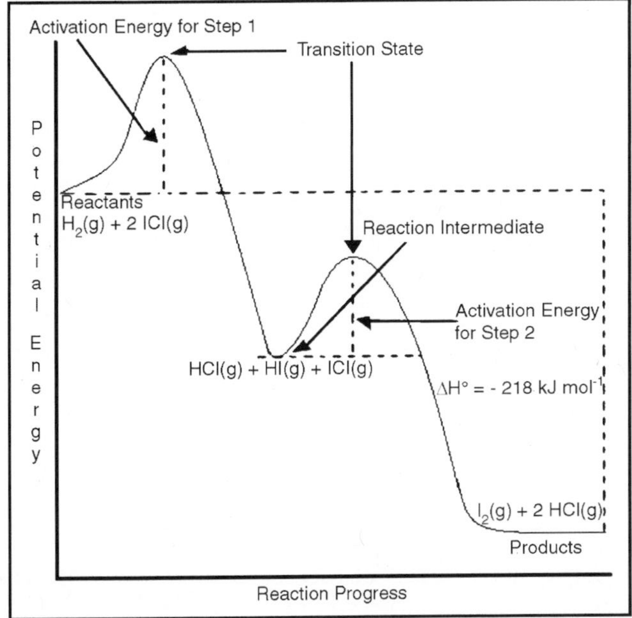

Mechanism Example with a Fast Reversible First Step then a Slow Step

Here is the equation of nitric oxide reacting with oxygen to produce nitrogen dioxide.

$$2\ NO_{(g)} + O_2 \rightarrow 2\ NO_{2\ (g)} \text{ rate of reaction} = k\ [NO_2]^2[O_2]$$

Since a termolecular mechanism, or a reaction occurring in a single step, is extremely rare, we can suggest a mechanism with two elementary steps that are bimolecular.

Step 1 (slow):

$$2\ NO_{\ (g)} \leftrightarrow N_2O_{2\ (g)}$$

Step 2 (fast):

$$N_2O2\ (g) + O_{2\ (g)} \rightarrow 2\ NO_{2\ (g)}$$

The proposed mechanism has the same stoichiometry as the overall reaction. Step 1 is a reversible process in which equilibrium is reached quickly. Step 2 is then the rate determinant, the step with a slow rate. The relative rate of the rate determinant is $k_3[N_2O2][O_2]$, which does not seem to match with the experimental rate of $k\ [NO_2]^2[O_2]$. However, N_2O_2 is actually a reaction intermediate, because it appears in both elementary steps but not in the equation of the overall reaction. Since an intermediate cannot be in the rate law for the overall reaction, we want to have a value that is equivalent to $[N_2O_2]$ to substitute into the rate equation. Since Step 1 is a fast, reversible reaction, we can make the assumption that:

$$k_1\ [NO]_2 = k_2\ [N_2O_2],$$

$$\text{so } [N_2O_2] = (k_1/k_2)\ [NO]_2$$

Through substitution, $k_3[N_2O2][O_2] = k_3(k_1/k_2)\ [NO]_2[O_2]$. The rate constant $k = k_3(k_1/k_2)$, so rate of reaction is $k\ [NO]_2[O_2]$, thus matching the experimental rate law. This proposed mechanism fulfills both requirements needed to be considered a possible mechanism.

ELEMENTARY REACTION STEPS - MOLECULARITY

We have made a big deal out of the fact that you cannot predict the rate law for a chemical reaction from the balanced equation for that reaction. We now introduce a new class of simple reactions, called elementary reaction steps, for which the rate law IS determined by the balanced reaction equation. We also introduce a new term, molecularity, which tells us the number of molecules involved in an elementary reaction step.

We emphasize that the reaction velocity for each elementary reaction step IS determined by the reaction step and, later on, we will string a sequence of elementary reaction steps together to form what is called amechanism. A reaction mechanism is a detailed (theoretical) description of how we think the chemical reaction proceeds. That is, it describes our thought about which molecule collides with which other molecule to form an intermediate product, which my go on to react with some other species, and so on, to produce the overall reaction. The elementary reaction steps must be balanced (as do all chemical reactions). We can usually tell the difference between an elementary reaction step and a balanced reaction by the fact that the elementary reaction step will have one or more rate constants associated with it, as we will see below.

It is probably easiest to describe the elementary reaction steps and their associated rate laws by just telling you what they are. As we proceed you will become aware of the fact that the rate law is completely determined by the elementary reaction step you write down. (The converse is not true. We will see in the examples below that several different elementary steps may give rise to the same rate law.)

Unimolecular Reaction Steps

The elementary reaction step,

$$A \xrightarrow{K_1} B,$$

is unimolecular because there is only one molecule reacting, that is, molecule "A" is reacting. This unimolecular reaction step implies the rate law,

$$\frac{d[B]}{dt} = k_1[A],$$

or, equivalently,

$$\frac{d[A]}{dt} = -k_1[A].$$

In words, these elementary reaction steps say that the molecule, A, spontaneously transforms into B at some rate k_1. The algebraic sign in front of k_1 tells whether you are gaining product or losing reactant depending on whether the concentration in the derivative is increasing or decreasing.

An elementary reaction step may be reversible or irreversible. An irreversible unimolecular step. Equation below is a reversible unimolecular step:

$$A \underset{k_{-1}}{\overset{k_1}{\rightleftarrows}} B.$$

This reversible unimolecular step implies the following rate laws,

$$\frac{d[B]}{dt} = +k_1[A] - k_2[B]$$

and/or,

$$\frac{d[A]}{dt} = -k_1[A] + k_2[B].$$

(Either one of these may be used, depending on whether we are trying to account for the disappearance of reactant, A, or the appearance of product, B, in our mechanism for a particular reaction.)

A unimolecular reaction step can have more than one product, for example,

$$A \xrightarrow{k_1} B + C \cdot$$

The unmimolecular process implies the same rate law as the reaction.

Bimolecular Reaction Steps

There are several varieties of bimolecular steps. For example,

$$A + B \xrightarrow{k_2} C$$

implies the rate law,

$$\frac{d[C]}{dt} = +k_2[A][B],$$

or,

$$\frac{d[A]}{dt} = -k_2[A][B],$$

and so on. We have given only one product, "C." We would get the same rate laws if there had been two or more products, for example as in,

$$A + B \xrightarrow{k_2} C + D \cdot$$

Reversible Bimolecular Steps

The bimolecular reaction,

$$A + B \underset{k_{-2}}{\overset{k_2}{\rightleftarrows}} C$$

implies the rate law,

$$\frac{d[C]}{dt} = +k_2[A][B] - k_2[C],$$

or it could be written as a rate of loss of A or B as we have seen above. The reversible bimolecular reaction,

$$A + B \underset{k_{-2}}{\overset{k_2}{\rightleftarrows}} C + D$$

implies,

$$\frac{d[C]}{dt} = +k_2[A][B] - k_2[C][D],$$

and its variants.

Termolecular Reaction Steps

Termolecular reaction steps require three molecules coming together at the same time. They are rare because three-body collisions in the gas phase are rare, but there are cases of termolecular reactions in the literature. There are many varieties of termolecular reactions and they may be reversible. A typical termolecular process might be,

$$A + B + C \xrightarrow{k_3} D + other\ products,$$

with its implied rate law,

$$\frac{d[D]}{dt} = +k_3[A][B][C].$$

A reaction mechanism is a combination of one or more elementary reaction steps which start with the appropriate reactants and end with the appropriate product(s). We will illustrate the construction of a reaction mechanism with the famous Lindemann mechanism for a class of gas-phase reactions.

FACTORS AFFECTING REACTION RATE; THE RATE LAW

The rate of a chemical reaction depends on a number of factors, some of which are listed below:

- The molecular structure(s) of the reactant(s)
- The medium in which the reaction is carried out (for example, does the reaction occur in the gas phase, or does it occur with reactants dissolved in a solvent?)
- Temperature
- The presence or absence of a catalyst
- The concentrations of the reacting species

The dependence of rate on molecular structure is the most obvious but least understood factor. This complex matter is a focus of intense laboratory research. At present, it is not possible to summarize this dependence in a neat, clean package. In fact, this incompletely understood factor is what prevents predictive ability in kinetics.

For molecules to react, they must directly interact; to directly interact, they must collide. This leads us to expect that rate should be proportional to the frequency of molecular collisions. In turn, the frequency of collisions should depend on the number of molecules per unit volume, which is the concentration. This concentration dependence of rate is indeed observed experimentally.

For most chemical reactions, the dependence of reaction rate on the concentrations of the reacting species is surprisingly simple. In almost all cases, the reaction rate is found to be proportional to the product of the concentrations of the reactants, each concentration raised to an integer power that is almost always 1 or 2. Thus for the general reaction, the rate law is expected to take the form:

$$\text{Rate} = -1/a\ \Delta[A]/dt = k\ [A]^x\ [B]^y$$

Equation is called the differential rate law, because it involves the ratio of differentials (derivative), D[*A*]/*dt*. The exponents, *x* and *y*, are integers, and for most reactions are either 1 or 2. It must be stressed that in most cases *x* and *y* are different from the stoichiometric coefficients, *a* and *b*; it is not valid to write the rate law for a reaction by raising the concentrations to the stoichiometric powers, unless the reaction is an elementary process.

We will have more to say about elementary processes in a subsequent section. The integer, *x*, is called the order of the reaction in reactant *A*; *y* is the order in *B*. The sum, $x + y$, is called the overall reaction order.

The reaction orders must be determined by experiment; they can not be written down simply by looking at the reaction stoichiometry. To illustrate this statement for a specific case, the reaction below, involving 31 moles of reactants, has the accompanying rate law. This very complex overall reaction has a simple rate law, which is first order in each reactant, and second order overall.

$$30C_2H_5OH + B_{10}H_{14} \rightarrow 10B(OC_2H_5)_3 + 22H_2(g)$$
$$\text{Rate} = k[C_2H_5OH][B_{10}H_{14}]$$

Finally, k (and in the expression just above) is the proportionality constant relating Rate to the concentration product. It is called the rate constant for the reaction. The rate constant contains implicit within it information about the dependence of rate on molecular structure, reaction medium, and temperature. A word about the units of the rate constant is in order. Generally, k has whatever units are required to give the rate units of concentration/time. Example.shows how units for the rate constant may be determined for various rate laws.

Example: For each rate law below, state the order of the reaction in each reactant, and determine the units of the rate constant, k.

- -d[A]/dt = k
- -d[A]/dt = k[A]
- -d[A]/dt = $k[A]^2$
- -d[A]/dt = k[A][B]
- -d[A]/dt = $k[A]^2[B]$

Solution:

Rate Law	Reaction Order	Units of k
-d[A]/dt = k	Zero order (no concentration dependence)	k has units of rate, Moles L^{-1} $time^{-1}$
-d[A]/dt = k[A]	First order in A	Moles/L-time = k(moles/L) so k has units of $time^{-1}$
-d[A]/dt = $k[A]^2$	Second order in A	Moles/L-time = $k(moles/L)^2$ so k has units L $moles^{-1}$ $time^{-1}$
-d[A]/dt = k[A][B]	First order in A First order in B Second order overall	Same as for the previous case
-d[A]/dt = $k[A]^2[B]$	Second order in A First order in B third order overall	Moles/L-time = $k(Moles/L)^3$ so k has units L^2moles^{-2} $time^{-1}$

Some Specific Examples of Rate Laws. Zero-order Reactions. Chemical reactions having overall order zero are rare. However, evaporation of a pure liquid, is an example of a physical process with a zero-order rate law. Recall that the rate of evaporation of a liquid having exposed surface area, A, is just,

$$Rate = k*A$$

As long as the exposed surface area remains constant, we may rewrite this as,

$$Rate = k', \text{ where } k' = k*A$$

Since this rate is independent of concentration of liquid, the process is zero order in liquid.

First-Order Reactions

Reactions and physical processes exhibiting first order rate laws are very common. Condensation is a first order physical process. The rate of condensation of vapour to liquid in a closed system is given by

$$Rate = k*P_{vap}$$

where P_{vap} is the vapour pressure of the substance. However, P is related to concentration in moles/L by the ideal gas law, leading to,

$$\text{Rate} = -d[\text{vapour}]/dt = k^*[\text{vapour}]$$

Molecular rearrangements (often called isomerization reactions) frequently follow first order rate laws.

Vision is critically dependent upon cis-retinal → trans-retinal where cis- and trans- retinal contain the same atoms, bonded together in the same way, but with somewhat different spatial arrangements. The rate of conversion of cis- to trans- form is given.

$$-d[\text{cis}]/dt = k[\text{cis}]$$

Similarly, trans- converts to cis- by a first order rate law with a different rate constant. Finally, nuclear decay processes are first order. Thus ^{238}U decays according to equation, which is governed by the rate law.

$$^{238}\text{U} \rightarrow {}^{234}\text{Th} + {}^{4}\text{He}$$

$$\text{Rate} = -d(\text{mass } ^{238}\text{U})/dt = k\ (\text{mass } ^{238}\text{U})$$

The rate constant for this process has the extremely small value, 1.54×10^{-10} year^{-1}.

The rate constant, k, for a first-order process has units of time^{-1}. It's reciprocal, 1/k, thus has units of time. The quantity, 1/k, is given the symbol t, and is called the time constant for the reaction. This is a measure of the time required for the reaction to take place. The time constant for decay of ^{238}U is $1/k = 6.49 \times 10^9$ years. The fact that this time constant is so large is the reason why ^{238}U is still found in the earth's crust, which is estimated to be about 4.5 billion years old.

Example: The time constant for the radioactive isotope, ^{226}Ra, is 2338 years. Calculate the number of atoms of ^{226}Ra that decay each minute in a 2.00 g sample of ^{226}Ra.

Solution: Nuclear decay is a first order process. We can therefore write the following rate law for decay of ^{226}Ra:

$$\text{Rate} = -d(^{226}\text{Ra atoms})/dt = k(\text{Number of } ^{226}\text{Ra atoms})$$

$$-dN/dt = kN$$

where N is the number of ^{226}Ra atoms in the sample, and $k = 1/t = 4.28 \times 10^{-4}$ year^{-1}.

$$N = (2.00\text{g}/226.025\text{ g/mole}) * 6.023 \times 10^{23}\text{ atoms/mole} = 5.33 \times 10^{21}\text{ atoms.}$$

$$-dN/dt = (4.28 \times 10^{-4}\text{ year}^{-1})(5.33 \times 10^{21}\text{ atoms}) = 2.28 \times 10^{18}\text{ atoms/year}$$

There are 5.256×10^5 minutes in 1 year. Therefore,

$-dN/dt = (2.28 \times 10^{18}\text{atoms/year})/(5.256 \times 10^5\text{ minutes/year})$

$= 4.34 \times 10^{12}$ atoms/minute

Second Order Processes

Many reactions exhibit second-order rate laws, either simple (second-order in a single reactant) or mixed (first-order in each of two reactants). Here we consider two specific reaction types that exhibit second-order kinetics.

Dimerization reactions, in which two molecules unite to form a larger molecule, obey simple second-order kinetics.

The dimerization of NO_2 is shown below:

$$: 2NO_2(g) \rightarrow N_2O_4(g)$$

$$\text{The rate law:: } -d[NO_2]/dt = k[NO_2]^2$$

where the second order rate constant is — $M^{-1}s^{-1}$ at °C. Other examples of dimerization processes are the formation of a Cl_2 molecule from two Cl atoms and the formation of one molecule of C_4H_8 from two molecules of C_2H_4. In general, a reaction of the general type is expected to obey simple second-order kinetics : $2\ A \rightarrow A_2$

Adduct formation processes, such as Lewis acid-base reactions and the formation of complexes involving enzymes and their substrates, are also second-order processes. Most such processes occur very rapidly. The role of enzyme-substrate complexes in enzyme catalysis will be discussed in a subsequent section.

"Net" Reaction Rate; A Prelude to the Next Section

In the next section we discuss experimental approaches to the determination of rate laws for reactions. To preface that discussion, we briefly discuss here the concept of net reaction rate. Except in some cases, we can experimentally measure only the net rate of a reaction, where the net rate is defined as:

Net reaction rate = rate of forward reaction – rate of reverse reaction

$$\text{Rate}_{net} = \text{Rate}_{forward} - \text{Rate}_{reverse}$$

The general reaction is reproduced below, this time acknowledging that in general the reverse process also occurs:

$$aA + bB \Leftrightarrow dD + fF$$

For this reaction we expect the forward rate to depend only on the concentrations of reactants, A and B, and the reverse rate to depend only on the concentrations of products, D and F:

$$\text{Rate}_{forward} = k_f[A]^x[B]^y$$

$$\text{Rate}_{reverse} = k_r[D]^p[F]^q$$

The net rate is therefore expected to be a potentially complex function of the concentrations of reactants AND products. The experimentally determined rate constant is a composite function of the forward and reverse rate constants, k_f and k_r.

$$\text{Rate}_{net} = k_f[A]^x[B]^y - k_r[D]^p[F]^q$$

Once reaction reaches equilibrium, the net rate will of course be zero. Applying this idea to, our familiar criterion for dynamic equilibrium:

$$\text{Rate}_{net} = 0 = \text{Rate}_f - \text{Rate}_r$$

$$\text{Rate}_f = \text{Rate}_r$$

There are a few situations in which the net rate is equal to the forward reaction rate:

- One of the products escapes from the reaction system, so that its concentration remains at or very near zero at all times. A gaseous product that escapes from a solution is an example of this situation.
- The reaction has a large value of K_{eq}, so that the reverse rate is negligible except near equilibrium, which is not achieved until almost all of the limiting reactant has been used.
- At the very beginning of a reaction in which only reactants are initially present. At time zero, the reverse rate must be zero because no products are present yet.

EXPERIMENTAL DETERMINATION OF THE RATE LAW THE INITIAL RATE METHOD

When applicable, the method of initial rates is a simple and clean way to obtain the rate law. The initial rate of a reaction is the instantaneous rate at t=0, when reactants are first mixed together and begin to react. In a plot of reactant concentration versus time, it is given by the slope of the tangent at t = 0. The slope of this type of plot is always steepest at the beginning; the initial rate is thus the fastest rate that we observe during the time course of the reaction process.

One major advantage of the initial rate method is that the reverse rate is zero, provided only reactants and NO products are initially present. Thus only the forward rate contributes to the initial rate, and complications resulting from the reverse reaction are avoided. The initial rate law depends only on concentrations of reactants. The method of initial rates is best illustrated by example.

Example: The initial rate was measured at several combinations of initial reactant concentration for the following reaction:

$$A + 2B + 3C \rightarrow D + E$$

The rate law is expected to have the following general form:

$$-d[A]/dt = k[A]^x[B]^y[C]^z$$

The following data were obtained. Use the data to determine the values of x, y and z.

Run Number	$[A]_o$, M	$[B]_o$, M	$[C]_o$, M	Initial Rate, M/s
1	0.05	0.10	0.10	6.00×10^{-3}
2	0.05	0.15	0.10	13.5×10^{-3}
3	0.05	0.15	0.05	13.5×10^{-3}
4	0.10	0.10	0.05	12.0×10^{-3}

Solution: The key to the initial rate method is to examine the dependence of rate on one reactant at a time, using data obtained when the other reactants do not vary in initial concentration. In the first two data entries above, the initial concentrations of both A and C are the same in both runs; only the initial concentration of B is changed.

These two data items can therefore be used to determine the order of the reaction in B. We see that when the initial concentration of B is increased from 0.10 to 0.15, an increase of a factor of 1.5, the rate increases from 6 to 13.5 x 10^{-3} moles/L-s, which is a factor of 2.25. We translate this information into equation form:

$$\text{Rate 2/Rate 1} = [B]_2^y/[B]_1^y = ([B]_2/[B]_1)^y$$
$$2.25 = (1.5)^y$$

We recognize that y must be 2. The reaction is therefore second order in B, and the rate law thus far is

$$-d[A]/dt = k[A]^x[C]^z[B]^2$$

In runs 2 and 3, the initial concentrations of A and B are the same, while that of C is changed. We see from these data that even though the concentration of C is only half as large in run 3 as in run 2, the rate is unaffected. We conclude that the rate does not depend on [C], hence that z = 0. The rate law to this point is

$$-d[A]/dt = k[A]^x[B]^2$$

There are no two kinetics runs in which only [A] is varied. However, runs 1 and 4 can be used to obtain the order in A because we have just determined that rate does not depend on [C]. Runs 1 and 4 indicate that the rate doubles when the initial concentration of A doubles. The order in A is thus 1, and the final rate law is

$$-d[A]/dt = k[A][B]^2$$

Once the rate law is known, the rate constant may be calculated from any one of the individual runs. Choosing the first run,

$$k = \text{Rate}/[A][B]^2 = 6.00 \times 10^{-3}/(0.05)(0.10)^2 = 12.0\ L^2/\text{mole}^2\text{s}$$

As this example shows, the initial rate method, when it is applicable, provides a clean, simple path to the rate law. The italicized qualifier is important, however; in practice, the method of initial rates is only rarely used, for several reasons. In most kinetics studies, the chemist examines the manner in which the concentration of a selected reagent varies with time. The resulting data are presented either in a table, or in a plot of concentration versus time.

The data obtained at the beginning of the reaction, where the rate is largest and the time interval is smallest, is usually less reliable than that obtained at later times, when the rate is less. It is undesirable to base an analysis on one's least reliable data. Second, and more important, the method of initial rates uses only a small fraction of the concentration time data; the rest is wasted.

Third, to apply the initial rate method as was done in the example above requires that full concentration-time profiles be collected for each run, in order to extract the slope of the concentration-time plot at t=0. Collection of this much data requires a lot of experimental effort.

Most of the resulting data is then ignored if the initial rate method is used. For these reasons and others, chemists have developed other methods for determining rate laws from data, which make use of all of the data collected. Some of these will be discussed now.

Integration of the Rate Law

It is possible to convert the differential rate law into an integrated form, in which concentration is expressed directly as a function of time. The concentration-time data obtained during a kinetics run in the lab may then be matched against the function. If the data conform to the function, it may be concluded that the reaction follows the corresponding differential rate law. Integration is very easy for rate laws of the form,

$$-d[A]/dt = k[A]^n$$

We will examine the first and second-order cases, with n = 1 and 2 respectively. The first order rate law may be rearranged to put the concentration terms on one side and time on the other side of the equals sign:

$$-d[A]/[A] = kdt$$

Both sides of the equation are then integrated over the limits, $[A] = [A]_o$ at t=0 and $[A] = [A]_t$ at some later time t:

$$\int_{[A]0}^{[A]t} d[A]/[A] = k\int_0^t dt$$

The integral of dx/x is ln x, and the integral of dx is x. Evaluating over the limits gives:

$$\ln ([A]_t/[A]_o) = -k\,t$$

This is the so-called integrated first order rate law. For a reaction following simple first order kinetics, a plot of ln $[A]_t$ versus t is linear, with slope -k. Thus if concentration-time data for a reaction are plotted according to equation, and linearity is obtained, it may be concluded that the reaction follows a simple first-order differential rate law. An alternative expression is obtained by exponentiating both sides:

$$[A]_t = [A]_o e^{-kt}$$

Here e = 2.718... is the base of natural logarithms. Substituting the reciprocal of the time constant, t, for the rate constant gives

$$[A]_t = [A]_o e^{-t/t}$$

Thus when an interval of time equal to the time constant has elapsed, the concentration of A will have decreased to 1/e (about 1/3) of its initial value.

With the integrated first-order rate law in hand, we are in a position to consider a quantity called the half-life, $t_{1/2}$, which is similar but not identical to the time constant. The half-life for a process is defined as the time required for the concentration of the limiting reactant to decrease to 1/2 of its initial value.

An expression for the half life is easily obtained from by realizing that $[A]_t = [A]_o/2$ at $t = t_{1/2}$. Making these substitutions leads to the following expression for the half-life of a first-order reaction:

$$t_{1/2} = \ln 2/k = 0.693/k$$

For n = 2, corresponding to a second-order process, rearrangement leads:

$$: -d[A]/[A]^2 = k\ dt$$

Integration of both sides of this equation results:

$$1[A] - 1/[A]_o = k\ t$$

For a reaction with a simple second order rate law, a plot of concentration-time data as 1/[A] versus t is linear, with slope k. Thus a reaction for which the concentration-time data produce a linear 1/[A] versus t plot may be concluded to follow a second-order rate law.

Example: Plot the concentration-time data in Example.according to equations to determine whether the hydrolysis of *t*-butyl bromide obeys a simple first- or second-order rate law.

Solution: The concentration-time data from Example.are reproduced below. Columns have been added for ln [*t*-BuBr] and 1/[*t*-BuBr]

time, s	[t-BuBr], moles/L	ln [t-BuBr]	1/[tBuBr]
0	0.100	–2.303	10
10	0.0876	–2.435	11.42
20	0.0768	–2.567	13.02
30	0.0672	–2.700	14.88
40	0.059	–2.830	16.95
50	0.0517	–2.962	19.34
60	0.0453	–3.094	22.08
80	0.0348	–3.358	28.74
100	0.0267	–3.623	37.45
120	0.0205	–3.887	48.78
180	0.0093	–4.678	107.53
240	0.0042	–5.473	238.10

The linearity of the first plot indicates that the reaction is first-order in *t*-BuBr. The rate constant, obtained from the slope is $1.32 \times 10^{-2}\ s^{-1}$ (recall that for a first order reaction, the units of the rate constant must be $time^{-1}$, resulting from the division of rate in conc/time by concentration).

Example: One product resulting from fission of ^{235}U in nuclear reactors is ^{137}Cs, a radioactive nucleus with a time constant of 43.3 years. How much time will it take for the radioactivity level of this waste to fall to 1per cent of its initial value, which is considered a safe level?

Solution: This problem is readily solved using equation :

$$\ln [A]/[A]_o = -kt = -t/t$$

$$t = -\ln(0.01)t = -\ln(0.01)(43.3) = 200 \text{ years}$$

The waste must be safely stored for 200 years before it will be safe.

Finite Difference Methods of Numerical Analysis

Another approach to determination of the rate law and rate constant from concentration-time data is to calculate the average rate over each experimental time interval. The average rates can then be plotted in turn against concentration, concentration squared, and so on until a linear plot is obtained.

This is a relatively simple way to determine whether a reaction is first-order or simple second order (that is, has a rate law of the form, Rate = $k[A]^2$). Estimating derivatives (slopes) and integrals (areas) using finite changes in variables is called the Method of Finite Differences. We illustrate this method using the data for hydrolysis of *t*-BuBr from Examples.

Example: Use finite difference methods to obtain the rate law and rate constant for the hydrolysis of *t*-butyl bromide according to equation.

Solution: The time and concentration data from Example are reproduced below. In addition, values of Dt, Δ[t-BuBr], and rate = Δ[t-BuBr]/Dt have been calculated for each pair of successive data points.

time, s	Δt, s	[t-BuBr],	D[t-BuBr], moles/L	Rate × 10^3 moles/L
0		0.100		
10	10	0.0876	–.0124	1.24
20	10	0.0768	–.0108	1.08
30	10	0.0672	–.0096	0.96
40	10	0.0590	–.0082	0.82
50	10	0.0517	–.0073	0.73
60	10	0.0453	–.0064	0.64
80	20	0.0348	–.0105	0.525
100	20	0.0267	–.0081	0.40
120	20	0.0205	–.0062	0.31
180	60	0.0093	–.0112	0.19
240	60	0.0042	–.0051	0.085

Thus the change in t-butyl bromide concentration over the first 10-second interval is the concentration at the 10 second mark minus the initial concentration at t = 0. The average rate over this interval is calculated by dividing the concentration change by the duration of the interval, Dt. Rate versus [t-BuBr] in a and Rate versus $[t\text{-BuBr}]^2$ in b.

The linearity confirms that the reaction is first order in t-butyl bromide. The slope of the plot gives $k = 1.36 \times 10^{-2}$ s^{-1}. This value is in satisfactory agreement with the value obtained from the ln[t-BuBr] versus time plot in Example. Even better agreement can be achieved by plotting average rate over a time interval versus the average [t-BuBr] over the same time interval. When the above data are plotted in this manner, the rate constant is found to be 1.33×10^{-2} s^{-1}.

Note that the method of finite differences, like the method based on integrated rate laws, uses the full range of concentration-time data, thus avoiding the major problem with the method of initial rates. For the method to be successful, however, it is important that the time intervals over which average rates are calculated be short relative to the total reaction time.

In the example above, the 10-second time intervals used in the early stages of the reaction, when concentration changes most rapidly, are small compared with the overall reaction time of 240 seconds. As a useful guideline, time intervals should not exceed 5per cent of the total reaction time; otherwise inaccurate rate constants will result from the finite difference approach. Electronic spreadsheets enable the calculations required for the finite difference method to be performed rapidly and conveniently using a computer. Further, the various data plots can be made quickly and professionally using the spreadsheet. With this modern tool, calculation by finite difference methods is routine. We will consider one additional example of the approach.

Example: The decomposition of N_2O occurs according to the following equation:

$$2N_2O(g) \rightarrow 2N_2(g) + O_2(g)$$

In a kinetics run, the concentration of N_2O was found to change with time as indicated in the first 2 columns of Table 15-1. The numbers in the remaining columns, calculated using an electronic spreadsheet, represent the average $[N_2O]$ over each time interval, squares of the N_2O concentrations, and the average rates of N_2O concentration change over the successive time intervals. The linearity of the plot and lack of intercept in b confirms that the reaction follows the rate law:

$$-d[N_2O]/dt = k[N_2O]^2$$

The rate constant, obtained from the slope of the plot, is 6.0×10^{-3} L/ mole-min.

Pseudo-Order Methods

Many chemical reactions obey a mixed second-order rate law of the type shown:

$$\text{Rate} = k[A][B]$$

The reaction is first order in each of two reactants. An example of a reaction governed by this rate law is given.

$$NH_4^+ + NO_2^- \rightarrow N_2(g) + 2H_2O$$

Suppose that a kinetics experiment is carried out by preparing a solution containing similar initial concentrations of the two reactants, NH_4^+ and NO_2^-, and monitoring the appearance of $N_2(g)$ as a function of time in order to measure the rate. Since the two reactants are present in similar concentrations, both substances undergo significant decreases in concentration over the course of the reaction. The rate of reaction reflects this simultaneous decrease in concentration of both reactants.

Consequently, a plot of Rate versus $[NH_4^+]$ (NH_4^+ = A) will be non-linear, even though the reaction is first order in NH_4^+, because the Rate also responds to the change in concentration of NO_2^- (B) significantly during reaction. Although it is possible to integrate obtain a rather complex equation that may then be used to plot the data, a simpler approach to the kinetics study is often possible.

In this approach, a kinetics run is performed on a solution that initially contains a substantial excess of one of the reactants, say B. If the initial concentration of B is more than 10 times larger than the initial concentration of A, the concentration of B will be nearly the same at the end of the reaction as at the beginning; consequently it may be considered to be constant during reaction. To illustrate this, consider a solution initially containing 0.002 M NH_4^+ and 0.10 M NO_2-. If reaction is allowed to proceed until all NH_4^+ is consumed, the concentration of NO_2– will have decreased by only 0.002 M to 0.098 M:

	NH_4^+ +	NO_2^- →	N_2 +	$2H_2O$
Initial	0.002	0.10	0	
Change	–0.002	–0.002	0.002	
Final	0	0.098	0.002	

Thus the concentration of NO_2- remains essentially constant while NH_4^+ is consumed, and the rate law simplifies as:

$$d[N_2]/dt = k[NH_4^+][NO_2-] = k'[NH_4^+]$$

Here the constancy of $[NO_2$-] has been recognized, and this concentration has been lumped into the rate constant to produce a new constant k'. The rate law now has a simple first-order form, even though the reaction is actually second-order. This simplified rate law, which may be integrated or subjected to finite difference analysis, is called a pseudo-first order rate law. The prefix, pseudo-, reminds us that the reaction is not truly first-order, but has only been made to appear that way by imposition of a large concentration of NO_2-.

If one carries out kinetic runs using several (relatively large) concentrations of [B], a pseudo-first order rate constant, k', will be obtained for each run. Of course, these will have different values depending on [B]. A plot of k' versus [B] is then made to obtain the true rate constant, k, from the slope.

Example: Pseudo-First Order Kinetics. For the reaction of NH_4^+ with NO_2-, several kinetics runs were carried out using the initial reactant concentrations in the first two columns of the table below. Plots of Rate versus $[NH_4^+]$ for each run were found to be linear, indicating that the reaction is first-order in NH_4^+. The first-order rate constants obtained from these plots varied with $[NO_2^-]$ as indicated in the third column of the table. What is the order of the reaction in NO_2-? What is the true rate constant for the reaction?

$[NH_4^+]$initial	$[NO_2-]$initial	Observed 1st order rate constant, k'
0.00896	0.197	7.06×10^{-5}
0.00904	0.395	1.42×10^{-4}
0.00523	0.593	2.13×10^{-4}

Solution: The order of reaction with respect to NH_4^+ is known to be 1, based on linear Rate vs. $[NH_4^+]$ plots. We thus focus on the relationship between the observed pseudo-first-order rate constants and the concentration of $[NO_2$–]. Note that for all three runs, the initial NO_2– concentration is much larger than the initial $[NH_4^+]$, guaranteeing that $[NO_2$–] remains essentially constant during the entire course of reaction. A linear plot of k' versus $[NO_2$–] raised to a power, n, indicates that the reaction is n-order in $[NO_2$–]. We can determine n by trying several plots, first with n = 1, then, if necessary, with n = 2, 3, and so on, until we obtain linearity.

Plots of k' versus $[NO_2^-]$ and $[NO_2^-]^2$. The top plot is linear, so the reaction is first order in $[NO_2^-]$. The overall second-order rate constant, $k = 3.6 \times 10^{-4}$ L/mole-s, is obtained from the slope.

The Experimental Measurement of Concentration

In order to carry out a kinetics study, it is necessary to have an experimental "handle" on at least one of the reactants or products; that is, the change in concentration of a reactant or product must be accompanied by a change in some measurable property of the system. For reactions involving gases, the total pressure of the reaction mixture changes as reaction proceeds, provided that there is a net change in the amount of gas in the system. Thus, the rate of reaction could be studied by monitoring the total pressure with time. Total pressure decreases as reaction proceeds because only 1 mole of gas forms for each 2 moles consumed.

$$2C_4H_6 \rightarrow C_8H_{12}$$

On the other hand, no change in total pressure accompanies because there is no change in moles of gas; reaction rate can not be studied by monitoring total pressure in this case.

$$H_2(g) + I_2(g) \rightarrow 2HI(g)$$

Rather than measuring system pressure at constant volume, one could measure system volume at constant pressure. In either case, the measured quantity is proportional to the number of moles of gas in the system at a particular time during reaction. Pressure (or volume) measurements are particularly convenient when only a single gaseous reactant or product is involved. In this case, the pressure (or volume) provides a direct measure of the amount of that reactant or product.

An alternative to pressure or volume measurements is to use gas chromatography to monitor the amounts of gaseous reactants and products as a function of time. The gaseous reaction mixture is periodically sampled with a syringe and injected into a gas chromatograph. In the ideal situation, it may be possible to simultaneously determine the amounts of all reactants and products at each sampling time.

For reactions carried out in solution, a number of methods are used to monitor concentration. The acidity of the solution changes during the course of reaction. Provided the reaction is slow enough, the kineticist could periodically "sample" an aliquot of reaction solution and carry out an acid-base titration using a suitable indicator to determine the $[H_3O^+]$ present at the known sampling time.

$$t\text{-BuCl} + 2H_2O \rightarrow t\text{-BuOH} + H_3O^+ + Cl^-$$

Alternately, since ions are produced but not consumed in the reaction, the kinetics could be monitored by measuring the electrical conductivity of the solution as a function of time. The measured conductivity values can then be related to the total concentration of ions present at that time. Perhaps the most common method for following the time course of reactions in solution is to monitor the absorption of light by the reacting solution. If one of the reactants (or products) has an absorption band in the *UV*-visible, infrared, or NMR region of the electromagnetic spectrum, the change with time of the amount of light absorbed is a measure of the rate of appearance or disappearance of the absorbing substance. The kinetics of reaction can easily be studied by this method.

$$Cr(H_2O)_6^{3+} + EDTA^{4-} \rightarrow Cr(EDTA)^- + 6H_2O$$

The product, $Cr(EDTA)^-$, has an absorption maximum at a wavelength of 542 nm in the visible region of the spectrum. As reaction proceeds, absorbance at this wavelength increases in direct proportion to the concentration of $Cr(EDTA)^-$, providing a "handle" on the reaction kinetics.

6

Elementary Reactions and Reaction Mechanisms

ELEMENTARY PROCESSES AND REACTION MECHANISM

A balanced chemical equation is intended to represent the net transformation of reactants to products that occurs when reaction takes place. It does not, and is not intended to, tell us anything about the details of the reaction pathway. In fact, most reactions occur via a series of relatively simple steps, each involving reaction of only a few (one, two, and very occasionally, three) molecules. Each of these simple steps is called an elementary process. The equation for an elementary process represents an actual collision event at the molecular level. The sequence of elementary processes by which a reaction is proposed to occur is called the mechanism for the reaction. For example, the reaction is thought to occur by the 2-step mechanism.

Overall:

$$2NO_2\,(g) + F_2\,(g) \rightarrow 2NO_2F\,(g)$$

Mechanism:

$$NO_2 + F_2 \rightarrow NO_2F + F$$
$$NO_2 + F \rightarrow NO_2F$$

with the first step proposed to be slow, the second one fast. Each of the two steps in the mechanism is understood to be an elementary process. We can make several important generalizations about the mechanism. First, each elementary process involves reaction (collision) of only two species, even though the overall reaction involves 3 moles of reactants. Second, F atoms, produced in the first step and consumed in the second, do not appear in the overall equation for the reaction. Such a species—first produced, then consumed—is called a reaction intermediate. Third, the elementary processes add to give the overall equation for the reaction. This must always be true.

It is important to realise that a mechanism for a reaction is a hypothesis about what goes on at the molecular level, made by the chemist based on her or his experimental observations. The overall equation for a reaction is NOT a hypothesis; it is an experimental fact, observable in the laboratory. The mechanism in all of its detail is seldom observable; it must be deduced from experiment, must be consistent with all known experimental facts about the reaction, and is subject to change if in conflict with a new experimental fact.

Types of Elementary Processes

There are three types of elementary process, named for their molecularity—the number of reactant molecules that they involve. A unimolecular process involves only one reactant molecule. Two examples of unimolecular processes are given.

$$O_3 \rightarrow O_2 + O$$
$$I_2 \rightarrow 2I$$

A bimolecular process involves two reactant molecules, which may be the same or different. Examples are given. Bimolecular elementary processes are by far the most common type.

$$NO_2 + F \rightarrow NO_2F$$
$$H_2O_2 + I^- \rightarrow H_2O + IO^-$$

A termolecular process requires the simultaneous collision of three reactant molecules. An example.

$$2NO + O_2 \rightarrow N_2O_4$$

Such three-body collisions are extremely improbable. Consequently, termolecular processes are rare. Elementary processes with molecularity greater than three do not occur, due to the overwhelming improbability of collisions involving four or more molecules. In contrast to the situation for the overall (net) reaction, the rate law for an elementary process may be written directly from its stoichiometry, because an elementary process is an actual, not a net, process. Thus for the elementary process,

$$A + B \rightarrow C + D$$

the rate law is $-D[A]/dt = k[A][B]$. Unimolecular processes have first-order rate laws. The rate law is

$$-D[O_3]/dt = k[O_3]$$

Bimolecular processes have rate laws that are second-order overall; and termolecular processes have overall third order rate laws.

$$-D[NO_2]/dt = k[NO_2][F]$$
$$-D[O_2]/dt = k[NO]^2[O_2]$$

In summary, an elementary process of molecularity n has a rate law of order *n*. The converse is NOT true: a first-order reaction is not necessarily either an elementary process or unimolecular.

The difference between order and molecularity is subtle but important. The concept of molecularity is appropriate only in the context of an elementary process, and indicates the number of reactant molecules involved in (undergoing collision) in the elementary step. The order of a reaction is the number to which the concentration of a reactant must be raised in the experimental rate law for the overall (net) reaction. This rate law, and the orders in it, do not necessarily correspond to a particular elementary step in the mechanism of the net reaction. We will say more about this shortly.

Rate Determining Step

A particular overall reaction consists of a sequence of elementary steps, which usually occur with different rates. Very often one of the elementary processes occurs much more slowly than do the others and limits the rate at which the overall reaction occurs. This slow step is then referred to as the rate-determining step (rds). The overall rate is equal to the rate of the rds. The fact that the overall reaction rate is limited by the rate of a particular elementary step is the reason why rate laws frequently do not correspond with the stoichiometry of the overall reaction.

A familiar example of an overall process occuring at a rate limited by a slow step. Theater filled to capacity, connected by a set of wide double doors to a smaller lobby, which is linked to the outside via a revolving door. When the movie is over, people begin to move from the theater to the lobby, and from there to the outside. We suppose that the double doors between theater and lobby are large enough so that 10 people per second can pass through; but that only 1 person per second can exit the revolving door. We then represent the overall process of leaving the theater in terms of 2 steps:

$$\text{theater} \rightarrow \text{lobby [10/s]}$$
$$\text{lobby} \rightarrow \text{outside [1/s]}$$
$$\text{Overall, theater} \rightarrow \text{outside [Rate = ?]}$$

Initially, people pass from theater to lobby at the rate of 10 per second, and begin to exit the revolving door one by one. However, the lobby quickly fills up, because it cannot be emptied nearly as rapidly as it was filled. The process "piles up" behind the slow step, which is passage through the revolving door. The maximum rate at which people can leave the theater is 1 per second, which is the rate of the slowest step in the process.

REACTION MECHANISM IN CHEMICAL CHANGE

In any chemical change, some bonds are broken and new ones are made. Quite often, these changes are too complicated to happen in one simple stage. Instead, the reaction may involve a series of small changes one after the other.

A reaction mechanism describes the one or more steps involved in the reaction in a way which makes it clear exactly how the various bonds are broken and made. The following example comes from organic chemistry. It doesn't matter in the least if it is unfamiliar to you!

This is a reaction between 2-bromo-2-methylpropane and the hydroxide ions from sodium hydroxide solution:

$$(CH_3)_3C\,Br + OH^- \rightarrow (CH_3)_3COH + Br^-$$

The overall reaction replaces the bromine atom in the organic compound by an OH group.

The first thing that happens is that the carbon-bromine bond in a small proportion of the organic compound breaks to give ions:

$$CH_3{-}C(CH_3)_2{-}Br \xrightleftharpoons{\text{Slow}} CH_3{-}C^+(CH_3)_2 + :Br^-$$

Carbon-bromine bonds are reasonably strong, so this is a slow change. If the ions hit each other again, the covalent bond will reform. The curly arrow in the equation shows the movement of a pair of electrons. If there is a high concentration of hydroxide ions present, the positive ion stands a high chance of hitting one of those. This step of the overall reaction will be very fast. A new covalent bond is made between the carbon and the oxygen, using one of the lone pairs on the oxygen atom.

$$CH_3{-}C^+(CH_3)_2 \quad :\bar{O}H \xrightarrow{\text{Fast}} CH_3{-}C(CH_3)_2{-}OH$$

Because carbon-oxygen bonds are strong, once the OH group has attached to the carbon atom, it tends to stay attached.

The mechanism shows that the reaction takes place in two steps and describes exactly how those steps happen in terms of bonds being broken or made. It also shows that the steps have different rates of reaction - one slow and one fast.

The Rate Determining Step

The overall rate of a reaction (the one which you would measure if you did some experiments) is controlled by the rate of the slowest step. In the example above, the hydroxide ion can't combine with the positive ion until that positive ion has been formed. The second step is in a sense waiting around for the first slow step to happen.

The slow step of a reaction is known as the *rate determining step*.

As long as there is a lot of difference between the rates of the various steps, when you measure the rate of a reaction, you are actually measuring the rate of the rate determining step.

Reaction Mechanisms and Orders of Reaction

The examples we use at this level are the very simple ones where the orders of reaction with respect to the various substances taking part are 0, 1 or 2. These tend to have the slow step of the reaction happening before any fast step(s).

To try to explain how fractional orders of reaction can arise is beyond the scope of UK A' level courses.

MOLECULARITY OF A REACTION

You may come across an older term known as the *molecularity* of a reaction. This has largely dropped out of UK A' level syllabuses, but if you meet it, it is important that you understand the difference between this and the *order* of a reaction. The terms were sometimes used carelessly as if they mean the same thing - they don't!

Order of Reaction

The important thing to realise is that this is something which can only be found by doing experiments. It gives you information about which concentrations affect the rate of the reaction. You cannot look at an equation for a reaction and deduce what the order of the reaction is going to be - you have to do some practical work! Having found the order of the reaction experimentally, you *may* be able to make suggestions about the mechanism for the reaction - at least in simple cases.

Molecularity of a Reaction

This starts at the other end! If you know the mechanism for a reaction, you can write down equations for a series of steps which make it up. Each of those steps has a molecularity. The molecularity of a step simply counts the number of species (molecules, ions, atoms or free radicals) taking part in that step. For example, going back to the mechanisms we've been looking at:

$$(CH_3)_3C{-}Br \xrightleftharpoons{\text{Slow}} (CH_3)_3C^+ + :Br^-$$

This step involves a single molecule breaking into ions. Because only one species is involved in the reaction, it has a molecularity of 1. It could be described as *unimolecular*.

The second step of this mechanism, involves two ions reacting together.

$$(CH_3)_3C^+ + :\overset{-}{O}H \xrightarrow{\text{Fast}} (CH_3)_3C{-}OH$$

This step has a molecularity of 2 - a *bimolecular* reaction.

The other reaction we looked at happened in a single step:

$$:\overset{-}{O}H + CH_3{-}\overset{\delta+}{C}H_2{-}\overset{\delta-}{Br} \longrightarrow CH_3{-}CH_2{-}OH + :Br^-$$

Because of the two species involved (one molecule and one ion), this reaction is also bimolecular.

Unless an *overall* reaction happens in one step (like this last one), you can't assign it a molecularity. You have to know the mechanism, and then each individual step has its own molecularity.

There's nothing the least bit complicated about the term*molecularity*. The only confusion is that you may sometimes find it used as if it meant the same as *order*. It doesn't!

Reaction Mechanisms and Orders of Reaction

Relating orders of reaction to mechanisms is relatively easy where the slow step is the first step of the reaction mechanism. It isn't so easy when it is one of the later steps. I want to look at this in a bit more detail in case your syllabus requires it.

We will revisit the simple case first where the slow step is the first step of the mechanism.

The First Step in the Mechanism

Suppose you had a reaction between A and B, and it turned out (from doing some experiments) to be first order with respect to both A and B. So the rate equation is:

$$\text{Rate} = k[A][B]$$

Which of these two mechanisms is consistent with this experimental finding?

Mechanism 1

$$A \xrightarrow{slow} C + D$$

$$B + C \xrightarrow{fast} E$$

Mechanism 2

$$A + B \xrightarrow{slow} D + E$$

Remember that in simple cases, where the slow step is the first step of the mechanism, the orders tell you what is taking part in the slow step.

In this case, the reaction is first order with respect to both A and B, so one molecule of each must be taking part in the slow step. That means that mechanism 2 is possible.

However, mechanism 1 must be wrong. One molecule of A is taking part in the slow step, but no B. The rate equation for that would be:

$$\text{Rate} = k[A]$$

THE REACTION CONDITIONS OF MECHANISM

The conditions under which some reaction occurs, including such variables as the temperature and concentrations of reactants, also are important in determining the course of the reaction. For reactions that have a high energy barrier between reactants and products, the rate is highly responsive to change in temperature, and such reactions become more likely at increased temperatures, so the minor products of a reaction often appear in larger proportions at higher temperatures.

Similarly, the concentration of reagents can be important to the course of a reaction, especially if two mechanisms are available that involve different numbers of molecules in the transition states. Higher concentrations of a particular reagent favour those mechanisms in which greater numbers of molecules are involved in the transition state. The pressure applied to the reacting system also may be significant, partly because it has an effect on concentration and partly because mechanisms involving closely associated transition states become more favourable at high pressures. The latter relationship comes about because associated transition states are those in which several molecules or ions are brought close together (and therefore take up less space), a situation that is encouraged by increased pressures.

ENERGY CHANGES INVOLVED IN REACTIONS

Collisions between molecules are rapid; therefore, reactions that occur spontaneously whenever the reagents collide are fast at ordinary concentrations. However, a reaction may be restricted in rate by its dependence on the occurrence of molecular collisions, because, for example, the reagents are present in such small amounts that reactions can only occur when they happen to encounter one another. Such a reaction is said to be diffusion-controlled, because it is dependent on the process of diffusion to bring the molecules together. In such cases the viscosity of the medium is relevant; the more viscous, or "thick," the medium, the more difficult the diffusion and the slower the reaction.

However, most reactions involve a rate-limiting energy barrier, and it is the nature of this barrier and of the molecular configuration at its top that determines the mechanism. Diagrams of energy changes during the course of reaction often are used to illustrate the energetic aspects of the reaction. An example of a possible energy diagram for a hypothetical one-stage process—the dissociation in a solution of a covalent molecule designated E–N, into its ions, E^+ and N^-.

In this diagram the energy is plotted against a reaction coordinate—a spatial relationship that varies smoothly during the course of the reaction—in this case the distance between the portions of the molecule designated E and N. At the left-hand energy minimum of the figure, the bond between E and N is fully formed; if energy is applied to excite the system in such a way that E and N are brought even closer together (region *a* of the figure), the atomic nuclei repel, with the result that energy rises steeply. Alternatively, excitation energy, such as thermal energy from collisions with other molecules, may stretch the bond, and the energy curve then moves into region *b*. The E–N bond is thereby weakened steadily until the transition state is reached. This point, as can be seen in the figure, has an energy peak on the reaction coordinate. At the same time, this point represents the minimum energy required to convert the reactants into the products.

The curve shown should be considered as only a planar section of a three-dimensional energy surface relating to the various possible spatial relations between the components of the reaction. The passage of the reactants from the initial state to the products then can be thought of as analogous to the climb from a valley (the initial state) through the lowest mountain pass leading to a second valley (the products).

Thus, although the transition state represents a peak on the single curve depicted, it really represents a secondary minimum (or pass) in the energy surface. From the top of this pass (the transition state), the molecule can only descend, losing energy by collision.

In doing so it may revert to the starting materials, or it may dissociate to give the products (region *c* in the figure). The products in the case of the reaction chosen are the ions E^+ and N", which are held together by electrostatic attraction as an ion pair at the right-hand minimum of the graph; beyond this point, further separation of the ions involves the consumption of energy (region *d*). In principle, the products may lie at higher or (as shown) lower energy than that of the initial state. The mechanism of a reaction such as this may be considered to be completely defined when the structures and energy properties of the starting materials, the products, and the transition state are known.

Ultimately, it should become possible to compute the properties of the molecules solely from the properties of their constituent atoms and also to deduce the transition states for any of the reactions these molecules may undergo. For a few simple situations, approaches already have been made to definitions of mechanism in this degree of detail. Systems involving several atoms, however, require a many-dimensional representation of the reaction course instead of the two-dimensional description shown in the figure. The problems of computation, and of testing theory against experiment, then become enormous. non-etheless, attempts have been made to deal with some simple reactions of systems involving up to about five atoms.

For a reaction involving several distinct stages, a more complicated description of the reaction course also is necessary. The figure gives an example of such a situation. This hypothetical reaction is reversible, with three successive intermediate complexes formed between the reactants. Unlike the case of the simpler situation above, the physical process that best approximates what is happening along the reaction coordinate changes from stage to stage across the diagram.

The highest point on the energy diagram corresponds to the energy of the transition state of the rate-limiting step in the reaction—that is, the slowest step in the reaction, the one that governs, or limits, the overall rate. The rate of reaction is independent of the nature and number of the intermediates that lie before this transition state on the reaction coordinate. Progress along the reaction coordinate cannot be identified with the time course of the reaction because any individual pair of reactants may reside for an appreciable time in the partially activated state represented by one of the intermediate complexes before final reaction is achieved.

Once the reactants have passed the rate-limiting transition state, they must lose energy (usually by collision with other molecules) to reach the final state. In a sense, the reaction coordinate in such a reaction may be thought of as representing the chemical course of the reaction (rather than a spatial or a time course).

If the diagram represents the only reaction of the system, then it is possible to apply the principle of microscopic reversibility, which states that the course taken by reverse action will be statistically identical with that taken by the forward reaction. This principle is not applicable to a reaction giving several different products not at equilibrium with one another, however.

REACTION MECHANISMS: NATURE OF REACTANTS, INTERMEDIATES, AND PRODUCTS

The stoichiometry of a reaction consists of the chemical formulas and relative molecular proportions of starting materials and products. Obviously these have a bearing on the mechanism of the reaction, for the overall reaction course must proceed from starting materials to the products.

The identification of a new chemical substance formed transiently in a reaction mixture, however, does not unambiguously imply that that substance is an intermediate in the reaction. In many cases a newly found material is only a temporary repository of the proportion of the reactants and ultimately produces the products by first reverting to the starting material.

The identification of the products of a reaction also helps to define the reaction course, because the mechanism in question clearly must account for their formation. Mechanistic theory has been greatly facilitated by the development of powerful methods of separation and purification based onchromatography (separation of compounds on the basis of their relative degrees of adsorption to certain solid substances, such as starch or silica) and also by modern methods available for the analysis of small quantities of materials. These spectroscopic procedures are often used, as is another instrumental method called polarimetry.

An important consideration with regard to the products of the reaction is whether the reaction is under kinetic or thermodynamic control. A reaction is said to be kinetically, rather than thermodynamically, controlled when the products are formed in proportions different from those that would prevail at equilibrium between the same products under the same conditions. Thermodynamic control leads to the equilibrium ratio of the products. Often, though not invariably, reactions under kinetic control give a greater amount of the thermodynamically less stable of two possible products; if thermodynamic control is then establzished, the products shift to their equilibrium proportions, which might give a misleading picture of the reaction course. Hence, inferences concerning the nature of the transition state can be drawn from the nature of the products only with a good deal of circumspection.

In determining the mechanism of a reaction, one of the major problems is to deduce the spatial or three-dimensional changes that occur to the molecules involved as they proceed from their initial state through the intermediate stages and transition states to the final products.

Knowledge about such changes can generally be deduced from knowledge of the stereochemistry (three-dimensional structures) of the starting materials, intermediates, and final products (provided these are obtained under kinetic control). Information of this kind is obtained by determinations of optical activity and analysis of the structures of the compounds by standard means.

In certain instances, information about the movement of atoms between molecules during the course of a reaction can be gained by using compounds containing isotopes of certain of the atoms. These isotopes behave much like the ordinary atoms they replace, but they can be identified by their behaviour. For example, in the hydrolysis of ethyl acetate (*see above* The reactants), it is crucial to a determination of the mechanism to be able to establish which of the two reactants (ethyl acetate or water) provides the oxygen atom that ends up in the product ethyl alcohol. In this case, the use of water labelled with oxygen-18 reveals that the oxygen atom in the alcohol comes from the ethyl acetate molecule.

ORDER OF REACTION, AND RATE CONSTANT

The term *order of reaction* can be applied to any elementary reaction considered in one direction only, and to certain composite reactions. For an elementary reaction occurring in one direction the order of reaction is equal to the molecularity, but it describes the kinetics not the mechanism. Thus, for the unimolecular elementary process,

$$EA \rightarrow E + A$$

the rate of reaction is proportional to the concentration of the reactant EA:

$$v = k\,[EA]$$

As the concentration [EA] is raised to the first power the order is unity and the reaction is said to be *first order*. The constant k is known as the *rate constant* of the reaction.

For the bimolecular elementary process,

$$E + A \rightarrow EA$$

the rate is proportional to the product of the reactant concentrations:

$$v = k\,[E]\,[A]$$

The reaction is first-order in E, first-order in A, but *second-order overall,* and k is again the rate constant. The orders for the individual reactants, unity for E and unity for A, are known as partial orders and the sum of all the *partial orders* of a reaction is the *overall order.* Second-order rate constants, such as k in Eqn, have the dimensions of reciprocal concentration multiplied by reciprocal time, whereas first-order rate constants, such as k in Eqn, have the dimensions of reciprocal time.

This difference in dimensions is not normally evident from the symbols used to represent rate constants, and care must therefore be taken to avoid making improper comparisons between rate constants of different orders. It is sometimes convenient to multiply a second-order rate constant by an appropriate concentration to produce a quantity with the dimensions of a first-order rate constant, *e.g., k* [A] from Eqn; such a product is called a *pseudo-first-order rate constant.*

The units of rate constants vary with the order of reaction in the same way that their dimensions vary. For a first-order rate constant, such as k in Eqn, the units are s^{-1}. For a second-order rate constant, such as k in Eqn, the units are $dm^3\ mol^{-1}\ s^{-1}$ or $L\ mol^{-1}\ s^{-1}$ or $M^{-1}\ s^{-1}$.

The concept of order of reaction (but not molecularity) can also be applied to certain reactions that occur by composite mechanisms, provided that the rate is proportional to reactant concentrations raised to powers (which need not be integral). However, this is rarely the case with enzyme-catalysed reactions and the concept of order cannot therefore be applied strictly to such reactions overall. non-etheless the individual steps of composite reactions have orders when considered in one direction. For processes that do not have a true order it is sometimes convenient to define an *apparent order* with respect to a reactant A as, $\frac{[A]}{v}\cdot\frac{dv}{d[A]}$

or $\frac{d\ln v}{d\ln[A]}$,

which is equivalent. For many enzyme-catalysed reactions the true order with respect to any substrate approximates to unity at very low concentrations and to zero at very high concentrations, but is not defined at intermediate concentrations. The apparent order, on the other hand, exists at any concentration.

It is sometimes useful to consider the rates of unidirectional elementary steps of a composite reaction in isolation. When the use of the term *rate of reaction* for such rates would cause ambiguity the term *chemical flux* or *chemiflux* may be used instead. For a full discussion of this terminology, see the IUPAC recommendations. In enzyme kinetics the need for an unambiguous terminology occurs mainly in discussions of the use of rates of transfer of isotopic labels as probes of the chemical fluxes in different parts of the composite reaction.

NUMBERING OF REACTIONS

The elementary reactions in a composite mechanism should be numbered as systematically as possible and in such a way that reverse processes are easily recognized. The reaction numbers should be used as subscripts to k, for rate constants, or v, for the individual rates (chemical fluxes). The preferred scheme for ordinary use is:

$$k_1, k_{-1}, k_2, k_{-2}....; v_1, v_{-1}, v_2, v_{-2}...$$

In this scheme positive subscripts may be prefixed by a + sign, *i.e.*, k_1 may be written as k_{+1}, if it helps avoid ambiguity resulting from confusion with scheme 3, or to emphasize that it refers to a step in the forward direction of the reaction. For some kinds of computer application and for theoretical discussions of enzyme mechanisms it is sometimes convenient to number the different forms of the enzyme rather than the elementary steps and then to denote the step from (*e.g.*) E_3 to E_4 as 34, *etc.* With this scheme the numbering of enzyme forms must be given explicity and the rate constants and rates listed above might become

$$k_{12}, k_{21}, k_{23}, k_{32}...; v_{12}, v_{21}, v_{23}, v_{32}...$$

If there are more than nine enzyme forms in the mechanism the subscripts should be separated by a comma, *e.g.*, $k_{10,11}$ but this can be omitted when it is not required for clarity.

The following scheme, in which odd subscripts refer to forward steps and even subscripts to reverse steps:

$$k_1, k_2, k_3, k_4....; v_1, v_2, v_3, v_4...$$

is less satisfactory, both because it conflicts with IUPAC Recommendations and because it makes it more difficult to recognize the forward and reverse rate constants for particular steps. It is unrealistic to expect any universal system of numbering rate constants to be equally satisfactory in all circumstances. For example, in a mechanism where enzyme forms in different states of protonation can undergo analogous reactions it may be clearer to assign the same numbers to the analogous steps and distinguish between them by the use of primes, *etc.*, rather than try to apply any of the above schemes in a rigid way. Regardless of what system is used, rate constants should never be referred to except in explicit relation to a mechanism or to a kinetic equation.

STEADY-STATE APPROXIMATION

If an intermediate is always present in amounts much less than those of the reactants (other than the enzyme) the rate of change of its concentration is much smaller than that of the reactants. This condition is ensured whenever, as is usually the case in enzyme-catalysed reactions, the concentration of substrate is much higher than that of the enzyme; it is not necessary for the amount of intermediate to be small compared with the amount of enzyme. For example, in the scheme,

$$E + A \underset{k_{-1}}{\overset{k_1}{\rightleftharpoons}} EA \xrightarrow{k_2} E + Z$$

if [EA] is always much less than [A] the following equation is obeyed to a good approximation:

$$\frac{d[EA]}{dt} = k_1[E][A] - k_{-1}[EA] - k_2[EA] = 0$$

7

Chemical Equilibrium

INTRODUCTION

Consider the following reaction: $N_{2(g)} + 3\ F_{2(g)} \rightarrow 2\ NF_{3(g)}$

The above equation implies that, if we mix 1 mole of N_2 and 3 moles of F_2 we would expect to produce 2 moles of NF_3. However, for many reactions in a close system like this one, a point will be reached where the concentration of products stop to increase with time even though there are reactants available in the mixture.

This observation is explained as follows: when N_2 and F_2 are initially mixed in a sealed container, molecular collisions result in the formation of NF_3. As time progresses and more products are formed, their molecules also collide with one another and cause the reverse reaction to occur. As reactant concentrations decrease and those of products increase, the rate of forward reaction decreases while that of the reverse reaction increases. At some point, the rates of opposing reactions become equal and the overall reactions come to a *state of equilibrium*.

At this point the concentrations of all species becomes constant until the reaction condition is altered. *Equilibrium* is a dynamic state; the reactions have not stopped, but both forward and reverse reactions continue to occur with the same rate. *Equilibrium* is *the exact balancing of two opposing processes*, in which all concentrations remain constant.

PRINCIPLES OF CHEMICAL EQUILIBRIUM

Equilibrium. In this chapter the dynamic equilibrium established in chemical reactions is explored, and its similarity to phase equilibrium is emphasized. The chapter culminates in a relationship between two signposts of spontaneity: the standard free energy of reaction and the equilibrium constant.

SPECIFIC CONCEPTS

Dynamic equilibrium is a balance of opposing processes occurring at equal rates.

- Chemical reactions achieve dynamic equilibrium similar to phase equilibrium and characterized by an equilibrium constant, K_{eq}, analogous to vapour pressure.

  Equilibrium as a balance of opposing rates. This state of balance gives the appearance of stasis at the macroscopic level, despite the continuous and dynamic activity at the microscopic level. Phase equilibrium is dynamic equilibrium. We summarize here the characteristics of dynamic equilibrium:
- It occurs only in closed systems.
- At least 2 states of a system must be simultaneously present (coexist).
- It is established only over a finite time interval, required to establish the equality of opposing rates.
- Values of macroscopic observables remain constant with time, but at the molecular level there is constant transit among the states present
- When subjected to a perturbation from outside, it responds to minimize the effect of the perturbation and returns to equilibrium, again over a finite time interval.
- It represents a compromise between the tendencies to minimize potential energy and maximize disorder.

In this, we extend the dynamic equilibrium concept to chemical reactions. We will see that, like a pure substance in a closed container, a chemical reaction in a closed system reaches a state of dynamic equilibrium, a state in which conversion of reactants to products is balanced by conversion of products to reactants. At the macroscopic scale, there is no change in the amount of a particular reactant or product with time. In contrast, at the molecular level there is a continuous reshuffling of atoms.

CHEMICAL EQUILIBRIUM

We introduce chemical equilibrium with a simple reaction, the dimerization of nitrogen dioxide to dinitrogen tetroxide, in equation:

$2\ NO_2 \rightarrow N_2O_4$

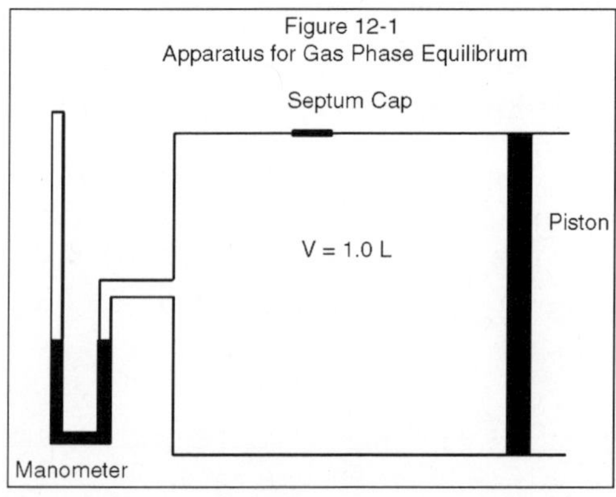

Figure 12-1
Apparatus for Gas Phase Equilibrum

As we have done several times before, we will study the reaction by carrying out several thought experiments designed to reveal the Behaviour of the reaction under various conditions. Throughout the following discussion, bear in mind that the experiments could be (and in fact have been) carried out in the Labouratory (although in a somewhat different manner than described).

Above figure shows an initially evacuated container of volume 1.0 L, with an attached manometre. The manometre allows the total gas pressure in the container to be measured. The temperature of the container is 298 K. The container is fitted with a piston that can be freely moved or locked in place. The following experiments are sequentially performed.

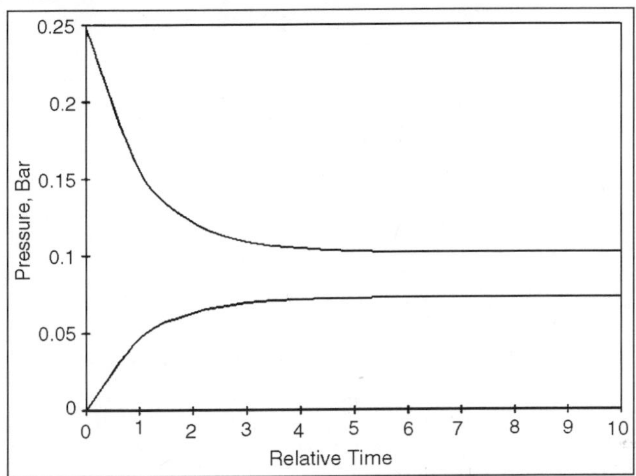

Expt 1. With the piston locked in place, 0.0460 g (0.0100 moles) of NO2 gas (red-brown in Colour, and toxic!) is injected into the container through a septum cap. The manometre quickly registers a pressure of 0.248 atm (188 mm of Hg). Then, over several minutes, the red-brown Colour fades and the pressure of the system falls, at first rapidly, then more and more slowly, until it levels off at a value of 0.175 atm.

Concept

The results of experiments on many different reactions over many years allow us to generalize as follows. For the generic reaction, in which the participating species are gases, the partial pressures at equilibrium obey the simple expressions, independent of the initial pressures of the gases.

$$a\,A + bB \rightarrow dD + fF$$

$$K_{eq} = P_D{}^d P_F{}^f / P_A{}^a P_B{}^b$$

In words, the equation says that if we divide the product of the pressures of the products, each raised to the power of its stoichiometric coefficient in the equation for the reaction, by a similar product of reactant pressures, the result is a constant, K_{eq}, that depends only on temperature. Every chemical reaction, simple or complex, obeys an equilibrium constant expression. This is a very simple result that makes possible calculations involving the direction and extent of reaction. If the value of K_{eq} for a reaction is large, the pressures of products will be large relative to those of reactants at equilibrium — the equilibrium lies to the right. If K_{eq} is small, pressures of products will be small relative to those of reactants at equilibrium — the equilibrium lies to the left.

Our conclusions are not restricted to gas phase reactions. An expression analogous as applied to reactions in solution, except that molarities must be used in place of pressures. In fact, when we recognize that gas pressure is proportional to n/V (molarity) through the ideal gas law, we realise the equivalence of expressions in terms of pressure and molar concentration. Equation below therefore applies to all reactions, in the gas phase, in solution, or in some combination:

$$K_{eq} = [D]^d[F]^f/[A]^a[B]^b$$

The brackets signify molarity. One caution is in order for gas phase reactions. The numerical value and the units of the equilibrium constant depend on whether pressures or molarities are used. If K_{eq} is specified in units of bar, pressures are used in calculations; if in moles/L, concentrations are used. If K_{eq} has no units, either pressure or molarity may be used.

LE CHATELIER'S PRINCIPLE

This principle applies as well to chemical equilibrium as to phase equilibrium. In fact, the principle is a qualitative manifestation of the quantitative characteristics of the equilibrium constant. Three stresses are particularly relevant to chemical equilibrium.

- Change in concentration (pressure) of a participant. For example, consider the reaction

$$N_2 + 3H_2 \rightarrow 2NH_3$$

Assume the reaction to be initially at equilibrium. Suppose that we add some NH_3 to the system. This imposes a stress on the system that will be relieved, according to Le Chatelier, by a shift that relieves the stress — *i.e.*, uses up NH_3.

The reaction shifts left until enough NH_3 is used and N_2 and H_2 produced to again make $Q = K_{eq}$. When the new equilibrium is reached, the pressures of all three gases will be larger than they were before the stress. Note that addition of NH_3 to the system does not change the value of K_{eq}. The same value applies before and after the stress.

Molecular View of Chemical Equilibrium

Consider now the general reaction in equations.

$$A + B \rightarrow 2C$$

$$K_{eq} = P_C^2/P_AP_B$$

Why do the stoichiometric coefficients appear as exponents in the expression for K_{eq}? Chemical equilibrium is like phase equilibrium; it results from a balance of opposing rates. In this case, the rates involved are those of the reaction in the forward and reverse directions:

$$\text{At equilibrium, } R_f = R_r$$

We assume now, and explore more fully later, that molecules must collide in order to react. This is certainly sensible, because it is not possible for atom rearrangements to occur unless the reacting atom aggregates are in close proximity.

The rate of a chemical reaction is thus expected to be proportional to the number of collisions between molecules per unit time: rate of rxn proportional to #collisions/time

For the forward and reverse processes, we write,

$$R_f = k_f\,(\text{\#collisions of A and B})$$

$$R_r = k_r'(\text{\#collisions of C and C})$$

The rate constants, k_f and k_r', are proportionality constants in the same way that k_E and k_C are. If there are N_A molecules of A and N_B of B in a container, a particular A molecule may collide with any of the N_B molecules of B. The total number of ways in which A and B may collide is thus $N_A \times N_B$.

Then,

$$R_f = k_f N_A N_B$$

N_C molecules of C can collide in a total of $N_C(N_C-1)/2$ ways. This arises because each molecule can collide with N_C-1 others. However, the simple product of $N_C$and N_C-1 double counts the collisions (*i.e.*, it includes collisions of molecule x with y and of y with x; however, there can be only one collision between two particular molecules). We divide by 2 to eliminate double counting.

Finally, because N_C is very large, $N_C - 1 = N_C$. It follows that:

$$R_r = k_r' N_C^2/2 = k_r N_C^2$$

At chemical equilibrium, then,

$$k_f N_A N_B = k_r N_C^2$$

Rearranging gives,

$$k_f/k_r = N_C^2/N_A N_B = K_{eq}$$

Since pressure is proportional to number of molecules, this equation is readily translated to pressure terms, giving the equilibrium constant expression in equation.

Notice that, as was true for phase equilibrium, the equilibrium constant is a ratio of rate constants for forward and reverse processes. Chemical equilibrium, like phase equilibrium, is dynamic rather than static. Although it appears at the macroscopic level that nothing is happening since concentrations remain constant, both forward and reverse reactions continue at equal rates at the molecular level.

FREE ENERGY AND EQUILIBRIUM

To this point we have encountered two quantities that indicate the extent to which a chemical reaction will proceed: its spontaneity. These are K_{eq} and the standard free energy change of reaction, $DG_R{}^o$. We now explore the relationship between these quantities. The sign of $DG_R{}^o$ indicates the spontaneity or non-spontaneity of a process in which reactants in their standard states are converted to products in their standard states. The magnitude of $DG_R{}^o$ tells us how far the standard state is from the equilibrium state. A process for which $DG_R{}^o$ is negative is spontaneous under standard conditions, and will proceed left to right to reach equilibrium.

At equilibrium, the reactants will be present at less than the standard state pressure (1 atm) or concentration (1 M), because they have been used up to some extent in achieving equilibrium. The products will be present at greater than the standard state pressure or concentration, because they have been formed in achieving equilibrium. In the equilibrium state, products will be present in relatively large amounts, reactants will be present in relatively small amounts, consistent with $K_{eq} > 1$.

Thus $DG_R{}^o < 0$ corresponds to $K_{eq} > 1$. Further, for a process with $DG_R{}^o = -50$ kJ, the standard state is much farther away from equilibrium than is the standard state for a process with $DG_R{}^o = -5$ kJ. The first reaction will have to proceed further to reach the equilibrium state than will the second reaction, and will thus have a larger K_{eq}. We can make very similar statements about processes for which $DG_R{}^o > 0$, except in reverse. The larger the magnitude of $DG_R{}^o$, the further the standard state is from the equilibrium position, and the further reaction will have to proceed to reach equilibrium from the standard state. For $DG_R{}^o > 0$, reaction must proceed from right to left towards equilibrium, decreasing the pressures or concentrations of products and increasing those of reactants. Thus $K_{eq} < 1$. The larger the magnitude of $DG_R{}^o$, the smaller K_{eq}. Finally, when $DG_R{}^o = 0$, the process is equally spontaneous in both forward and reverse directions; the standard state is the equilibrium state.

Supplement Applications Consider the Reaction Below

$$2NO_2(g) \rightarrow N_2O_4(g) \quad K_{eq} = 7.0 \text{ atm}^{-1}$$

If you mix 0.12 atm NO_2 and 0.43 atm N_2O_4 in a closed bulb, will any reaction occur? In which direction? (you must provide convincing support for your answers.)

What will the partial pressures of NO_2 and N_2O_4 be once equilibrium is attained? Consider the reaction below:

$$2NOBr(g) \rightarrow 2NO(g) + Br_2(g)$$

Over a time interval between t_0 and t_1, the reaction is at equilibrium at some temperature, with constant partial pressures of reactants and products as At time t_1, the volume of the reaction system is instantly halved. Show on the graphs the immediate effect of the volume decrease on the partial pressures of species, and the manner in which these pressures change as equilibrium is Re-established. Your graphs should be as quantitatively correct as possible. You are studying the equilibrium $P_4(g)\ 2P_2(g)$. In a previously evacuated vessel, you place a quantity of diatomic phosphorus gas.

What is the value of the reaction quotient Q? Which direction will reaction proceed to achieve equilibrium? You allow the system to come to equilibrium and then decide to tamper with it. In which direction would the equilibrium shift if you add $P_4(g)$? Increase the pressure of the system (at const T)? Increase the volume of the system (at const T)? Increase the T of the system (at const P)? What do you know about a system at equilibrium when you know that the value of the equilibrium constant is 0.34 at a particular temperature?

EQUILIBRIUM CONSTANT

For the hypothetical chemical reaction,

$$a\ A + b\ B \rightleftarrows c\ C + d\ D$$

the equilibrium constant is defined as,

$$K_c = \frac{[C]c[D]d}{[A]a[B]b}$$

The notation [A] signifies the molar concentration of species A. An alternative expression for the equilibrium constant involves partial pressures.

$$K_p = \frac{P_C c\ P_D d}{P_A a\ P_B b}$$

Note that the expression for the equilibrium constant includes only solutes and gases; pure solids and liquids do not appear in the expression. For example, the equilibrium expression for the reaction,

$$CaH_2\ (s) + 2\ H_2O\ (g) \rightleftarrows Ca(OH)_2\ (s) + 2\ H_2\ (g)$$

is,

$$K_c = \frac{[H_2]2}{[H_2o]2}$$

Observe that the gas-phase species H_2O and H_2 appear in the expression but the solids CaH_2 and $Ca(OH)_2$ do not appear.

The equilibrium constant is most readily determined by allowing a reaction to reach equilibrium, measuring the concentrations of the various solution-phase or gas-phase reactants and products, and substituting these values into the Law of Mass Action.

Calculations Involving Reaction Stoichiometry

The one species existed in the gas phase; all other species were solids. The pressure of the system could thus be used to directly calculate the equilibrium constant, because the system pressure was the same as the partial pressure of the gas-phase species. For most reactions, however, the calculations are more complicated.

Consider the following reactions, which is an important atmospheric reaction.

$$O_2\ (g) + 2\ SO_2\ (g) \rightleftarrows 2\ SO_3\ (g)$$

Suppose the system initially contains only oxygen and sulfur dioxide. As the reaction progresses, sulfur trioxide is formed and oxygen and sulfur dioxide are consumed. The pressure of the system is the sum of the partial pressures of the components of the gas, and each of these partial pressures differs from the initial values.

$$P = P_{O2} + P_{SO2} + P_{SO3}$$

How can one take these changes into account and determine each individual partial pressure from the system pressure?

A useful concept for this purpose is the extent of reaction, which will be given the symbol x for this experiment. In reality the reaction occurs on the molecular scale, with individual molecules of oxygen, sulfur dioxide, and sulfur trioxide being involved. Chemists prefer to think in terms of moles of material, however. Each unit of x (the extent of reaction) corresponds to the number of molecules of each reactant and product indicated by the respective stoichiometric coefficients. Thus when $x = 1$, one mole of oxygen and two moles of sulfur dioxide have been consumed and two moles of sulfur trioxide have been formed. The sense of the change can be reversed. When $x = -1$, one mole of oxygen and two moles of sulfur dioxide have been formed and two moles of sulfur trioxide have been consumed.

The extent of reaction allows the change in amount of one species to be related to the change in the amount of a different species. The most straightforward way to illustrate this relationship is through the use of a table such as that shown below.

	O_2	SO_2	SO_3
Initial	P_1	P_2	P_3
Change	$-x$	$-2x$	$+2x$
Equilibrium	$P_1 - x$	$P_2 - 2x$	$P_3 + 2x$

Strictly speaking, the table should be constructed in terms of moles of each species. The equilibrium expression, however, requires partial pressures or molar concentrations. So long as each species exists in the same phase (and thus the volume is the same for each species), the moles can be replaces by partial pressure or molar concentration. In this example the extent of reaction is expressed in terms of partial pressure.

The first row of the table contains the initial partial pressures of each species. These values are called "analytical partial pressures" and are the partial pressures of each species the experimenter put in the system. The second row simply reflects the stoichiometry of the reaction and employs the extent of reaction to show how the change in amount of one reactant or product is linked to the change in the amounts of the other reactants and products. The bottom row is simply the sum of the first two rows and contains the actual equilibrium partial pressures of each species. In this context, the extent of reaction, x, is the value necessary to reach equilibrium. At equilibrium, the values in the bottom row of the table may be substituted into the equilibrium expression to evaluate K_P.

$$P_{O2} = P_1 - x$$
$$P_{SO2} = P_2 - 2x$$
$$P_{SO3} = P_3 + 2x$$

$$K_p = \frac{P_{SO_3}^2}{P_{O2}\,P_{SO_2}^2} = \frac{(P_3 + 2x)^2}{(P_1 - x)(P_2 - 2x)^2}$$

Can you calculate the value of x if you are given the pressure P and the three initial partial pressures P_1, P_2, and P_3?

Can you calculate the individual partial pressures and the equilibrium constant, if you are given x and the three initial partial pressures P_1, P_2, and P_3?

Will this strategy for determining K_P from P work for the reaction shown below?

$$CO\ (g) + H_2O\ (g) \rightleftarrows CO_2\ (g) + H_2\ (g)$$

Le Châtelier's Principle

The first step in the conversion of coal into gasoline is the steam reforming reaction.

$$C\ (s) + H_2O\ (g) \rightleftarrows CO\ (g) + H_2\ (g)$$

Suppose that 0.0500 mole each of carbon, water, carbon monoxide, and hydrogen are placed in a 10.0 L glass bulb, which is then heated to 1000 K and the steam reforming reaction is allowed to reach equilibrium. This system is depicted in the experiment shown below.

The question to be answered in this experiment is:

How does a change in the analytical amount of a reactant or product affect the equilibrium amounts of reactants and products in the system?

The distinction between analytical amount and equilibrium amount is especially important in this experiment. The analytical amount of material, such as carbon, is the amount physically added to the system. The equilibrium amount is the amount of material that actually exists in the system at equilibrium. The two amounts are generally not the same, because the reaction will consume or produce the material in order to reach equilibrium.

In this experiment, for example, the analytical amount of each reactant and product is 0.0500 mole, but the equilibrium amounts of carbon and water are 0.0093 mole and the equilibrium amounts of carbon monoxide and hydrogen are 0.0907 mole. One consequence of the dynamic nature of the system is that the amount of a compound placed in a system is rarely the amount that actually exists in the system.

Le Châtelier's Principle: Effect of a Change in Volume

When a stress is brought to bear on a system at equilibrium, the system will react in the direction that serves to relieve the stress.

In the previous experiments Le Châtelier's Principle was employed to predict the behaviour of a system when the amount of a reactant or product is changed or when the temperature is changed. In this experiment the stress brought upon the system is a change in volume.

What is the consequence of changing the volume of a gas?

The fundamental effect of a change in volume is to simultaneously change the partial pressures (and molar concentrations) of all reactants and products by the same relative amount. If the volume of the gas is reduced, the concentrations of all reactants and products are simultaneously increased, which corresponds with an increase in the total concentration of species in the gas. Using Le Châtelier's Principle, the system should react in a way to oppose this stress; that is, the system should react to reduce the total concentrations.

Note that species that do not participate in a reaction are unaffected by the position of the equilibrium for the reaction. Thus adding an unreactive species, such as nitrogen gas or helium, has no influence upon the reaction, because such species are neither reactants nor products in the reaction. (Strictly speaking, this is true only for systems that behave ideally.)

In addition, the only species that are affected by a change in volume are gas-phase or solution-phase species. (For a solution, the volume change would be accomplished by adding pure solvent or removing pure solvent by evaporation.) Such species are the only type whose concentrations are altered by a volume change.

To illustrate this effect, consider the steam reforming reaction used in previous experiments.

$$C\ (s) + H_2O\ (g) \rightleftarrows CO\ (g) + H_2\ (g)$$

In this experiment the volume of the system can be changed and the corresponding changes in equilibrium amounts of carbon, water, carbon monoxide, and hydrogen can be observed.

Effect of a Change in Temperature

When a stress is brought to bear on a system at equilibrium, the system will react in the direction that serves to relieve the stress.

A change in the analytical amount of a reactant or product produces a stoichiometric stress on the system. A change in temperature also constitutes a stress on the system, but what is the nature of this stress? Unlike changes in the analytical amounts of reactants and products, a change in temperature produces a change in the equilibrium constant itself. To understand the effect of a temperature change on the value of the equilibrium constant, it is necessary to understand how temperature affects a reaction.

An increase in temperature is the result of the flow of heat into the system. Conversely a flow of heat out of the system reduces the temperature. A reaction itself can be a source or sink for heat. The heat flow for a reaction is characterized by the Enthalpy of Reaction or the Heat of Reaction, ΔH_{rxn}.

If $\Delta H_{rxn} > 0$, the reaction is said to be endothermic, which means that the reaction draws heat from its surroundings as the reaction occurs.

If $\Delta H_{rxn} < 0$, the reaction is said to be exothermic, which means that the reaction releases heat into its surroundings as the reaction occurs.

Suppose an *exothermic* reaction is at equilibrium, and then the temperature of the system is increased. The increase in temperature corresponds to introducing heat into the system. This influx of heat pushes the system away from equilibrium and nature restores equilibrium by removing some of this additional heat. Because the reaction is exothermic, the reaction *produces* heat when it proceeds in the forward direction. Le Châtelier's Principle states that the system will react to *remove* the added heat, thus the reaction must proceed in the reverse direction, converting products back to reactants. Conversely if the temperature of the system were decreased (heat removed from the system), the system would react in a direction that opposed the removal of the heat. The forward reaction would thus occur to release heat in an attempt to offset the heat that was removed from the system.

Bear in mind that the system is being held at a specific temperature, so the net reaction (which may produce or release heat) does not actually change the temperature of the system. That extra heat is removed or replaced as necessary to maintain the temperature.

The steam reforming reaction studied in the previous experiment is an example of an endothermic reaction.

$$\mathrm{C\ (s) + H_2O\ (g) \rightleftarrows CO\ (g) + H_2\ (g)}$$
$$\Delta H_{rxn} = +\ 131.3\ \mathrm{kJ\ mole^{-1}}$$

As an endothermic reaction, the steam reforming reaction draws heat out of the system as it occurs. Although it is not strictly correct to include *heat* in a chemical equation, which shows stoichiometric relations between physical substances, one could loosely represent the endothermic nature of the reaction by writing *heat* as a reactant:

$$\textit{heat} + \mathrm{C\ (s) + H_2O\ (g) \rightleftarrows CO\ (g) + H_2\ (g)}$$

The advantage of this construct is that it permits the appplication of Le Châtelier's Principle through the same reasoning as employed when the analytical amount of a substance is changed.

The temperature-dependence of the equilibrium constant, K_P in this case, is determined by the enthalpy of reaction.

PREDICTING EQUILIBRIUM CONCENTRATIONS

This is by far the most common kind of equilibrium problem you will encounter: starting with an arbitrary number of moles of each component, how many moles of each will be present when the system comes to equilibrium?

The principal source of confusion and error for beginners relates to the need to determine the values of several unknowns (a concentration or pressure for each component) from a single equation, the equilibrium expression. The key to this is to make use of the stoichiometric relationships between the various components, which usually allow us to express the equilibrium composition in terms of a single variable. The easiest and most error-free way of doing this is adopt a systematic approach in which you create and fill in a small table as shown in the following problem example. You then substitute the equilibrium values into the equilibrium constant expression, and solve it for the unknown.

This very often involves solving a quadratic or higher-order equation. Quadratics can of course be solved by using the familiar quadratic formula, but it is often easier to use an algebraic or graphical approximation, and for higher-order equations this is the only practical approach. There is almost never any need to get an exact answer, since the equilibrium constants you start with are rarely known all that precisely anyway.

EFFECTS OF DILUTION ON EQUILIBRIUM

In the section that introduced the LeChâtelier principle, it was mentioned that diluting a weak acid such as acetic acid CH_3COOH ("HAc") will shift the dissociation equilibrium to the right:

$$HAc + H_2O \rightarrow H_3O^+ + Ac^-$$

Thus a 0.10*M* solution of acetic acid is 1.3per cent ionized, while in a 0.01M solution, 4.3per cent of the HAc molecules will be dissociated. This is because as the solution becomes more dilute, the product $[H_3O^+][Ac^-]$ decreases more rapidly than does the [HAc] term. At the same time the concentration of H_2O becomes greater, but because it is so large to start with (about 55.5M), any effect this might have is negligible, which is why no $[H_2O]$ term appears in the equilibrium expression.

For a reaction such as $CH_3COOH(l) + C_2H_5OH(l) \rightarrow CH_3COOC_2H_5(l) + H_2O(l)$ (in which the water concentration does change), dilution will have no effect on the equilibrium; the situation is analogous to the way the pressure dependence of a gas-phase reaction depends on the number of moles of gaseous components on either side of the equation.

Phase Distribution Equilibria

It often happens that two immiscible liquid phases are in contact, one of which contains a solute. How will the solute tend to distribute itself between the two phases? One's first thought might be that some of the solute will migrate from one phase into the other until it is distributed equally between the two phases, since this would correspond to the maximum dispersion (randomness) of the solute. This, however, does not take into the account the differing solubilities the solute might have in the two liquids; if such a difference does exist, the solute will preferentially migrate into the phase in which it is more soluble.

For a solute S distributed between two phases a and b the process S_a = $S_b$is defined by the distribution law,

$$K_{ab} = \frac{[S]_a}{[S]_b}$$

in which $K_{a,b}$ is the *distribution ratio* (also called the distribution coefficient) and $[S]_i$ is the solubility of the solute in the phase.

SPREADSHEETS IN EQUILIBRIUM CALCULATIONS

A systematic approach to the calculation of equilibrium pressures in gas phase reactions. However, we have been careful to choose examples that generate no higher than a second-order polynomial — a quadratic equation — which is readily solved. There are many chemical reactions that generate polynomials of order 3 and higher when our general method is applied. Such polynomials cannot be solved analytically. That is, there is no analog of the quadratic formula that gives the roots of a polynomial of order > 3. Consider the very important reaction

$$N_2(g) + 3H_2(g) \rightarrow 2NH_3(g)$$

for which the equilibrium constant expression is :

$$K_{eq} = P_{NH_3}{}^2 / P_{N_2} P_{H_2}{}^3$$

Application of our approach to this reaction generates an equation in x^4 (from the denominator of K_{eq}) which cannot be solved analytically.

However, the availability of personal computers and modern Spreadsheet programmes enables such problems to be easily solved. We consider a specific example:

	$N_2(g)$ +	$3H_2(g) \rightarrow$	$2NH_3(g)$	$K_{eq} = 15.0$ atm^{-1}
initial	10.5	0.38	0.24	

For the initial conditions, $Q = 0.10$, so reaction will proceed right to equilibrium. The change and equilibrium lines then follow:

Change	$-x$	$-3x$	$2x$
equal	$10.5 - x$	$0.38 - 3x$	$0.24 + 2x$

Now, instead of substituting into the K_{eq} expression as we normally do (which would generate a polynomial of order 4), we set limits on the value of x;

i.e., we figure out the minimum and maximum values that x can have. The minimum is clearly zero, which would correspond to no reaction. The maximum value is calculated by assuming that reaction proceeds to the right until the limiting reagent is used up. The limiting reagent is clearly H_2, so the maximum possible x value is 0.38/3 = 0.127 atm. The numerical range in which x must lie is thus $0 < x < 0.127$. We now explore this range with a trial-and-error procedure that goes as follows:

- Vary x systematically between 0 and 0.127
- Calculate Q for each x
- Compare Q with K_{eq}
- When $Q = K$ ($Q - K = 0$, or $Q/K = 1$),

we have found x.

Table shows the variation of Q with x as x is incremented in steps of 0.01.

Table. Q as a Function of x for Reaction 12-20

x	Q
0	0.1
0.01	
•	
•	
0.09	12.7 (Q < K)
0.10	36.3 (Q > K)
•	
•	
0.127	infinite

Incrementation of x in steps of 0.01 allows us to greatly narrow the range of possible x values to $0.09 < x < 0.10$. We now "zoom" in on this range by incrementing x by 0.001:

0.09	12.7
0.091	14.0(Q < K)
0.092	15.4(Q > K)
•	
•	
•	
0.10	36.3

We have now limited x to the range between 0.091 and 0.092. If desired, we could continue this zooming process to obtain x to any desired degree of precision. However, since the initial hydrogen pressure is given to only 2 significant figures, the value $x = 0.092$ atm is adequately precise (0.092 is chosen rather than 0.091 because 15.4 is closer to K_{eq} than is 14.0).

The trial and error calculation illustrated above can be carried out manually. However, it is much faster and more reliable to use a spreadsheet. Further, once the spreadsheet is set up for a particular reaction, it can be used for any specified initial conditions. Figure shows a spreadsheet set-up for the ammonia reaction. The cell entries are as follows:

First few lines	explanatory text
$B_{12} - B_{14}$	initial pressures
B_{15}	initial Q: B_{14} ^ 2/(B_{12})(B_{13} ^ 3)
E_{15}	minimum x value (0)
G_{15}	maximum x value, calculated as $P^{\circ}_{H_2}/3$
A_{17-19}	reminders of relationship between x and pressures
B_{21}	increment of x (Del x)
Row 25 ff	x values, calculated concentrations, and Q/K comparison
A_{26}	new x calculated by adding del x (B21) to old value of x in A_{25}. $ tells spreadsheet not to change the row and column.
B_{26}	new value of P_{N_2} for x in A_{26}.
C_{26}, D_{26}	P_{H_2}, P_{NH_3}
E_{26}	Value of Q/K

To get subsequent entries in columns A – E, copy row 26. Once the value of x is narrowed down, define a new "del x" for zooming. Copy the table headings and proceed as in the first table. The spreadsheet method can be used for any equilibrium problem, involving any power of x. It is therefore very general and powerful.

ADDITIONAL EQUILIBRIA INVOLVING TWO PHASES

In our discussion of the phases of a single pure substance, we found that equilibrium involving two phases is achieved when the rate of passage from phase 1 to phase 2 is balanced by an equal rate of passage from phase 2 to phase 1:

$$\text{Rate}_{1\rightarrow 2} = \text{Rate}_{2\rightarrow 1}$$

The rate of passage from phase 1 to phase 2 is directly proportional to the concentration of substance in phase 1 and to the surface area, SA, of contact between the two phases. A similar statement holds for the rate of passage from phase 2 to 1:

$$\text{Rate}_{1\rightarrow 2} = k_{1\rightarrow 2}{*}\text{SA}{*}[\text{substance}]_1$$
$$\text{Rate}_{2\rightarrow 1} = k_{2\rightarrow 1}{*}\text{SA}{*}[\text{substance}]_2,$$

When these two rates become equal, equilibrium exists, and the following relationship characterizes it:

$$\text{Rate}_{1\rightarrow 2} = \text{Rate}_{2\rightarrow 1}$$
$$k_{1\rightarrow 2}{*}\text{SA}{*}[\text{substance}]_1 = k_{2\rightarrow 1}{*}\text{SA}{*}[\text{substance}]_2$$
$$K_{eq} = k_{1\rightarrow 2}/k_{2\rightarrow 1} = [\text{substance}]_2/[\text{substance}]_1$$

K_{eq} is an equilibrium constant, and is seen to be the ratio of rate constants for opposing processes. We now consider a number of other two-phase equilibrium situations for which the above considerations apply.

Distribution of a Solute Between Two Immiscible Liquids in Contact. When two immiscible liquids such as water and chloroform ($CHCl_3$) are placed in contact, they mutually exclude each other, forming two liquid layers.

The less dense liquid floats on the denser one, and a boundary surface is clearly visible. It is interesting to think about what will happen if a third substance, solid or liquid, is added to the system consisting of two immiscible liquids, *A* and *B*.

If the third substance, *C*, is soluble in one of the liquids (say *A*) and insoluble in the other (*B*), it will dissolve exclusively in the *A* layer.

More interesting is the situation in which *C* is soluble in both liquids to at least some extent. In this case, we find that *C* partitions, or distributes, itself between the two liquid layers.

We further find that the concentration of the third substance is higher in the liquid with which its intermolecular forces are most compatible. The ratio of the concentration of *C* in liquid *B* to that in liquid *A* is found to attain a constant value K_d, defined as in:

$$K_d = [C]_B / [C]_A$$

This constant ratio is established no matter how the 3-component mixture is prepared. Three such ways follow:

A and *B* are first brought into contact. *C* is then added, and the container is shaken.

C is first dissolved in *A*. *A* (containing *C*) and *B* are then brought into contact, and the container is shaken.

C is first dissolved in *B*. *A* and *B* (containing *C*) are then brought into contact, and the container is shaken.

In addition, the ratio is Re-established if the mixture is perturbed in any of the following ways:

More *A* is added.
More *B* is added.
More *C* is added.

We can understand this behaviour in terms of the diagram in Figure, which shows layers of *A* and *B* in contact at a boundary, and molecules of *C* initially dissolved in liquid *A*.

The boundary between the two liquids provides a "doorway" from one to the other through which *C* molecules, which have reasonable interactions with both *A* and *B* molecules, can pass. Thus molecules of *C* will tend to move across the boundary from liquid *A*, where they are initially dissolved, into liquid *B*, in which there are initially no molecules of *C*. In other words, they begin to distribute themselves between the two liquids. After some time, enough *C* molecules will have passed into liquid *B* that the rates of passage of *C* from *A* to *B* and back will be the same. This is a situation of dynamic equilibrium, represented as in first equation. The equality of rates is shown in second equation.

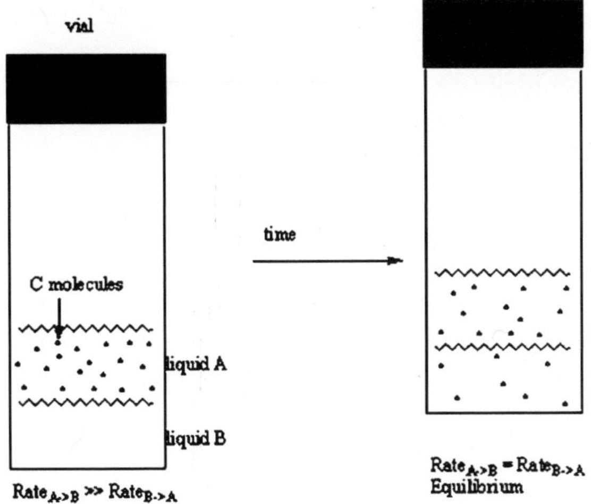

Fig. Distribution of a Solution between two Solvents

$$C \text{ dissolved in } A \Leftrightarrow C \text{ dissolved in } B$$

$$\text{Rate}_{A \to B} = \text{Rate}_{B \to A}$$

The rate of passage from A to B is directly proportional to the concentration of C in liquid A, and to the surface area over which A and B are in contact.

$$\text{Rate}_{A \to B} = k_{A \to B} {}^* SA {}^* [C]_A$$

Similarly,

$$\text{Rate}_{B \to A} = k_{B \to A} {}^* SA {}^* [C]_B$$

Equating the two rates at equilibrium gives

$$[C]_B / [C]_A = k_{A \to B} / k_{B \to A} = K_d$$

K_d is an equilibrium constant, the ratio of rate constants for opposing processes.

Once the equilibrium situation is attained, there will be no further change in the concentration of C in the two layers. In other words, the constant ratio will be attained. From the magnitude of K_d, we may easily deduce which liquid C prefers. If $K_d < 1$, C prefers A; if > 1, C prefers B.

In summary, distribution of a substance between two liquid phases is an example of dynamic equilibrium, a situation in which two opposing processes occur at equal rates.

Example: A vial contains 1 mL of water and 1 mL of 1,1,2,2-tetrachloroethane, which are immiscible. To the vial is added 0.1 g of acetone, which is soluble in both water and TCE. When the equilibrium distribution of acetone between the two solvents has been achieved, it is found that $K_d = [\text{acetone}]_{H2O}/[\text{acetone}]_{TCE} = 2.8$.

- What mass of acetone is present in each liquid layer at equilibrium?
- What is the molarity of acetone in each layer at equilibrium?

  To the vial is now added 1 mL of TCE containing 0.05 g of acetone, and the vial is shaken until equilibrium is once again attained.

- At equilibrium, what masses of acetone will be present in the water and TCE layers?
- Show graphically how the concentrations in the 2 layers change with time as a result of the addition.

Solution:

- At equilibrium, the ratio of the concentration of acetone in water to that in TCE must be 2.8. Because the two solvents are present in equal volumes, the ratio of the moles acetone in water to moles acetone in TCE must also be 2.8. Further, the total moles of acetone present is 0.1 g/58.08 g/mole = 0.00172 moles.
  moles acetone in water/moles acetone in TCE = 2.8. Thus
  moles acetone in water = 2.8*moles acetone in TCE
  moles acetone in water + moles acetone in TCE
  = 0.00172
  2.8 moles acetone in TCE + moles acetone in TCE
  = 0.00172
  Thus moles acetone in TCE = 0.00172/3.8
  $= 4.53 \times 10^{-4}$
  and moles acetone in water $= 2.8 \times 4.53 \times 10^{-4}$
  $= 1.27 \times 10^{-3}$ moles
  Masses may now be calculated from moles:
  mass acetone in water $= 1.27 \times 10^{-3}$ moles $\times$ 58.08 g/mole = 0.074 g
  mass acetone in TCE = 0.10 g 0.074 g = 0.026 g
- Molarities may be easily calculated from the moles acetone and the volume of each layer:
  Molarity acetone in water $= 1.27 \times 10^{-3}$ moles/0.001 L
  = 1.27 M
  Similarly, molarity acetone in TCE = 0.453 M
- To address this, we introduce an approach that is generally useful for situations in which a system proceeds from specified concentration conditions to equilibrium. The approach begins with an equation representing the process:

$$\text{Acetone}_{\text{TCE}} \Leftrightarrow \text{Acetone}_{\text{H2O}}$$

Beneath each substance in the equation we write the moles of that substance that we start with. Thus we put moles of acetone present in each phase prior to addition of 0.05 g of acetone in 1 mL TCE. These are equilibrium values, so we write "original equilibrium" to the left of these amounts:

	$\text{Acetone}_{\text{TCE}}$	$\Leftrightarrow$	$\text{Acetone}_{\text{H2O}}$
original equilibrium:	4.53×10^{-4} moles		1.27×10^{-3} moles

Beneath the "original equilibrium" line of information, we enter the amount of acetone added to the TCE. This is 0.05 g, or 8.60×10^{-4} moles.

	$Acetone_{TCE}$	⇔	$Acetone_{H2O}$
original equilibrium:	4.53×10^{-4} moles		1.27×10^{-3} moles
add:	8.60×10^{-4}moles		0 moles

The amounts in the start and add lines are now added to give the "initial" values, from which the system will proceed towards a new equilibrium:

	$Acetone_{TCE}$	⇔	$Acetone_{H2O}$
original equilibrium:	4.53×10^{-4} moles		1.27×10^{-3} moles
add:	8.60×10^{-4} moles		0 moles
initial:	13.13×10^{-4} moles	1.27×10^{-3} moles	

It is unlikely that the system is still in equilibrium following addition of extra acetone and TCE. Assuming this, there must be movement of acetone from one solvent to the other in order to restore the equilibrium concentration ratio to 2.8. We will suppose that "x" moles of acetone leaves the TCE layer and enters the water layer as equilibrium is restored. This information is entered in a "change" line:

	$Acetone_{TCE}$	⇔	$Acetone_{H2O}$
original equilibrium:	4.53×10^{-4} moles		1.27×10^{-3} moles
add:	8.60×10^{-4} moles		0 moles
initial:	13.13×10^{-4} moles		1.27×10^{-3} moles
change:	$-x$ moles		x moles

Note that the change in amount of acetone in TCE is taken as negative, consistent with our assumption that acetone will move from the TCE layer to the water layer. Finally, the amounts present at the new equilibrium are calculated by adding the change to the initial value:

	$Acetone_{TCE}$	⇔	$Acetone_{H2O}$
original equilibrium:	4.53×10^{-4} moles		1.27×10^{-3} moles
add:	8.60×10^{-4} moles		0 moles
initial:	13.13×10^{-4} moles		1.27×10^{-3} moles
change:	$-x$ moles		x moles
new equilibrium:	$1.313 \times 10^{-3} - x$ moles	$1.27 \times 10^{-3} + x$ moles	

We now have expressions for the moles of acetone in each solvent after the new equilibrium is established, and can convert these to concentrations by dividing by the volumes of the solvents:

$[acetone]_{H2O} = (1.27 \times 10^{-3} + x)$ moles/0.001 L

$[acetone]_{TCE} = (1.313 \times 10^{-3} - x)$ moles/0.002 L

Substituting these into the K_d expression produces

$[(1.27 \times 10^{-3} + x)/0.001]/[(1.313 \times 10^{-3} - x)/0.002] = 2.8$

$[(1.27 \times 10^{-3} + x)]/[(1.313 \times 10^{-3} - x)] = 1.4$

$x = 2.37 \times 10^{-4}$ moles acetone

Substituting this value into the expressions in the "new equilibrium" line gives

moles acetone in TCE = 0.001075; mass = 0.062 g

moles acetone in water = 0.001505; mass = 0.087 g

- Initially, the vial contains 1 mL of water and 1 mL of TCE, with acetone concentrations of 1.27 and 0.453 M, respectively. Then 1 mL of TCE containing 0.05 g acetone is added, and the system returns to equilibrium. At equilibrium, the vial contains 1 mL of water and 2 mL of TCE, with acetone concentrations of 1.50 and 0.538 M, respectively (these values are calculated from known moles acetone in each layer, and known volume of each layer). Thus from time zero to the time at which the addition of acetone in TCE takes place, concentrations are level at 1.27 and 0.453 M in the water and TCE layers respectively. At a time sufficiently long following the addition, concentrations are again level at 1.50 and 0.538 M, respectively. These two set of conditions are shown.

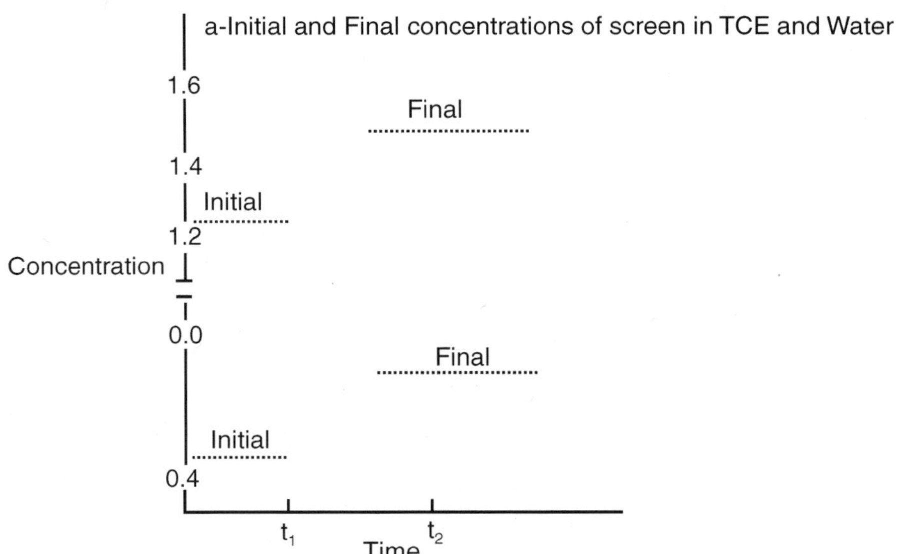

Fig. Adjustment of Distribution of Acetone betweenWater andTCE

It remains for us to determine what happens to concentrations during the time interval between t_1, when TCE addition takes place, and t_2, by which equilibrium has once again been attained. We can easily determine the concentrations of acetone in the two layers immediately following TCE addition, but before any redistribution of acetone takes place. First, the concentration of acetone in the water layer is the same as before TCE addition, at 1.27 M. The concentration of acetone in the TCE layer jumps from 0.453 M just before addition to 0.657 M immediately following addition.

The latter value is calculated by dividing the total number of moles of acetone in the TCE layer by the volume of the TCE layer after addition, 0.002 L. From the post-addition values, the concentrations in the two layers must smoothly evolve according to typical dynamics curves as the acetone redistributes itself to reattain equilibrium. Thus the concentration-time curves are completed as shown in Figure.

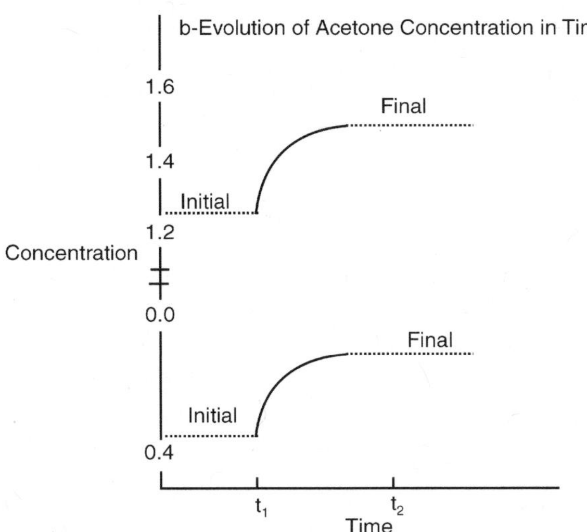

Fig. Adjustment of Distribution of Acetone between Water and TCE

Solutions of Molecular Solids in Liquids. A molecular solid is one in which covalent molecules are arranged in a lattice held together by intermolecular forces exerted by one molecule on its neighbours. When such a solid is placed in contact with a solvent in which it dissolves, molecules at the surfaces of crystals in contact with the solvent can escape the potential well of the crystal if they possess sufficient kinetic energy, and can enter the potential well of the solvent. The rate of this escape from the solid phase is proportional to the surface area of the solid and to the "concentration" of substance in the molecular solid, which is constant. Thus

$$\text{Rate}_{\text{solid}\to\text{soln}} = k_{\text{solid}\to\text{soln}} * SA$$

Similarly, dissolved molecules collide with the surface of the solid and return to the solid phase at a rate given in equation:

$$\text{Rate}_{\text{soln}\to\text{solid}} = k_{\text{soln}\to\text{solid}} * SA * [\text{substance}]$$

At equilibrium, when the two opposing rates are equal :

$$K_{\text{eq}} = k_{\text{solid}\to\text{soln}}/k_{\text{soln}\to\text{solid}} = [\text{substance}]$$

Thus the equilibrium constant is simply the concentration of substance dissolved in the solvent. As we might expect from our work with phase diagrams, a change in pressure has little effect on the equilibrium established between a pure solid and a saturated liquid solution of the solid.

$$B(s) \to B(aq)$$

Both the solid and solution (liquid) phases are condensed, hence incompressible—their volumes are essentially unaffected by pressure.

Unless there is a change in volume with pressure, there can be no change in concentration (the number of molecules per unit volume); and unless there is a concentration change in a phase, there can be no change in the rate at which molecules or ions leave the phase as long as T is constant (we have seen that rate processes are related directly to concentration). All phase equilibria are dynamic equilibria involving a balance of opposing rates. Unless one of the rates is affected by pressure, a change in P will have no effect on the position of equilibrium.

Since neither the forward nor the reverse rate of an equilibrium involving two condensed phases is affected by pressure, this type of equilibrium is unperturbed when P is varied. The equilibrium between a solid and its saturated solution is of this type, where in this case the two balanced rates are the rate of dissolution and the rate of crystallization.

The equilibrium between s and l phases of a pure substance is very similar. The temperature at which such an equilibrium exists is unaffected by even huge changes in P. Thus the solubility of a solid in a liquid should be independent of pressure.

Example: The solubility of a molecular solid, $A(s)$, in water is 9.0 g per 100 mL of water at 20°C. 6.4 g of $A(s)$ is added to 23.6 mL H_2O. What fraction of $A(s)$ dissolves?

To the resulting solution is added 14.4 mL of water. What is the concentration of $A(s)$ when equilibrium is reached, in grams $A(s)$ per 100 mL water?

Solution: The mass of A that will dissolve in 23.6 mL H_2O is mass A = 9.0 g A/100 mL water) * 23.6 mL water = 2.12 g A.

The fraction of A that dissolves is 2.12/6.4 = 0.33.

The equilibrium constant for the process

$$A(s) \Leftrightarrow A(aq)$$

is simply the concentration of $A(aq)$ in a saturated solution. Thus K_{eq} = 9.0 g/100 mL. Before addition of extra water, the system is at equilibrium and satisfies this K_{eq}. Immediately after addition of water, the concentration of dissolved A is 2.12 g A/(23.6 + 14.4) mL H_2O = 5.58 g A/100 mL water. According to Le Chatelier, the process above must shift to the right to offset the dilution of A caused by addition of water.

Shift will continue either until all solid A is dissolved, or until the solution is once again saturated. To saturate 38 mL water requires (9.0 g A/100 mL water) * 38 mL = 3.42 g A, which is less than the total initial amount of A present. Thus the solution will achieve equilibrium with 3.42 g A dissolved and 6.4 – 3.42 = 3.0 g remaining in the solid phase.

Solutions of Gases in Liquids. The solubility of a gas in a liquid is markedly affected by the pressure at which the gas is maintained over the liquid. This is a consequence of the effect of pressure on the volume, hence the concentration, of a gas. We begin with a quantity of water at 25°C in a cylinder fitted with a piston and a gas inlet.

Initially, the space above the water is free of all gas except water vapour, which exerts the characteristic vapour pressure at 25°C. We now fill the space above the water with a gas, $B(g)$, to some pressure P_{B1} by opening the gas inlet valve, while holding the piston fixed in place, until the desired pressure is reached.

As soon as the gas is admitted to the cylinder, gas molecules begin to strike the walls of the cylinder and the surface of the water with a frequency that depends on the concentration (pressure) of gas molecules in the space above the liquid.

Some of the molecules that strike the water surface enter the liquid—i.e., they dissolve. The rate of dissolution is directly proportional to the pressure of the gas and to the surface area, *SA*, of the water, as indicated in equation.

$$\text{Rate}_{\text{dissolution}} = k_1 * SA * P_{B1}$$

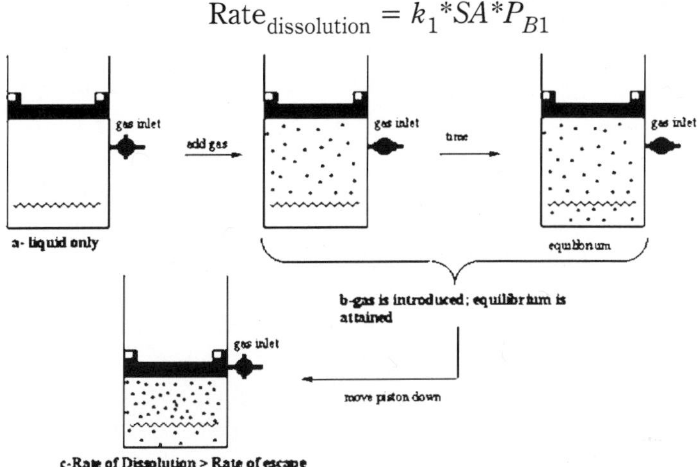

Fig. Equilirium between a Gas and its Solution in a Liquid Solvent.

Here k_1 is a proportionality constant (a rate constant). Gas molecules that dissolve move randomly in the liquid phase, and at any given time, some of them will be moving towards the surface. Those with sufficient energy will escape the liquid and Re-enter the gas phase. The rate of escape is directly proportional to the surface area of the liquid, and to the number of molecules travelling toward the surface, which in turn is directly related to the concentration of dissolved gas. Thus

$$\text{Rate}_{\text{escape}} = k_2 * SA * [B]$$

where k_2 is also a rate constant. Immediately after the gas is introduced to the cylinder, the rate of escape is 0 because there is no dissolved gas, but the rate of dissolution is greater than 0 because $P > 0$. The initial imbalance of rates leads to a buildup of dissolved molecules, an increase in $[B]$, and a corresponding increase in $\text{Rate}_{\text{escape}}$, until $\text{Rate}_{\text{dissolution}} = \text{Rate}_{\text{escape}}$. At this point, dynamic equilibrium between the gas and solution phases has been established. The dynamic equilibrium between a gas and its solution is expressed as in equation.

$$B(g) \Leftrightarrow B(\text{soln})$$

where the double arrow shows the opposing equal rates. Equating the right sides of equation and rearranging gives

$$[B]/P_{B1} = k_1 / k_2 = K$$

Equilibrium is characterized by a particular ratio of the concentration of dissolved B to the pressure of B above the solution.

Now consider quickly moving the piston down so that the volume above the liquid is halved. This doubles the pressure of the gas above the solution, and immediately doubles $R_{dissolution}$ because the number of collisions of gas molecules with the surface is now twice as great. However, R_{escape} is initially unaffected. The immediate effect of the pressure increase is thus to make $R_{dissolution} > R_{escape}$.

This imbalance results in a net flow of molecules from the gas to the liquid phase, causing a gradual buildup in [B], until the two opposing rates again become equal. The new equilibrium is established with higher values of both P and [B], but their ratio, given by k_1 / k_2, is the same as before the pressure increase. Thus the concentration of dissolved gas—its solubility—is directly proportional to the pressure of gas above the solution. We express this in mathematical terms by writing equation somewhat differently : $[B] = KP_B$.

Equation, and the verbal statement that it summarizes, is known as Henry's Law. K is called the Henry's Law constant. The values of K for dissolution of several gases in water are given in Table. For a gas pressure of 1 atm, these constants are numerically the same as the molar solubilities of the gases.

Table. Henrys Law Constants (Solubilities) of several gases in water at 20°C, with 1 atm gas pressure

Gas	*K(M/atm), x 10^3*
N_2	0.69
CO	1.04
O_2	1.38
Ar	1.50
Kr	2.79

Example: A corked bottle of champagne is allowed to warm from 0°C to 25°C over a 1-hour period. The pressure of CO_2 in the bottle was 2.5 atm at 0°C. What is the concentration of dissolved CO_2 at 0°C? What is the pressure of CO_2 in the bottle at 25°C? You may assume that, The Henry's Law constant for CO_2 is 1.5 times greater at 0°C than at 25°C;

The Henry's Law Constant for CO_2 is 3.1×10^{-2} moles/L-atm at 25°C;

The volume of champagne in the bottle is exactly 1.00 L;

The volume of the gas space over the liquid in the bottle is exactly 50.0 mL.

Solution: We start at 0°C because we know the pressure of CO_2 and can calculate the Henry's Law constant at this temperature. The concentration of dissolved CO_2 at 0°C can be calculated from Henry's Law as follows:

$$[CO_2] = K_{eq}P_B$$
$$= 1.5 \times 3.1\times 10^{-2} \times 2.5 = 11.6 \times 10^{-2} \text{ moles/L}$$

We have multiplied by 1.5 to obtain the Henry's Law constant appropriate at the lower temperature. Thus at 0°C we have 11.6×10^{-2} moles/L * 1.00 L = 0.116 moles CO_2 dissolved in the liquid, and $n = PV/RT = (2.5)(0.050)/(0.08206)(273) = 0.00558$ moles CO_2 in the gas space above the liquid. The total moles CO_2 in the bottle is therefore 0.116 + 0.00558 = 0.1216 moles. This total must remain constant as the champagne warms.

At 25°C the ratio of pressure to concentration must satisfy the Henry's Law equilibrium constant, which is 3.1×10^{-2} at the higher temperature. Thus

$$[CO_2]/P_{CO2} = 3.1 \times 10^{-2}$$

So that we can make use of conservation of total moles CO_2 during the warming process, we should convert the ratio on the left to a mole ratio. This is easy using the volumes of the two phases:

$[CO_2]$ = moles CO_2 dissolved/volume of solution
= moles CO_2 dissolved/1.00 L
= moles CO_2 dissolved
P_{CO2} = moles CO_2 in gas * RT/V
= moles CO_2 in gas * 489

Substituting for $[CO_2]$ and P_{CO2} and rearranging gives moles CO_2 dissolved/moles CO_2 in gas = $3.1 \times 10^{-2} \times 489 = 15.16$.

Knowing both the ratio and the sum of the moles dissolved and in the gas allows us to calculate both:

moles CO_2 dissolved = 0.114
moles CO_2 in gas = 0.00752

Finally we convert back to pressure and concentration:

$[CO_2]$ = 0.114 moles/1.00 L = 0.114 M
$P_{CO2} = nRT/V = (0.00752)(0.08206)(298)/0.050 = 3.68$ atm

In summary, warming causes a slight decrease in concentration of dissolved gas from 0.116 to 0.114 M, and an increase in pressure of gas in the head space from 2.5 to 3.68 atm.

Solutions. of Ionic Solids in Liquids; The Solubility Product, K_{sp}. The Solubility Rules. Almost all salts are strong electrolytes; they dissolve as separate positive and negative ions. However, it is easy to show experimentally that although some salts are very soluble (dissolve readily) in water, others dissolve to only a small extent.

For example 34.7 g of sodium chloride, NaCl, dissolves in 100 mL of water at 10°C, compared with only 0.000089 g of silver chloride, AgCl. Chemists have developed empirical rules that can be used to decide whether or not a salt is soluble. These are presented below. For application of these rules, we consider a salt to be soluble if its solubility exceeds 0.01 moles per litre of solution. Salts that dissolve to a lesser extent than 0.01 M are insoluble.

Solubility Rules

A salt is soluble if it contains:

- A group 1 metal
- NO_3-, ClO_3-, ClO_4-, or CH_3COO- (acetate). Exceptions are $AgCH_3COO$ and $KClO_4$.
- Cl-, Br, or I^-. Exceptions for Ag^+, Pb^{2+}, Hg_2^{2+}.
- SO_4^{2-}. Exceptions are the group 2 metals, Ag^+, and Pb^{2+}.

A salt is insoluble if it contains:

- O^{2-} or OH-. Exceptions are the group 1 and 2 cations.
- CO_3^{2-}, PO_4^{3-}, S^{2-}, and SO_3^{2-}. Exceptions are the group 1 cations and NH_4^+.

Example: Use the solubility rules to predict solubilities of NaCl, Fe_2S_3, $(NH_4)_2S$, and $CaCO_3$.

Solution:

NaCl. According to rule 1, this is soluble.

Fe_2S_3. According to rule 7, this is insoluble.

$(NH_4)_2S$. According to rule 7, this is soluble.

$CaCO_3$. According to rule 7, this is insoluble.

The solubility rules can be used to predict whether a precipitate (solid) will form when solutions of two electrolytes are mixed.

Example: A solution of sodium chloride is mixed with a solution of silver nitrate. Will a precipitate form?

Solution: According to the solubility rules, NaCl and $AgNO_3$ are soluble salts. This makes it possible to prepare a fairly concentrated solution of each salt. Because salts are strong electrolytes, these solutions contain the constitutent ions of the salts: Na^+(aq), Cl^-(aq) and Ag^+(aq), NO_3-, respectively. Mixing the solutions produces a solution containing all four ions. Is there a cation-anion combination that will produce an insoluble salt? Rule 3 says that AgCl is insoluble. A precipitate of AgCl will form when the two solutions are mixed, as shown below.

$Na^+(aq) + Cl^-(aq) + Ag^+(aq) + NO_3\text{-}(aq) \rightarrow Na^+(aq) + NO_3\text{-}(aq) + AgCl(s)$

Eliminating spectator ions gives the net ionic equation:

$$Ag^+(aq) + Cl^-(aq) \rightarrow AgCl(s)$$

The solubility rules, based on empirical observation, are useful for predicting the outcome of mixing solutions of electrolytes. This is applied in chemical analysis, in which selective precipitation of an ion from solution can enable us to determine the amount of the ion that was initially present by weighing the precipitate.

Although many ionic salts show substantial solubility in water, a number of simple and common salts fail to dissolve. These are called slightly soluble salts.

Combinations of OH^-, CO_3^{2-}, and S^{2-} with many cations, particularly those of the transition metals, are examples of such salts. We now want to examine the quantitative aspects of their dissolution, using the principles of equilibrium learned thus far.

Consider adding a quantity of the salt $MX(s)$ (M^{p+} is the cation, X^{p-} is the anion) to a beaker of water.

Assume that we know how much of the salt will dissolve in the water, and that the amount added is more than this. What happens at the molecular level after the salt is placed in contact with the water?

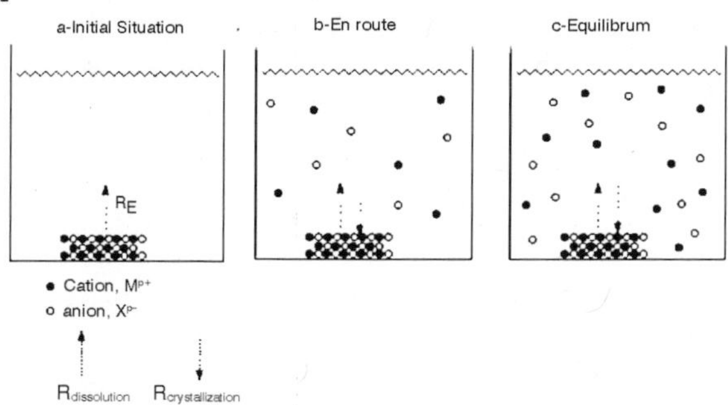

Fig. Equilibrium between an Ionic Solid and its Ions in Solution

The initial situation, immediately after addition of the salt to the water, but before dissolution of the solid. Upon contact with water, ions on the surface of the crystals begin to escape into the liquid at a rate that depends upon the forces of attraction between these ions and other ions in the crystal on one hand, and the ions and water molecules on the other.

The rate of dissolution is proportional to the surface area, *SA*, of the solid salt, and the concentration of salt, which is constant in the solid.

$$\text{Rate}_{\text{dissolution}} = \text{k}_{\text{dissolution}} * SA$$

The rate of dissolution remains essentially constant as long as solid is present and the temperature is maintained constant. As dissolution proceeds, the concentrations of the ions in the liquid phase gradually build up. These ions move randomly in the liquid phase, occasionally colliding with the surface of the solid, where some of them stick—i.e., they crystallize.

As their number builds, so does the rate at which they collide with the surface of the solid and crystallize. The rate of crystallization is proportional to the exposed surface area of solid, and to the product of the concentrations of positive and negative ions. The product is required because positive and negative ions must collide at the surface in order for crystallization to occur.

$$\text{Rate}_{\text{crystallization}} = k_{\text{crystallization}} * SA * [M^{p+}][X^{p-}]$$

A situation in which dissolution and crystallization occur with unequal rates. Eventually, rates of dissolution and crystallization become equal, and there is no further net dissolution of solid. At this point, dynamic equilibrium has been established between the solid and its ions in solution. The equilibrium is represented in equation, where the double arrow shows the two opposing rates.

$$MX(s) \Leftrightarrow M^{p+}(aq) + X^{p-}(aq)$$

The equalization of the rates of dissolution and precipitation over time. The equilibrium in equation is certainly a phase equilibrium, because it involves two phases: a solid phase, represented by the salt, and a liquid phase, the solution. Equating the expressions for the rates of dissolution and crystallization, the expression for the equilibrium constant is obtained:

$$K_{eq} = K_{sp} = k_{\text{dissolution}} / k_{\text{crystallization}} = [M^{p+}][X^{p-}]$$

We give a special symbol to the equilibrium constant for dissolution of a slightly soluble salt: K_{sp}. The subscript, *sp*, stands for "solubility product' because it is the product of ion concentrations, which are related to the molar solubility of the salt. The molar solubility is the number of moles of salt that dissolve per litre of solution. The solubility product expression for the slightly soluble salt, bismuth sulfide, which dissolves according to

$$Bi_2S_3(s) \Leftrightarrow 2Bi^{3+} + 3S^{2-}, \text{ is } K_{sp} = [Bi^{3+}]^2[S^{2-}]^3.$$

The value of K_{sp} is 2×10^{-70} at 298 K, indicating that the concentrations of ions in equilibrium with solid bismuth sulfide are very small.

Before considering calculations using solubility product equilibria, we define a few terms used to describe solutions of salts. Consider that we have a beaker containing one litre of pure water. To this we add a quantity of a salt, $M_nX_m(s)$. Suppose first that the quantity of solid added is less than the amount that will dissolve in one litre of water. We will see a gradual disappearance of solid salt as dissolution occurs, until all of the solid is gone. The resulting solution, containing less than the maximum possible amount of dissolved salt, is said to be unsaturated. In such a solution, the product $[M^{p+}]^n [X^{q-}]^m$ is less than K_{sp} for the salt.

Now suppose that we add more solid to the beaker. Since the solution is unsaturated, we expect that some of the added solid will dissolve, until the solution contains the maximum possible amount. This amount depends on the solubility of the salt in water at the water temperature. We can identify this situation visually, because we will see a quantity of undissolved solid on the bottom of the beaker that does not change in size with time. This indicates that there is no longer any net dissolution of solid. At the molecular level, of course, a dynamic conversion between solid salt and dissolved ions is taking place, in which salt is dissolving and ions are combining to form solid at equal rates.

At the macroscopic level, as is true for any equilibrium, the dynamic aspects of the microscopic balance are not apparent, and it appears as if nothing is happening. A solution containing the maximum possible amount of dissolved solid at a particular temperature is said to be saturated. In a saturated solution, it must be true that the product of ion concentrations, each concentration raised to the appropriate power, is equal to K_{sp}. Finally, suppose that we are able somehow to cause more than the maximum possible amount of salt to dissolve. In such a solution, which is said to be supersaturated, the product of ion concentrations exceeds K_{sp}, and the system is not at equilibrium. Eventually, we expect precipitation (solid formation) to occur until the product of ion concentrations becomes equal to K_{sp}, at which point equilibrium will have been established.

How do we prepare a supersaturated solution of a salt? Suppose that the salt becomes more soluble in water as the temperature of the water increases. Both sodium acetate, $NaC_2H_3O_2$, and alum, $KAl(SO_4)_2$, qualify, as do many other salts. We consider sodium acetate. The solubility of sodium acetate in water at 25°C is 125 g per 100 mL of water. At 100°C (the boiling point of water) the solubility increases to 170 g per 100 mL of water. If we weigh an amount of sodium acetate that is greater than 125 g but less than 170 g, add it to 100 mL of water, and warm the mixture with stirring to near 100°C, all of the sodium acetate should dissolve. (In order to observe supersaturation, it is necessary to remove all solid material from the solution while it is hot.

This can be done by careful filtration of the hot solution.) We now carefully cover the solution to exclude dust and dirt, and allow it to slowly cool to room temperature. With luck, all of the sodium acetate stays dissolved, even though the equilibrium solubility at 25°C is exceeded. This is a supersaturated solution. How can a substance remain dissolved when it should precipitate? Why does the salt not crystallize from solution as the system cools, so as to always maintain the solubility product equilibrium? The answer is interesting—and simple. Formation of crystals of solid from ions or molecules dissolved in a solvent is similar to the freezing process, in which solid forms from liquid:

$$l \Leftrightarrow s$$

For this process, $\Delta PE < 0$, favoring formation of solid; but $\Delta S < 0$ (because solid is more ordered than liquid), opposing formation of solid. Supersaturation is thus analogous to supercooling of liquids. For crystallization of a salt to begin, it is necessary that positive and negative ions, moving randomly in solution, coalesce in the proper arrangement for a crystal to begin to form.

Because this process is opposed by entropy, it often takes some time, during which the solution can hold more dissolved material than it should. Once the beginnings of a crystal form, however, this serves as a nucleus, or template, on which other ions can deposit, and crystallization proceeds rapidly. Thus we can initiate crystallization of sodium acetate from the supersaturated solution by adding a small crystal of sodium acetate as a template.

Crystallization then proceeds until the K_{sp} is satisfied (i.e., until the solution is saturated). (Often, addition of any small solid particles to a supersaturated solution promotes crystallization. That is why filtration to remove dust is necessary.) Generally, the more complicated and/or unsymmetrical the structures of the ions of the salt, the more difficult it is for the ions to coalesce in the correct way to initiate crystallization. Thus we expect salts of complex ions to more readily form supersaturated solutions than salts of simple ions. In agreement with this guideline, it is very difficult to prepare a supersaturated solution of NaCl; whereas preparation of supersaturated sodium acetate, which contains an anion of complex structure, is readily accomplished.

Example: Determine the molar solubility of $CaCO_3$ in water at 25°C.

Solution: In this example, we illustrate what might be called the standard approach to the determination of salt solubility using K_{sp}. After carrying this through in the usual way, we will compare the result with the known solubility of $CaCO_3$ to demonstrate that K_{sp} calculations must in most cases be taken with a grain of salt (no pun intended).

We write the equation for dissolution of calcium carbonate, and underneath the species we write the initial concentrations, the changes in concentrations that take place during reaction, and the concentrations that prevail at equilibrium. This is done below.

	$CaCO_3(s) \rightarrow$ $K_{sp} = [Ca^{2+}][CO_3^{2-}]$	$Ca^{2+}(aq)$ + $= 5 \times 10^{-9}$ M^2	CO_3^{2-}
initial	*xs*	0	0
change	$-x$	x	x
equilibrium	*xs*	x	x

(Note: because $CaCO_3$ is a pure solid, it is omitted from the K_{sp} expression. In order to establish equilibrium with its ions, it is necessary only that a sufficient amount of it be present. Thus we have written under its formula that we have an excess (*xs*) initially, and still have an excess at equilibrium.)

Substituting, $x^2 = 5 \times 10^{-9}$, giving $x = 7 \times 10^{-5}$ M.

Thus according to our calculation, 7×10^{-5} moles (0.007 g) of $CaCO_3$ will dissolve in one litre of water. The solubility of calcium carbonate is 0.0015 g per 100 mL of water, or 0.015 g per litre, for one crystalline form; and 0.014 g per litre for another. The solubility thus depends on the crystalline form adopted by the solid, so already we have a complication not addressed by our standard calculation above. Further, we see that at best our calculated result is wrong by a factor of about 2. The solubility of $CaCO_3$ is actually two times greater than we would expect. The major reason for the discrepancy is that the carbonate anion is substantially basic, and reacts with water in the hydrolysis process below:

$$CO_3^{2-} + H_2O \rightarrow HCO_3^- + OH^-$$

This process causes the concentration of the carbonate dianion to decrease, shifting the solubility equilibrium, to the right by Le Chatelier's Principle. Reaction therefore enhances the solubility of $CaCO_3$. It is the rule rather than the exception that the solubilities of slightly soluble salts are influenced by reactions that the ions produced by dissolution may subsequently undergo. For this reason, the standard approach to K_{sp} calculations gives at best only a gross estimate of concentrations, and at worst, a completely inaccurate result.

The equilibria among phases of a single pure substance, the equilibrium vapour pressure has emerged as a quantity of great importance. It measures the "escaping tendency" of the molecules in the liquid (or solid) phase by directly reflecting at the macroscopic scale the magnitudes of the forces of attraction between molecules.

By measuring its temperature dependence, we can obtain the depth of the liquid potential well and can measure the change in disorder that occurs when the liquid vaporizes. By measuring its decrease, or consequences thereof, in solutions, we can determine the amount and even the molecular formula of a dissolved solute. This very simple macroscopic property of liquids, then, provides us with an abundance of information about happenings in the liquid at the molecular level.

More generally, we have seen that dynamic equilibrium, a concept of immense importance in chemistry, has a number of defining characteristics:

- It occurs only in closed systems.
- At least 2 states of a system must be simultaneously present (coexist).
- It is established only over a finite time interval, required to establish the equality of opposing rates.
- Values of macroscopic observables remain constant with time, but at the molecular level there is constant transit among the states present.
- When subjected to a perturbation from outside, it responds to minimize the effect of the perturbation and returns to equilibrium, again over a finite time interval.
- It represents a compromise between tendencies to minimize potential energy and maximize disorder.

The word, "equilibrium," is used somewhat casually by scientists. It is useful to distinguish between equilibrium in general and dynamic equilibrium as we have characterized it here. A pure substance subjected to a pressure and a temperature such that only the liquid phase of the substance may exist is certainly at equilibrium in the sense that no change in the system will occur in the absence of a perturbation. However, this equilibrium is not dynamic, because only one state of the system is present, so opposing processes do not occur. At least 2 states of a system must coexist in order for a situation of balanced opposing processes to be possible.

THE EQUILIBRIUM CONSTANT EXPRESSIONS

According to the *law of mass action*, equilibrium reactions or processes are written as follows:

$$wA + xB \rightleftarrows yC + zD$$

where A, B, C, and D are the chemical species, and w, x, y, and z are their respective coefficients. The equilibrium constant expression is given as,

$$K = \frac{[C]^y.[D]^z}{[A]^w.[B]^x}; \quad (K \text{ is called the equilibrium constant})$$

At constant temperature, *K* has fixed numerical value. For example, the reaction:

$2SO_{2(g)} + O_{2(g)} \rightleftarrows 2SO_{3(g)}$ has reach equilibrium constant,

$$K = \frac{[SO_3]^2}{[SO_2]^2[O_2]}$$

$[SO_2]$, $[O_2]$, and $[SO_3]$ are molar concentrations of individual substances at equilibrium. For example, if at a certain temperature an equilibrium mixture contains 1.50 *M* SO_2, 1.25 *M* O_2, and 3.50 *M* SO_3, the equilibrium constant for the reaction at that temperature is,

$$K = \frac{[SO_3]^2}{[SO_2]^2[O_2]} = \frac{(3.50\ M)^2}{(1.50\ M)^2(1.25\ M)} = 4.36$$

A given reaction has one equilibrium constant at a given temperature, but it has an infinite number of sets of concentrations, called the equilibrium positions. For example, at the same temperature, the above reaction may have a different set of concentrations and another equilibrium position, such as, $[SO_2]$ = 0.590 M. $[O_2]$ = 0.045 M, and $[SO_3]$ = 0.260 M. The equilibrium constant for the reaction is

$$K = \frac{[SO_3]^2}{[SO_2]^2[O_2]} = \frac{(0.260\ M)^2}{(0.590\ M)^2(0.045\ M)} = 4.32$$

These two values of *K* are within experimental error. Thus, a given reaction may have many equilibrium positions, but only one equilibrium constant at a given temperature. The equilibrium constant, *K*, only changes if temperature changes. For exothermic reactions, the value of *K* decreases when temperature increases, while for endothermic reactions, *K* increases as temperature increases.

FACTORS AFFECTING THE EQUILIBRIUM POSITIONS AND EQUILIBRIUM CONSTANTS

The numerical value of the equilibrium constant, K, enable us to say something about the equilibrium position or the extent of a reaction when it reaches equilibrium at a particular temperature. A large *K* values imply that a considerable amount of products have been formed at equilibrium. In fact, if *K* is very, very large, we can simply assume that the reaction has gone to completion when equilibrium is reached. While a small value of *K* means that very little products have been formed at equilibrium.

The equilibrium position is affected by factors such as changes in concentrations, pressure, and temperature of the reaction mixture. Le Chatelier's principle states that, when a change is imposed on a system at equilibrium, the system will react and shifts its equilibrium position in a direction that tends to reduce that change.

Effect of Changes in Concentration

Consider the reaction: $N_{2(g)} + 3H_{2(g)} \rightleftarrows 2\,NH_{3(g)}$;
Suppose that a system at equilibrium contains the following concentrations:
$[N_2] = 0.399$ M, $[H_2] = 1.197$ M, $[NH_3] = 0.203$ M
What will happen if 1.000 mol/L of N_2 is added into the system at equilibrium?

- $[N_2]$ increases and according to the Le Chatelier's principle, the system will try to decrease this concentration. A net forward reaction occurs as a result until a new equilibrium concentrations is established.

In general, increasing the concentration of any of the reactants will cause the equilibrium to shift in the direction of the formation of products, because by doing so, it decreases the concentration of the reactant and reduces the stress.

Effect of Changing Volume and Pressure

In reactions involving gaseous substances, compressing the mixture into smaller volumes causes the pressure to increase (Boyle's Law). According to the Le Chatelier's principle, if a system at equilibrium experiences a change in pressure, a net reaction that tends to reduce the overall pressure will occur until a new equilibrium position is established.

- Consider the reaction: $N_{2(g)} + 3\,H_{2(g)} \rightleftarrows 2NH_{3(g)}$;

In this reaction, the forward reaction decreases the total pressure because the forward reaction results in fewer number of molecules in the mixture. Whereas the reverse reaction increases the total number of molecules, hence increases the total pressure.

If the mixture is compressed into a smaller vessel, the pressure will increase and a net forward reaction occurs to reduce the pressure and reach a new equilibrium position. (We say that increasing the pressure by compression causes an equilibrium shift to the right.) The new equilibrium position will contain a higher concentration of ammonia. Thus, the production of ammonia from nitrogen and hydrogen gases favors a high-pressure condition.

On the other hand, if the mixture (at equilibrium) is transferred into a larger vessel, the total gas pressure will drop, causing a stress, and a net reaction in the reverse direction will occur to increase the pressure and reach a new equilibrium position.

- Consider the reaction: $CH_{4(g)} + H_2O_{(g)} \rightleftarrows CO_{(g)} + 3\,H_{2(g)}$,

In this reaction, the forward reaction causes the number of gaseous molecules to increase. If the equilibrium mixture is compressed into a smaller vessel, a net reaction in the reverse direction will occur until a new equilibrium position with lower pressure is established. Reducing the pressure by transferring the equilibrium mixture into a larger vessel will result in a net forward reaction to achieve a new equilibrium position. Thus, the reaction of methane gas with steam to produce CO and H_2 gas favors a low-pressure condition.

- Consider the following reaction: $CO_{(g)} + H_2O_{(g)} \rightleftarrows CO_{2(g)} + H_{2(g)}$,

There is no net change in the total number of molecules. Compressing or expanding the volume of the reaction vessel will not affect the equilibrium position because the change imposes equal stress on both reactants and products.

Compressing or expanding the volume of the reaction mixture will affect equilibrium position only if the total number of gaseous molecules of products is different from that of reactants.

Effects of Temperature on Equilibrium

- Consider the following exothermic reaction, in which heat is a product:

$$2\ SO_{2(g)} + O_{2(g)} \rightleftarrows 2\ SO_{3(g)};\ \Delta H^{o} = -180\ \text{kJ}$$

If the temperature is raised (more heat added), the system will experience a stress and a net reaction that absorbs heat will occur to reach a new equilibrium position, which is in the reverse direction for exergonic reactions. If the temperature is lowered (heat is removed), the system will also experience a stress, and a net forward reaction that produces heat occurs to reach a new equilibrium position. All exothermic reactions favour a low temperature condition.

- For the following endothermic reaction:

$$CH_{4(g)} + H_2O_{(g)} \rightleftarrows CO_{(g)} + 3\ H_{2(g)};\ \Delta H^{o} = +205\ \text{kJ}$$

increasing the temperature will cause a net forward reaction to a new equilibrium position, and lowering the temperature causes a net reverse reaction (produces heat). That is, endothermic reactions favour high temperature condition.

PRESSURES CAN EXPRESS CONCENTRATIONS

Although we commonly write equilibrium quotients and equilibrium constants in terms of molar concentrations, any concentration-like term can be used, including mole fraction and molality. Sometimes the symbols K_c, K_x, and K_m are used to denote these forms of the equilibrium constant. Bear in mind that the *numerical values* of K's and Q's expressed in these different ways will not generally be the same.

Most of the equilibria we deal with in this course occur in liquid solutions and gaseous mixtures. We can express K_c values in terms of moles per litre for both, but when dealing with gases it is often more convenient to use partial pressures. These two measures of concentration are of course directly proportional:

$$C = \frac{n}{V} = \frac{\frac{PV}{RT}}{V} = \frac{P}{RT}$$

so for a reaction A*(g)* → B*(g)* we can write the equilibrium constant as,

$$Kp = \frac{P_B}{P_A}$$

All of these forms of the equilibrium constant are only approximately correct, working best at low concentrations or pressures. The only equilibrium constant that is truly "constant" (except that it still varies with the temperature!) is expressed in terms of *activities*, which you can think of as "effective concentrations" that allow for interactions between molecules. In practice, this distinction only becomes important for equilibria involving gases at very high pressures (such as are often encountered in chemical engineering) and in ionic solutions more concentrated than about 0.001 *M*. We will not deal much with activities in this course.

For a reaction such as $CO_2(g) + OH^-(aq) \rightarrow HCO_3^-(aq)$ that involves both gaseous and dissolved components, a "hybrid" equilibrium constant is commonly used:

$$K = \frac{\left[HCO_3^-\right]}{P_{CO2}\left[OH^-\right]}$$

Clearly, it is essential to be sure of the units when you see an equilibrium constant represented simply by "*K*".

Converting between K_p and K_c

It is sometimes necessary to convert between equilibrium constants expressed in different units. The most common case involves pressure- and concentration equilibrium constants.

The ideal gas law relates the partial pressure of a gas to the number of moles and its volume:

$$PV = nRT$$

Concentrations are expressed in moles/unit volume n/V, so by rearranging the above equation we obtain the explicit relation of pressure to concentration:

$$P = (n/V)RT$$

Conversely, $c = (n/V) = P/RT$

... so a concentration [A] can be expressed as P_A(RT).

For a reaction of the form 2 A = B + 3 C, we can write

$$K_p = \frac{(P_C)^2}{(P_A)\ (P_B)^3} = \frac{([C]RT)^2}{([A]RT)\ ([B]RT)^3} = \underbrace{\boxed{\frac{[C]^2}{[A]\ [B]^3}}}_{K_c}(RT)^{\underbrace{\boxed{-2}}_{\Delta n_g}}$$

Assuming that all of the components are gases, the difference (moles of gas in products) – (moles of gas in reactants) = Δn_g is given by

$$K_p = K_c (RT)^{\Delta ng}$$

UNCHANGING CONCENTRATIONS

Substances whose concentrations undergo no significant change in a chemical reaction do not appear in equilibrium constant expressions. How can the concentration of a reactant or product *not*change when a reaction involving that substance takes place? There are two general cases to consider.

The Substance is also the Solvent

This happens all the time in acid-base chemistry. Thus for the hydrolysis of the cyanide ion

$$CN^- + H_2O \rightarrow HCN + OH^-$$

we write,

$$\frac{[HCN][OH^-]}{[CN^-]}$$

in which no $[H_2O]$ term appears. The justification for this omission is that water is both the solvent and reactant, but only the tiny portion that acts as a reactant would ordinarily go in the equilibrium expression. The amount of water consumed in the reaction is so minute (because K is very small) that any change in the concentration of H_2O from that of pure water (55.6 mol L^{-1}) will be negligible. Similarly, for the "dissociation" of water $H_2O = H^+ + OH^-$the equilibrium constant is expressed as the "ion product"$K_w = [H^+][OH^-]$.

But... be careful about throwing away H_2O whenever you see it. In the esterification reaction,

$$CH_3COOH + C_2H_5OH \rightarrow CH_3COOC_2H_5 + H_2O$$

that we discussed in a previous section, a $[H_2O]$ term must be present in the equilibrium expression if the reaction is assumed to be between the two liquids acetic acid and ethanol. If, on the other hand, the reaction takes place between a dilute aqueous solution of the acid and the alcohol, then the $[H_2O]$ term would not be included.

The Substance is a Solid or a Pure Liquid Phase

This is most frequently seen in solubility equilibria, but there are many other reactions in which solids are directly involved:

$$CaF_2(s) \rightarrow Ca^{2+}(aq) + 2F^{\rightarrow}(aq)$$

$$Fe_3O_4(s) + 4\,H_2(g) \rightarrow 4\,H_2O(g) + 3Fe(s)$$

These are *heterogeneous reactions* (meaning reactions in which some components are in different phases), and the argument here is that concentration is only meaningful when applied to a substance within a single phase.

Thus the term [CaF_2] would refer to the "concentration of calcium fluoride within the solid CaF_2", which is a constant depending on the molar mass of CaF_2 and the density of that solid. The concentrations of the two ions will be independent of the quantity of solid CaF_2 in contact with the water; in other words, the system can be in equilibrium as long as any CaF_2 at all is present.

Throwing out the constant-concentration terms can lead to some rather sparse-looking equilibrium expressions. For example, the equilibrium expression for each of the processes shown in the following table consists solely of a single term involving the partial pressure of a gas:

1)	$CaCO_3(s) \rightarrow CaO(s) + CO_2(g)$	$K_p = P_{CO2}$	Thermal decomposition of limestone, a first step in the manufacture of cement.
2)	$Na_2SO_4{\cdot}10\ H_2O(s)$ $Na_2SO_4(s) + 10\ H_2O(g)$	$K_p = P_{H2O}{}^{10}$	Sodium sulfate decahydrate is a solid in which H_2O molecules (*"waters of hydration"*) are incorporated into the crystal structure.)
3)	$I_2(s) \rightarrow I_2(g)$	$K_p = P_{I2}$	sublimation of solid iodine; this is the source of the purple vapor you can see above solid iodine in a closed container.
4)	$H_2O(l)$! $H_2O(g)$	$K_p = P_{H2O}$	Vaporization of water. When the partial pressure of water vapor in the air is equal to *K*, the relative humidity is 100per cent.

The last two processes 3 and 4 represent changes of state (*phase changes*) which can be treated exactly the same as chemical reactions.

In each of the heterogeneous processes shown in the table, the reactants and products can be in equilibrium (that is, permanently coexist) only when the partial pressure of the gaseous product has the value consistent with the indicated K_p. Bear in mind also that these K_p 's all increase with the temperature.

VALUES OF EQUILIBRIUM CONSTANTS

The numerical value of a quantity in terms of what it means in a practical sense is an essential part of developing a working understanding of Chemistry. This is particularly the case for equilibrium constants, whose values span the entire range of the positive numbers.

Although there is no explicit rule, for most practical purposes you can say that equilibrium constants within the range of roughly 0.01 to 100 indicate that a chemically significant amount of all components of the reaction system will be present in an equilibrium mixture and that the reaction will be *incomplete* or "*reversible*".

As an equilibrium constant approaches the limits of zero or infinity, the reaction can be increasingly characterized as a one-way process; we say it is "*complete*" or "*irreversible*".

The latter term must of course not be taken literally; the Le Châtelier principle still applies (especially insofar as temperature is concerned), but addition or removal of reactants or products will have less effect.

Although it is by no means a general rule, it frequently happens that reactions having very large equilibrium constants are *kinetically hindered*, often to the extent that the reaction essentially does not take place. The examples in the following table are intended to show that numbers (values of *K*), no matter how dull they may look, do have practical consequences!

Reaction	*K*	Remarks
$N_2(g) + O_2(g) \rightarrow 2 NO(g)$	5×10^{-31} at 25°C, 0.0013 at 2100°C	These two very different values of Killustrate very nicely why reducing combustion-chamber temperatures in automobile engines is environmentally friendly.
$3 H_2(g) + N_2(g) \rightarrow 2 NH_3(g)$	7×10^{5} at 25°C, 56 at 1300°C	See the discussion of this reaction in the section on the Haber process.
$H_2(g) \rightarrow 2 H(g)$	10^{-36} at 25°C, 6×10^{-5} at 5000°	Dissociation of any stable molecule into its atoms is endothermic. This means that all molecules will decompose at sufficiently high temperatures.
$H_2O(g) \rightarrow H_2(g) + ½ O_2(g)$	8×10^{-41} at 25°C	You won't find water a very good source of oxygen gas at ordinary temperatures!
$CH_3COOH(l) \rightarrow 2 H_2O(l) + 2 C(s)$	$K_c = 10^{13}$ at 25°C	This tells us that acetic acid has a great tendency to decompose to carbon, but nobody has ever found graphite (or diamonds!) forming in a bottle of vinegar. A good example of a super kinetically-hindered reaction!

Units Equilibrium Constants

The equilibrium expression for the synthesis of ammonia,

$$3 H_2(g) + N_2(g) \rightarrow 2 NH_3(g)$$

can be expressed as,

$$K_p = \frac{P^2_{NH_3}}{P_{N_2} P^{.3}_{H_2}}$$

or,

$$K_c = \frac{[NH_3]^2}{[N_2][N_2]^3}$$

so K_p and Q_p for this process would appear to have units of atm^{-1}, and K_c and Q_c would be expressed in mol^{-2} L^2. And yet these quantities are often represented as being dimensionless. Which is correct? The answer is that both forms are acceptable. There are some situations (which you will encounter later) in which *K*'s must be considered dimensionless, but in simply quoting the value of an equilibrium constant it is permissible to include the units, and this may even be useful in order to remove any doubt about the units of the individual terms in equilibrium expressions containing both pressure and concentration terms. In carrying out your own calculations, however, there is rarely any real need to show the units.

Strictly speaking, equilibrium expressions do not have units because the concentration or pressure terms that go into them are really *ratios* having the forms (n mol L^{-1})/(1 mol L^{-1}) or (n atm)/(1 atm) in which the unit quantity in the denominator refers to the *standard state* of the substance; thus the units always cancel out. (But first-year students are not expected to know this!)

For substances that are liquids or solids, the standard state is just the concentration of the substance within the liquid or solid, so for something like CaF*(s)*, the term going into the equilibrium expression is $[CaF_2]/[CaF_2]$ which cancels to unity; this is the reason we don't need to include terms for solid or liquid phases in equilibrium expressions.

The subject of standard states would take us beyond where we need to be at this point in the course, so we will simply say that the concept is made necessary by the fact that energy, which ultimately governs chemical change, is always relative to some arbitrarily defined zero value which, for chemical substances, is the standard state.

It is important to remember that an equilibrium quotient or constant is always tied to a specific chemical equation, and if we write the equation in reverse or multiply its coefficients by a common factor, the value of *Q* or *K* will change.

The rules are very simple:

- Writing the equation in reverse will invert the equilibrium expression;
- Multiplying the coefficients by a common factor will raise *Q* or *K* to the corresponding power.

Here are some of the possibilities for the reaction involving the equilibrium between gaseous water and its elements:

$2\,H_2 + O_2 \rightarrow 2\,H_2O$	$10\,H_2 + 5\,O_2 \rightarrow 10\,H_2O$	$H_2 + \tfrac{1}{2}\,O_2 \rightarrow H_2O$	$H_2O \rightarrow H_2 + \tfrac{1}{2}\,O_2$
$K_p = \dfrac{P^2_{H_2O}}{P^2_{H_2}P_{O_2}}$	$K_p = \dfrac{P^{10}_{H_2O}}{P^{10}_{H_2}P^5_{O_2}}$	$K_p = \dfrac{P_{H_2O}}{P_{H_2}P^{1/2}_{O_2}}$	$K_p = \dfrac{P_{H_2}P^{1/2}_{O_2}}{P_{H_2O}}$

Heterogeneous Reactions

Heterogeneous reactions are those involving more than one phase. Some examples:

$Fe(s) + O_2(g) \rightarrow FeO_2(s)$	air-oxidation of metallic iron (formation of rust)
$CaF_2(s) \rightarrow Ca(aq) + F^+(aq)$	dissolution of calcium fluoride in water
$H_2O(s) \rightarrow H_2O(g)$	sublimation of ice (a *phase change*)
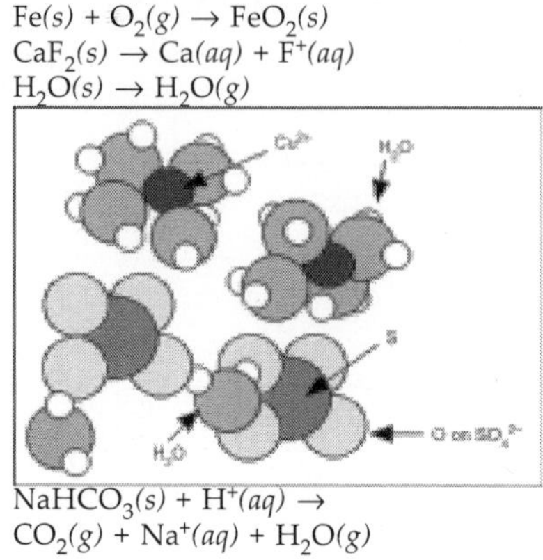	
$NaHCO_3(s) + H^+(aq) \rightarrow$ $CO_2(g) + Na^+(aq) + H_2O(g)$	formation of carbon dioxide gas from sodium bicarbonate when water is added to baking powder (the hydrogen ions come from tartaric acid, the other component of baking powder.)

The Vapor Pressure of Solid Hydrates

A particularly interesting type of heterogeneous reaction is one in which a solid is in equilibrium with a gas. The sublimation of ice illustrated in the above table is a very common example. The equilibrium constant for this process is simply the partial pressure of water vapor in equilibrium with the solid— the *vapor pressure* of the ice.

Many common inorganic salts form solids which incorporate water molecules into their crystal structures. These water molecules are usually held rather loosely and can escape as water vapor. Copper(II) sulfate, for example forms a pentahydrate in which four of the water molecules are coordinated to the Cu^{2+} ion while the fifth is hydrogen-bonded to SO_4^{2-}. This latter water is more tightly bound, so that the pentahydrate loses water in two stages on heating:

$$CuSO_4 5H_2O \xrightarrow{140°C} CuSO_4 5H_2O \xrightarrow{400°C} CuSO_4$$

These dehydration steps are carried out at the temperatures indicated above, but at any temperature, some moisture can escape from a hydrate. For the complete dehydration of the pentahydrate we can define an equilibrium constant:

$$CuSO_4{\cdot}5H_2O(s) \rightarrow CuSO_4(s) + 5\ H_2O(g)\ K_p = 1.14{\times}10^{10}$$

The vapor pressure of the hydrate (for this reaction) is the partial pressure of water vapor at which the two solids can coexist indefinitely; its value is $K_p^{1/5}$ atm. If a hydrate is exposed to air in which the partial pressure of water vapor is less than its vapor pressure, the reaction will proceed to the right and the hydrate will lose moisture. Vapor pressures always increase with temperature, so any of these compounds can be dehydrated by heating.

Loss of water usually causes a breakdown in the structure of the crystal; this is commonly seen with sodium sulfate, whose vapor pressure is sufficiently large that it can exceed the partial pressure of water vapor in the air when the relative humidity is low. What one sees is that the well-formed crystals of the decahydrate undergo deterioration into a powdery form, a phenomenon known as *efflorescence*.

When a solid is able to take up moisture from the air, it is described as *hygroscopic*. A small number of anhydrous solids that have low vapor pressures not only take up atmospheric moisture on even the driest of days, but will become wet as water molecules are adsorbed onto their surfaces; this is most commonly observed with sodium hydroxide and calcium chloride. With these solids, the concentrated solution that results continues to draw in water from the air so that the entire crystal eventually dissolves into a puddle of its own making; solids exhibiting this behaviour are said to be *deliquescent*.

Name	Formula	Vapor Pressure, Torr	
		25°C	30°C
sodium sulfate decahydrate	$Na_2SO_4 \cdot 10H_2O$	19.2	25.3
copper(II) sulfate pentahydrate	$CuSO_4 \cdot 5H_2O$	7.8	12.5
calcium chloride monohydrate	$CaCl_2 \cdot H_2O$	3.1	5.1
(water)	H_2O	23.5	31.6

8

Enzymes Kinetics

ENZYME

Enzymes are macromolecules that help accelerate (catalyze) chemical reactions in biological systems. This is usually done by accelerating reactions by lowering the transition state or decreasing the activation energy.

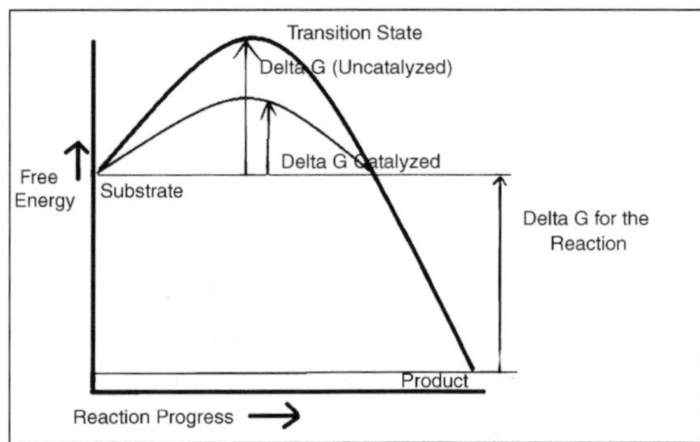

Fig. Enzyme Decreases Activation Energy.

Some biological reactions in the absence of enzymes may be as much as a million times slower. Virtually all enzymes are proteins, though the converse is not true and other molecules such as RNA can also catalyze reactions. The most remarkable characteristics of enzymes are their ability to accelerate chemical reactions and their specificity for a particular substrate. Enzymes take advantage of the full range of intermolecular forces (van der waals interactions, polar interactions, hydrophobic interactions and hydrogen bonding) to bring substrates together in most optimal orientation so that reaction will occur. Also, enzymes can be inhibited by specific molecules by called competitive, uncompetitive, and non-competitive inhibitors.

Catalysis happens at the active site of the enzyme. It contains the residues that directly participate in the making and breaking of bonds. These residues are called the catalytic groups. Although enzymes differ widely in structure, specificity, and mode of catalysis a number of generalizations concerning their active sites can be made:

- The active site is a three dimensional cleft or crevice formed by groups that come from different parts of the amino acid sequence - residues far apart in the amino acid sequence may interact more strongly than adjacent residues in the sequence.
- The active site takes up a relatively small part of the total volume of an enzyme. Most of the amino acid residues in an enzyme are not in contact with the substrate, which raises the question of why enzymes are so big. Nearly all enzymes are made up of more than 100 amino acid residues. The "extra" amino acids serve as a scaffold to create the three dimensional active site from the amino acids that are far apart in the primary structure. In many proteins the remaining amino acids also constitute regulatory sites, sites of interaction with other proteins, or channels to bring the substrate to the active sites.
- Active sites are unique microenvironments. In all enzymes of known structure, substrate molecules are bound to a cleft or crevice. Water is usually excluded unless it is a reactant. The non-polar microenvironment of the cleft enhances the binding of substrates as well as catalysis. Nevertheless, the cleft may also contain polar residues. Certain of these polar residues acquire special properties essential for substrate binding or catalyse.
- Substrates are bound to enzymes by multiple weak interactions. Stated above
- The specificity of binding depends on the precise defined arrangement of atoms in the active site. Because the enzyme and the substrate interact by means of short-range forces that require close contact, a substrate must have a matching shape to fit into the site. However, the active site of some enzymes assume a shape that is complementary to that of the substrate only after the substrate is bound. This process of dynamic recognition is called induced fit.

Enzymes are highly specific and may require cofactors for catalysis. A cofactor is a non-protein chemical compound bound to a protein; there are 2 types of cofactors: Metals and organic/metalloorganic (which are derived from vitamins). An example of a metal cofactor is zinc and the enzyme, carbonic anhydrase, tightly binds the zinc at the active site. The process involves binding water to carbon dioxide and deprotonating it into carbonic acid. Then the carbonic acid becomes a bicarbonate ion due to the displacement of water.

Catalysts can fasten the reaction speed by lowering the activation energy (not the transition state) of the process. The active site is a location on the enzyme which has complementary shape to the substrate. It is also where the amino acids with a complementary charge, polarity and shape to the ligand are. The enzyme function and catalysis result from the ability to stabilize the transition state in a chemical reaction. The transition state is the highest energy species in a reaction. It is a transitory molecular structure that is no longer the substrate but is not yet the product. It is the most seldom occupied species along the reaction pathway. The difference in free energy between the transition state and the substrate is called the Gibbs free energy of activation or simply the activation energy.

Thus we can see the key to how enzymes operate: Enzymes accelerate reactions by decreasing the activation energy. The combination of substrate and enzyme creates a reaction pathway whose transition state is lower than that of the reaction in the absence of the enzyme. Because the activation energy is lower more substrate molecules have the energy required to reach the transition state.

It is important to note that enzymes have evolved specifically to recognize the transition states of chemical reactions. Therefore, enzymes do not bind to any reactive species before the species have actually begun to react; enzymes only recognize and bind the *transition states* of such species.

In fact, if enzymes were to bind to the reactants of a reaction "on sight", or immediately, this would result in an even higher activation energy than before! For this reason, enzymes recognize only the transition state and bind to reactive species only when this high-energy state has been achieved. The fact that enzymes can recognize structures as specific and short-lived as transition states is a testament to their incredible specificity and efficiency.

Each enzyme is optimized for a particular reaction transition state. This ensures that enzymes will not compete with each other and hinder cellular reactions instead of help them. Enzyme inhibition occurs when the activity of a given enzyme is disrupted or interrupted in some fashion. Inhibitors can be molecules that have a similar shape, structure, or charge to the substrate in question so that the active site of an enzyme will "mistake" the inhibitor for the substrate.

This affects the affinity of the enzyme for the substrate, as well as the rate of the overall reaction. Several types of inhibition can occur in the cell; more detailed explanations on these can be found in the corresponding sections.Because of the active sites, enzymes are highly specific catalysts. These catalysts are governed by the ability to lower the free energy of thermodynamics to overcome transition states.

The Michaelis-Menten Model describes the kinetic properties of many enzymes. The interaction between the substrate and the enzyme helps accelerate the reaction, and the specificity of enzymes result in minimal side reactions.

It is of great importance to note that an enzyme cannot alter the laws of thermodynamics and consequently cannot alter the equilibrium of the reaction. The amount of product formed for a reaction utilizing an enzyme is always equal to the amount of product form of the same reaction occurring in the same reaction mixture without the enzyme. The enzyme just allows the reaction to reach its equilibrium faster. The equilibrium position is a function only of the free-energy difference between reactants and products.

- Enzymes only alter reaction rate, not the reaction equilibrium. Enzyme cannot alter the laws of thermodynamics; therefore, it cannot alter the equilibrium of a chemical reaction. Enzyme is present, the amount of products form faster compared with enzyme is absent. Enzyme is only accelerating the reaction rate, not shipping the position of equilibrium (free energy, delta G)

GENERAL ENZYMATIC KINETICS

Enzymes are catalysts, most are proteins, that bind temporarily to one or more of the reactants of the reaction they catalyze. In doing so, they lower the amount of activation energy needed and thus speed up the reaction.

Enzymes are Agents of Metabolic Function. A given Enzyme is quite selective, both in the substances with which it interacts and in the reaction it catalyzes. The substances upon which an enzyme acts are called Substrates. The selective qualities of an enzyme are recognized as its specificity.The specific site on the enzyme where substrates bind and where catalysis occurs is called the Active site

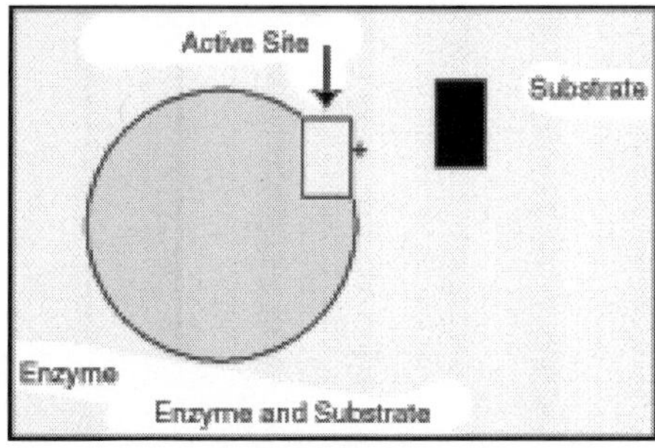

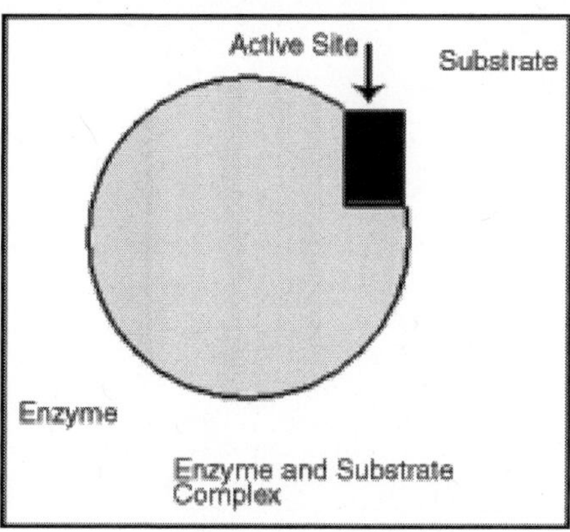

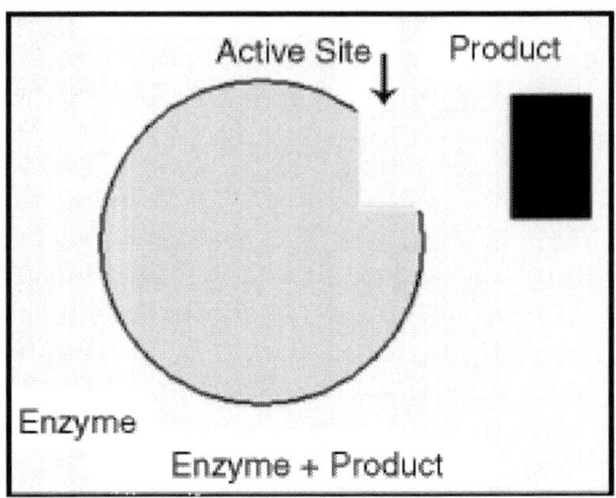

Catalysts lower the Free energy of Activation for a reaction given by the Arrhenius Equation: $k = Ae^{-Ea/RT}$

Can the rate of an Enzyme-catalyzed Reaction be defined in a mathematical way?

Enzymes Kinetics exhibit the following rates of chemical reactions:

- Zero-Order Reactions
- First-Order Reactions

We seek to determine the maximum reaction velocity that an enzyme can attain and its binding affinity for substrates and inhibitors.

When working with enzyme catalyzed reactions, generally the enzyme is less than the total substrate, so that it is a valid assumption that the concentration of free substrate concentration is equal to the Total Amount of Substrate

$$[E] << [S]_{total}$$
$$[S]_{free} = [S]_{total}$$

We will use this assumption throughout the discussion of enzyme kinetics.

Most frequently one takes the phrase "enzyme kinetics" refers to the analysis of the "steady state" kinetics of enzymatic reactions. The term "steady state" has a specific meaning. The steady state is the phase of a reaction in which reactive intermediates are both formed and decomposed at the same rate so that their concentrations are essentially constant.

Analogies are always helpful for an intuitive understanding.

A good analogy is that an animal population can be in a Steady-state Approximation

This occurs when the rate of birth and death are equal.

The "reactive intermediates" (the live animals in the population) are then in the steady state, with the most "stable" states being those of pre-birth and death.

Consider the following chemical reaction:

$$A + B \rightleftharpoons AB^* \rightleftharpoons C + D$$

If the AB^* complex is a high energy species relative to the $A + B$ and $C + D$ ground states then, as a good approximation, as soon as it is formed it will either go to $A + B$ or $C + D$. This means that it's concentration will be very low and nearly constant over a very large part of the reaction.

The idea of steady state kinetics applied to enzyme catalyzed reactions is that the enzyme is generally present only in very low concentrations relative to the total substrate concentration, and that the reactive intermediate corresponding to AB^* above is the E^*S complex.

Basic Equations of Enzyme Kinetics

$$E + S \rightleftharpoons E^*\,S \rightleftharpoons E + P$$

The typical mechanism for enzyme kinetics:

Enzyme + Substrate goes to the Enzyme Substrate Complex, and further yields Enzyme and Substrate!

If $[S]_{\text{total}} >> [E]_{total}$, as is usually the case, then $[S]$ in the E^*S complex is a very small portion or fraction of $[S]_{total}$.

This is a necessary condition for the steady state assumption to be valid: the reactive intermediate {{ math.formula("ES") }} must be at a very low and constant concentration compared to the reactants {{ math.formula("S") }}.

We generally miss the pre-steady state part of enzymatic reactions when we mix enzyme catalyzed reactions by hand in a cuvette and stick it in a spectrophotometer to measure it. By the time we start taking data, the reaction is already in the steady state because rate constants for enzyme catalyzed reactions are generally large.

In 1913, Michaelis and Menten published the idea that enzymes and substrates formed reasonably stable complexes with their substrates that then subsequently undergo a reaction.

The Big, Big Equation for Enzyme Kinetics

The Michealis-Menten Equation,

$$E + S \rightleftharpoons E^*S \rightleftharpoons E + P$$

The mechanism involves constants: K_M (Michealis Constant) and k_{cat} (turnover number)

$$v_o = \left(\frac{dP}{dt}\right)_o = \frac{V\max[S]}{K_m + [S]}$$

ENZYME INHIBITION

Enzymes are proteins that speed up the rate of a reaction by providing an alternate route to overcoming the activation energy. The graph below shows the path of a reaction both with and without the presence of an enzyme.

The black line indicates the path a reaction would take in the absence of an enzyme while the yellow line indicates the path a reaction would take with an enzyme.

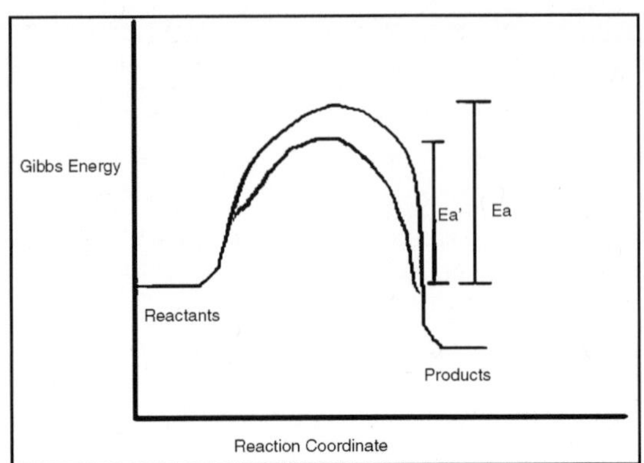

The general reaction of an enzyme is E+S ↔ ES ↔ E + P and the velocity of the reaction is given by the Michaelis-Menten equation:

$$v = vmax[S]km + [S]$$

I is the concentration of the inhibitor and K_i is the dissociation constant of the E_I complex. A Lineweaver-Burke plot of 1/V vs. 1/S shows us how the Km and the V_{max} change as inhibitor concentration changes. The red line corresponds to the reaction with no inhibitor present while the black lines are with various inhibitor concentration. The black arrow shows an increasing inhibitor concentration. We see from the graph that a competitive inhibitor changes the apparent value of K_m but does not affect the Vmax.

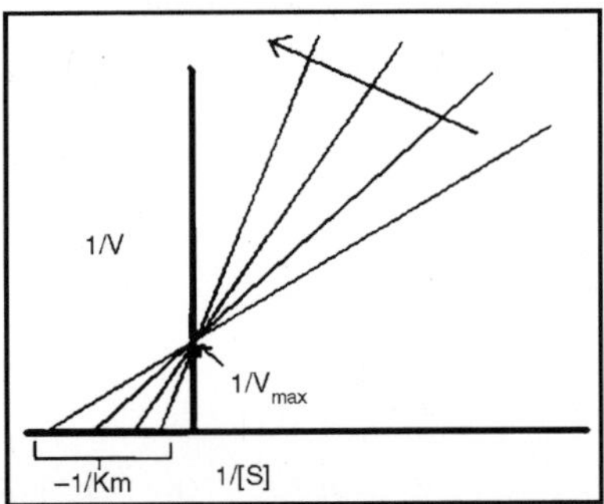

Non-Competitive Inhibition

In non-competitive inhibition, the inhibitor binds to the enzyme at a location other than the active site in such a way that the inhibitor and substrate can simultaneously be attached to the enzyme. The substrate and the inhibitor have no effect on the binding of the other and can bind and unbind the enzyme in either order. The inhibitor and the substrate are not binding to the same site. The general reaction for this is E+S ↔ ES ↔ E + P, except at the same time E+I ↔ EI, ES+I ↔ EIS, and EI+S ↔ EIS. The Michaelis-Menten equation for this becomes:

$$v = vmax[S]km(1+[I]ki)+[S](1+[I]ki)$$

In uncompetitive inhibition, the inhibitor binds to the enzyme only after the substrate has bound. The general reaction for this is E+S $\leftrightarrow$ ES $\leftrightarrow$ E + P, except at the same time ES+I $\leftrightarrow$ ESI. The Michaelis Menten equation for this becomes:

$$v=vmax[S]km+[S](1+[I]ki)$$

I is the concentration of uncompetitive inhibitor and Ki is the dissociation constant of the ESI complex. The Lineweaver-Burke plot below shows that for uncompetitive inhibition both the apparent K_m and apparent V_{max} decrease with increase in inhibitor concentration.

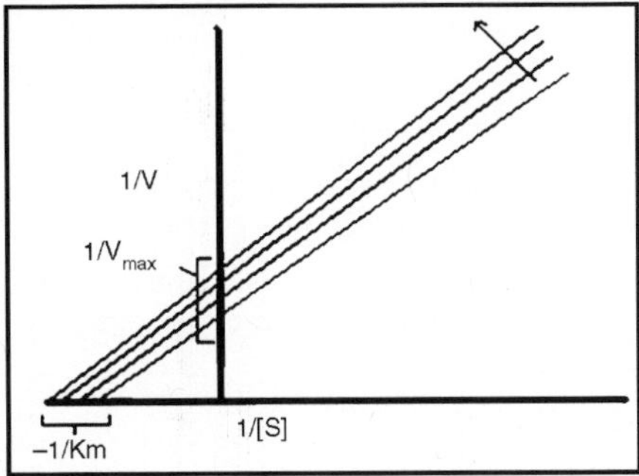

Mixed Type Inhibition

Mixed type inhibition is similar to non-competitive inhibition except that binding of the substrate or the inhibitor affect the enzyme's binding affinity for the other. The change in binding affinity is included in the chemical equation by the term ki. For mixed type inhibition ki>1, which means that binding affinity for the substrate is decreased when the inhibitor is present.

$$\begin{array}{lcl} E+S \xrightarrow{ks} ES \xrightarrow{kp} E+P \\ \uparrow ki \qquad\qquad \uparrow aki \\ EI+S \xleftrightarrow{aks} EIS \end{array}$$

Note that it is possible to completely deactivate the enzye activity by saturating it with I because the EIS complex does not form product.The velocity equation for mixed type is:

$$v=vmax[S]ks(1+[I]ki)+[S](1+[I]aki)$$

As usual, the basic velocity graph is not very useful, so we take the double reciprocal of the velocity equation. The Lineweaver-Burke equation for mixed type is:

$$\frac{1}{\text{v}} = \frac{ks}{\text{v max}}\left(1 + \frac{[I]}{ki}\right)\frac{1}{[S]} + \frac{1}{\text{v max}}\left(1 + \frac{[I]}{aki}\right)$$

Due to the ? factor, we cannot solve for all of the kinetic constants from the Lineweaver-Burke plot alone. To solve this problem, we make re-plots of the slope and the y-intercept at varying concentrations of I.

$$y - \text{int}\,ercept = \frac{1}{\text{v max}\bullet aki}\frac{1}{[I]} + \frac{1}{\text{v max}}$$

$$slope = \frac{ks}{\text{v max}\bullet ki}\frac{1}{[I]} + \frac{ks}{\text{v max}}$$

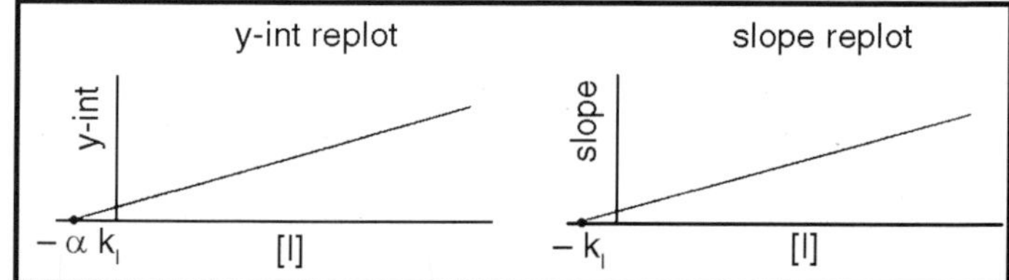

p style=”text-align: justify;”>Determining the kinetic constants

- *Ks:* Can be found from the original double reciprocal plot ([I]=0 line)
- *vmax:* Can be found in the original double reciprocal plot ([I]=0 line)
- *Ki:* Can be found from the slope of the slope re-plot
- *?:* Can be found from the slope of the y-intercept re-plot.

MICHAELIS-MENTEN KINETICS OF A TWO-SUBSTRATE REACTION

For a two-substrate reaction in the absence of products Eqn (13) simplifies to the following equation:

$$v = \frac{[E]_O}{\frac{1}{k_O} + \frac{1}{k_A[A]} + \frac{1}{k_B[B]} + \frac{1}{k_{AB}[A][B]}}$$

If the concentration of one substrate, known as the *constant substrate,* is held constant, while that of the other, known as the *variable substrate,* is varied, the rate is of the form of the Michaelis-Menten equation in terms of the variable substrate, because Eqn (14) can be rearranged to,

$$v = \frac{k_O^{app} k_A^{app} [E]_O [A]}{k_O^{app} + k_A^{app} [A]}$$

(cf. Eqn 10), where,

$$k_O^{app} = \frac{k_O k_B [B]}{k_O + k_B [B]}$$

is known as the apparent *catalytic constant,* and

$$k_A^{app} = \frac{k_A k_{AB} [B]}{k_A + k_{AB} [B]}$$

is known as the *apparent specificity constant* for A. It follows from Eqns (16) and (17) that $k_O{}^{app} = k_0$ and $k_A{}^{app} = k_A$ when [B] is extrapolated to an infinite value. This relationship provides the basis for defining the*catalytic constant* and the specificity constants in reactions with more than one substrate: in general, the catalytic constant of an enzyme is the value of $v/[E]_0$ obtained by extrapolating all substrate concentrations to infinity; for any substrate A the *specificity constant* is the apparent value when all other substrate concentrations are extrapolated to infinity.

Eqn (14) may also be rearranged into a form resembling Eqn (11), as follows:

$$v = \frac{V[A][B]}{K_{iA} K_{mB} + K_{mb}[A] + K_{mA}[B] + [A][B]}$$

in which $V = k_0[E]_0$ is the *limiting rate,* which may also, subject to the reservations noted in section 4.1, be called the maximum rate or maximum velocity and symbolized as *V*max; *K*mA = k_0/k_A is the Michaelis constant for A; $K_{mB} = k_0/k_B$ is the Michaelis constant for B; and $K_{iA} = k_B/k_{AB}$ is the *inhihition constant* for A. In some mechanisms K_{iA} is equal to the true dissociation constant for the EA complex: when this is the case the alternative symbol K_{iA} and the name *substrate-dissociation constant* for A (cf. section 3.2) may be used. If Eqn(18) is interpreted operationally rather than as the equation for a particular mechanism it is arbitrary whether the constant in the denominator is written with $K_{iA}K_{mB}$ (as shown) or as $K_{mA}K_{iB}$, where $K_{iB} = k_A/k_{AB}$. However, for some mechanisms only one of the two ratios k_B/k_{AB} and k_A/k_{AB} has a simple mechanistic interpretation and this may dictate which inhibition constant it is appropriate to define.

The term inhibition constant and the symbol K_{iA} derive from the fact that the quantity concerned is related to (and in the limiting cases equal to) the inhibition constant K_{ic} or K_{iu} (as defined below in Section 6.4) measured in experiments where the substrate is treated as an inhibitor of the reverse reaction. However, the relationships are not always simple and quantities such as K_{iA} in Eqn (18) can be and nearly always are defined and measured without any reference to inhibition experiments. For these reasons some members of the panel feel that the symbolism and terminology suggested are not completely satisfactory. No alternative system has so far gained wide support, however.

An *apparent Michaelis constant* for A (and similarly for B) may be defined by dividing Eqn (16) by Eqn (17):

$$K_{mA}^{app} = k_O^{app} / k_A^{app} = \frac{K_{iA}K_{mB} + K_{mA}[B]}{K_{mB} + [B]}$$

This equation provides the basis for defining the Michaelis constant for any substrate in a reaction with more than one substrate: the *Michaelis constant* for A, K_{mA}, is the value of the apparent Michaelis constant for A when the concentrations of all substrates except A are extrapolated to infinity. This definition applies to reactions with any numbers of substrates, as also does the definition of the limiting rate V as k_0 $[E]_0$, but in other respects it becomes very cumbersome to define constants resembling K_{iA} for reactions with more than two substrates. The symbolism of Eqn (13) (or the alternative in terms of Dalziel coefficients) is readily extended to reactions with three or more substrates, however.

LIMITING KINETICS OF ENZYME-CATALYSED REACTIONS

At very low concentrations of substrate many enzyme-catalysed reactions display approximately second-order kinetics, with rate given by the following equation:

$$v = k_A\,[E]_0\,[A]$$

in which the symbol k_A (or, in general, k_R for a reactant R) is the apparent second-order rate constant or *specificity constant* and $[E]_0$, which may also be written as $[E]_t$ or $[E]_{stoich}$, is the total or stoichiometric concentration of catalytic centres. (This corresponds to the total enzyme concentration only if there is a single catalytic centre per molecule.) The rationale for the subscript 0 is that the total enzyme concentration is normally the concentration at the instant of mixing, i.e. at time zero. Conversely, at very high substrate concentrations the same reactions commonly display approximately first-order kinetics (zero-order with respect to substrate):

$$v = k_0\,[E]_0$$

in which k_0, which may also be written as k_{cat} is the apparent first-order rate constant. Although these limiting types of behaviour are not universally observed, they are more common than *Michaelis-Menten kinetics* and provide a basis for classifying inhibitory and other effects independently of the need for Michaelis-Menten kinetics.

The apparent second-order rate constants k_A and k_B of competing substrates A and B determine the partitioning between competing reactions, regardless of whether the substrate concentrations are very small or not, and it is for this reason that the name *specificity constant* is proposed for this parameter of enzymic catalysis. The apparent first-order rate constant k_0 is a measure of the catalytic potential of the enzyme and is called the*catalytic constant.*

The quantity $k_0[E]_0$ is given the symbol *V* and the name *limiting rate.* It is particularly useful when k_0 cannot be calculated because the total catalytic-centre concentration is unknown, as in studies of enzymes of unknown purity, sub-unit structure and molecular mass. The symbol V_{max} and the names *maximum rate* and *maximum velocity* are also in widespread use although under normal circumstances there is no finite substrate concentration at which $v = V$ and hence no maximum in the mathematical sense. The form V_{max} is convenient in speech as it avoids the need for a cumbersome distinction between 'capital *V*' and 'lower case *v*'. When a true maximum does occur (as insubstrate inhibition; Section 4.3) the symbol v_{max} (not V_{max}) and the name maximum rate may be used for the true maximum value of *v* but care should be taken to avoid confusion with the limiting rate.

MICHAELIS-MENTEN KINETICS

Sometimes the relationship between the rate of an enzyme-catalysed reaction and the substrate concentration takes the form,

$$v = \frac{V[A]}{K_{mA} + [A]}$$

where *V* and K_{mA} are constants at a given temperature and a given enzyme concentration. The reaction is then said to display *Michaelis-Menten kinetics.* (The term *hyperbolic kinetics* is also sometimes used because a plot of *v* against [A] has the form ot a rectangular hyperbola through the origin with asymptotes $v = V$ and $[A] = -K_{mA}$. This term, and others that imply the use of particular kinds of plot. should be used with care to avoid ambiguity, as they can be misleading if used out of context.) The constant *V* is the *limiting rate,* with the same meaning as in Section 4.1. The second constant K_{mA} is known as the *Michaelis constant* for A; the alternative name *Michaelis concentration* may also be used and has the advantage of emphasizing that the quantity concerned has the dimensions of a concentration and is not, in general, an equilibrium constant. When only one substrate is being considered the qualifier A may be omitted, so that the symbol becomes K_m. When the qualifier is included its location is a matter of typographical convenience; no particular significance attaches to such variants as K_m^A or K_{mA}. The Michaelis constant (or Michaelis concentration) is the substrate concentration at which $v = 0.5\ V$, and its usual unit is mol dm^{-3}, which may be written as mol L^{-1} or M. The term Michaelis constant and the symbol K_m should not be used when Michaelis-Menten kinetics are not obeyed.

For a reaction obeying Michaelis-Menten kinetics the rate in the limit of very low substrate concentrations is $v = V[A]/K_{mA}$, and comparison with Eqn

(8) shows that $V/K_{mA} = k_A\,[E]_0$. In the limit of very high substrate concentrations $v = V$, and comparison with Eqn (9) gives $V = k_0\,[E]_0$. The Michaelis constant K_{mA} is therefore k_0/k_A, and Eqn (10) can be written as,

$$v = \frac{K_O k_A [E]_O [A]}{k_O + k_A [A]}$$

An indefinitely large number of mechanisms generate Michaelis-Menten kinetics, and still more generate limiting behaviour of the kind described in Section 4.1. Consequently there is no general definition of any of the kinetic parameters k_A, k_0, V and K_{mA} in terms of the rate constants for the elementary steps of a particular mechanism.

NON-MICHAELIS-MENTEN KINETICS

When the kinetic behaviour does not conform to Eqn (10) or Eqn (11) the reaction is said to exhibit *non-Michaelis-Menten kinetics.* If the Michaelis-Menten equation is obeyed approximately over a restricted range of substrate concentrations it may be convenient to regard this behaviour as a deviation from this equation rather than as an unrelated phenomenon. For example, a reaction may obey an equation of the following form,

$$v = \frac{V'[A]}{K'_{mA} + [A] + [A]^2 / K_{iA}}$$

in which the constants V', K'_{mA} and K_{iA} are used for illustration without any implication of universal or standard definitions. If K_{iA} is large compared with *K*mA the behaviour predicted by Eqn (12) will approximate to that predicted by Eqn (11), with V and K_{mA} replaced by V' and K_{mA}, in the lower range of substrate concentrations. However, with Eqn (12) the rate passes through a maximum as the concentration increases, and there is said to be *inhibition by substrate,* and the constant K_{iA}, which has the dimensions of a concentration, is called the *substrate inhibition constant.*

When more complex kinds of non-Michaelis-Menten behaviour occur it is usually unhelpful to use terminology and symbolism suggestive of the Michaelis-Menten equation; instead the approach discussed in Section 10 is appropriate. In all cases it is advisable to avoid the term Michaelis constant and the symbol K_m when the Michaelis-Menten equation is not obeyed, because it is defined as a parameter of that equation. The symbol $[A]_{0.5}$ or $[A]_{1/2}$, not K_{mA}, may be used for the value of [A] at which $v = 0.5\,V$.

KINETIC ANALYSIS

The observed velocities predicted by the Michaelis–Menten equation can be used to directly model the time course disappearance of substrate and the production of product through incorporation of the Michaelis–Menten equation into the equation for first order chemical kinetics. This can only be achieved however if one recognises the problem associated with the use of Euler's number in the description of first order chemical kinetics. i.e. e^{-k} is a split constant that introduces a systematic error into calculations and can be rewritten as a single constant which represents the remaining substrate after each time period.

$$[S] = [S]0\ (1 - k)^t$$

$$[S] = [S]_0\ (1 - v/[S]_0)^t$$

$$[S] = [S]_0\ (1 - (V_{max}\ [S]0/(K_M + [S]_0)/[S]_0))^t$$

In 1983 Stuart Beal (and also independently Santiago Schnell and Claudio Mendoza in 1997) derived a closed form solution for the time course kinetics analysis of the Michaelis-Menten mechanism. The solution, known as the Schnell-Mendoza equation, has the form:

$$\frac{[S]}{K_M} = W\left[\frac{[S]_0}{K_M}.\exp\left(\frac{[S]_0}{K_M} - \frac{V\max}{K_M}t\right)\right]$$

where W[] is the Lambert-W function.

Linear Plots of the Michaelis–Menten Equation

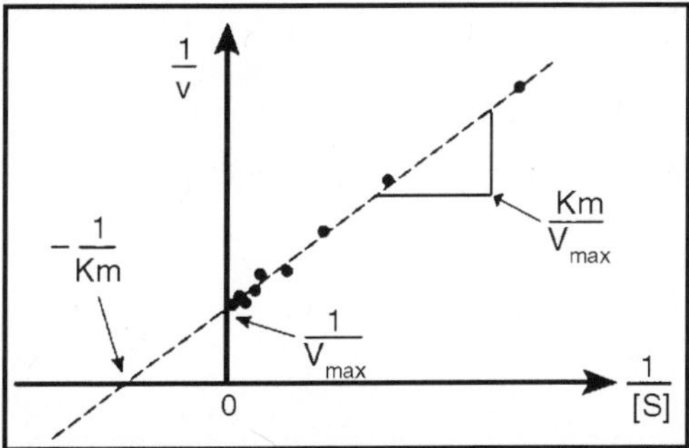

Lineweaver–Burk or double-reciprocal plot of kinetic data, showing the significance of the axis intercepts and gradient.

The plot of v versus [S] above is not linear; although initially linear at low [S], it bends over to saturate at high [S]. Before the modern era of non-linear curve-fitting on computers, this non-linearity could make it difficult to estimate K_M and V_{max} accurately. Therefore, several researchers developed linearisations of the Michaelis–Menten equation, such as the Lineweaver–Burk plot, the Eadie–Hofstee diagram and the Hanes–Woolf plot. All of these linear representations can be useful for visualising data, but none should be used to determine kinetic parameters, as computer software is readily available that allows for more accurate determination bynonlinear regression methods.

The Lineweaver–Burk plot or double reciprocal plot is a common way of illustrating kinetic data. This is produced by taking the reciprocal of both sides of the Michaelis–Menten equation. As shown on the right, this is a linear form of the Michaelis–Menten equation and produces a straight line with the equation

$y = mx + c$ with a y-intercept equivalent to $1/V_{max}$ and an x-intercept of the graph representing $-1/K_M$.

$$\frac{1}{v} = \frac{K_M}{V\max[S]} + \frac{1}{V\max}$$

Naturally, no experimental values can be taken at negative 1/[S]; the lower limiting value 1/[S] = 0 (the y-intercept) corresponds to an infinite substrate concentration, where $1/v = 1/V_{max}$ as shown at the right; thus, the x-intercept is an extrapolation of the experimental data taken at positive concentrations. More generally, the Lineweaver–Burk plot skews the importance of measurements taken at low substrate concentrations and, thus, can yield inaccurate estimates of V_{max} and K_M. A more accurate linear plotting method is the Eadie-Hofstee plot. In this case, v is plotted against v/[S]. In the third common linear representation, the Hanes-Woolf plot, [S]/v is plotted against [S]. In general, data normalisation can help diminish the amount of experimental work and can increase the reliability of the output, and is suitable for both graphical and numerical analysis.

Practical Significance of Kinetic Constants

The study of enzyme kinetics is important for two basic reasons. Firstly, it helps explain how enzymes work, and secondly, it helps predict how enzymes behave in living organisms. The kinetic constants defined above, K_M and V_{max}, are critical to attempts to understand how enzymes work together to control metabolism.

Making these predictions is not trivial, even for simple systems. For example, oxaloacetate is formed by malate dehydrogenase within the mitochondrion. Oxaloacetate can then be consumed by citrate synthase, phosphoenolpyruvate carboxykinase or aspartate aminotransferase, feeding into the citric acid cycle, gluconeogenesis or aspartic acidbiosynthesis, respectively. Being able to predict how much oxaloacetate goes into which pathway requires knowledge of the concentration of oxaloacetate as well as the concentration and kinetics of each of these enzymes. This aim of predicting the behaviour of metabolic pathways reaches its most complex expression in the synthesis of huge amounts of kinetic and gene expression data into mathematical models of entire organisms. Alternatively, one useful simplification of the metabolic modelling problem is to ignore the underlying enzyme kinetics and only rely on information about the reaction network's stoichiometry, a technique called flux balance analysis.

Michaelis–Menten Kinetics with Intermediate

One could also consider the less simple case,

$$E + S \underset{k_{-1}}{\overset{k_1}{\rightleftarrows}} ES \xrightarrow{k_2} EI \xrightarrow{k_3} E + P$$

where a complex with the enzyme and an intermediate exists and the intermediate is converted into product in a second step. In this case we have a very similar equation,

$$v_0 = k_{cat}\frac{[S][E]_0}{K'_M + [S]}$$

but the constants are different,

$$K'_M \stackrel{def}{=} \frac{k_3}{k_2 + k_3} K_M = \frac{k_3}{k_2 + k_3} \cdot \frac{k_2 + k_{-1}}{k_1}$$

$$k_{cat} \stackrel{def}{=} \frac{k_3 k_2}{k_2 + k_3}$$

We see that for the limiting case $k_3 \gg k_2$, thus when the last step from EI to $E + P$ is much faster than the previous step, we get again the original equation. Mathematically we have then $K'_M \approx K_M$ and $k_{cat} \approx k_2$.

Multi-substrate Reactions

Multi-substrate reactions follow complex rate equations that describe how the substrates bind and in what sequence. The analysis of these reactions is much simpler if the concentration of substrate A is kept constant and substrate B varied. Under these conditions, the enzyme behaves just like a single-substrate enzyme and a plot of v by [S] gives apparent K_M and V_{max} constants for substrate B.

If a set of these measurements is performed at different fixed concentrations of A, these data can be used to work out what the mechanism of the reaction is. For an enzyme that takes two substrates A and B and turns them into two products P and Q, there are two types of mechanism: ternary complex and ping–pong.

Ternary-complex Mechanisms

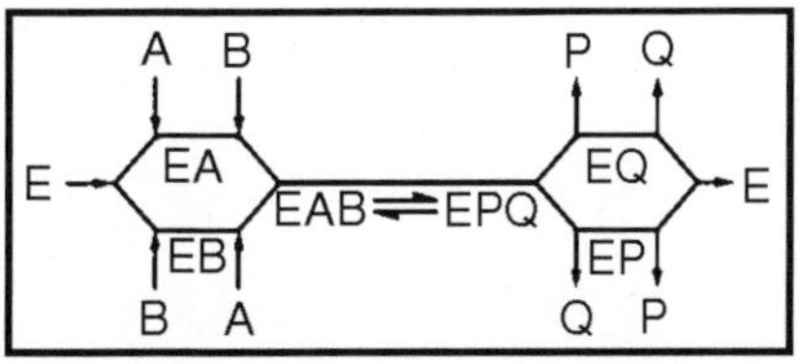

Random-order ternary-complex mechanism for an enzyme reaction. The reaction path is shown as a line and enzyme intermediates containing substrates A and B or products P and Q are written below the line.

In these enzymes, both substrates bind to the enzyme at the same time to produce an EAB ternary complex. The order of binding can either be random (in a random mechanism) or substrates have to bind in a particular sequence (in an ordered mechanism). When a set of v by [S] curves (fixed A, varying B) from an enzyme with a ternary-complex mechanism are plotted in a Lineweaver–Burk plot, the set of lines produced will intersect. Enzymes with ternary-complex mechanisms include glutathione *S*-transferase, dihydrofolate reductase and DNA polymerase. The following links show short animations of the ternary-complex mechanisms of the enzymes dihydrofolate reductase[β] and DNA polymerase[γ].

Ping–Pong Mechanisms

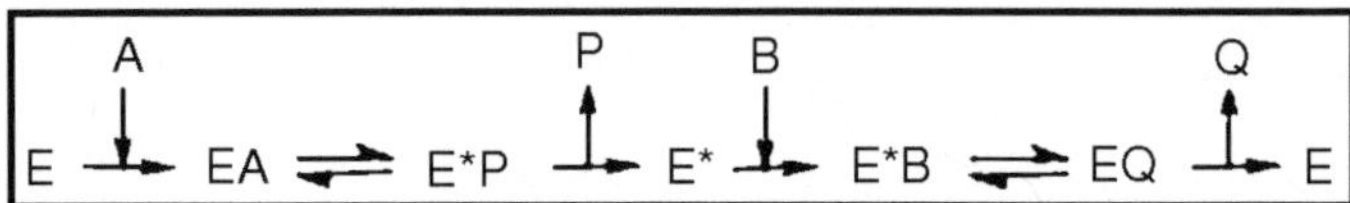

Ping–pong mechanism for an enzyme reaction. Intermediates contain substrates A and B or products P and Q.

As shown on the right, enzymes with a ping-pong mechanism can exist in two states, E and a chemically modified form of the enzyme E*; this modified enzyme is known as an intermediate. In such mechanisms, substrate A binds, changes the enzyme to E* by, for example, transferring a chemical group to the active site, and is then released. Only after the first substrate is released can substrate B bind and react with the modified enzyme, regenerating the unmodified E form.

When a set of v by [S] curves (fixed A, varying B) from an enzyme with a ping–pong mechanism are plotted in a Lineweaver–Burk plot, a set of parallel lines will be produced. This is called a secondary plot.

Enzymes with ping–pong mechanisms include some oxidoreductases such as thioredoxin peroxidase, transferases such as acylneuraminate cytidylyltransferase and serine proteases such as trypsin andchymotrypsin. Serine proteases are a very common and diverse family of enzymes, including digestive enzymes (trypsin, chymotrypsin, and elastase), several enzymes of the blood clotting cascade and many others.

In these serine proteases, the E* intermediate is an acyl-enzyme species formed by the attack of an active site serine residue on a peptide bond in a protein substrate. A short animation showing the mechanism of chymotrypsin is linked here.[δ]

ENZYME-SUBSTRATE INTERACTIONS

The favored model of enzyme-substrate interaction is known as the induced-fit model. This model proposes that the initial interaction between enzyme and substrate is relatively weak, but that these weak interactions rapidly induce conformational changes in the enzyme that strengthen binding and bring catalytic sites close to substrate bonds to be altered. After binding takes place, one or more mechanisms of catalysis generate transition-state complexes and reaction products. The possible mechanisms of catalysis are four in number:

1. *Catalysis by Bond Strain:* In this form of catalysis, the induced structural rearrangements that take place with the binding of substrate and enzyme ultimately produce strained substrate bonds, which more easily attain the transition state. The new conformation often forces substrate atoms and bulky catalytic groups, such as aspartate and glutamate, into conformations that strain existing substrate bonds.
2. *Catalysis by Proximity and Orientation:* Enzyme-substrate interactions orient reactive groups and bring them into proximity with one another. In addition to inducing strain, groups such as aspartate are frequently chemically reactive as well, and their proximity and orientation towards the substrate thus favors their participation in catalysis.
3. *Catalysis Involving Proton Donors (Acids) and Acceptors (Bases):* Other mechanisms also contribute significantly to the completion of catalytic events initiated by a strain mechanism, for example, the use of glutamate as a general acid catalyst (proton donor).
4. *Covalent Catalysis:* In catalysis that takes place by covalent mechanisms, the substrate is oriented to active sites on the enzymes in such a way that a covalent intermediate forms between the enzyme or coenzyme and the substrate. One of the best-known examples of this mechanism is that involving proteolysis by serine proteases, which include both digestive enzymes (trypsin, chymotrypsin, and elastase) and several enzymes of the blood clotting cascade. These proteases contain an active site serine whose R group hydroxyl forms a covalent bond with a carbonyl carbon of a peptide bond, thereby causing hydrolysis of the peptide bond.

ALLOSTERIC ENZYMES

In addition to simple enzymes that interact only with substrates and inhibitors, there is a class of enzymes that bind small, physiologically important molecules and modulate activity in ways other than those described above. These are known as allosteric enzymes; the small regulatory molecules to which they bind are known as effectors. Allosteric effectors bring about catalytic modification by binding to the enzyme at distinct allosteric sites, well removed from the catalytic site, and causing conformational changes that are transmitted through the bulk of the protein to the catalytically active site(s).

The hallmark of effectors is that when they bind to enzymes, they alter the catalytic properties of an enzyme's active site. Those that increase catalytic activity are known as positive effectors. Effectors that reduce or inhibit catalytic activity are negative effectors.

Most allosteric enzymes are oligomeric (consisting of multiple subunits); generally they are located at or near branch points in metabolic pathways, where they are influential in directing substrates along one or another of the available metabolic paths.

The effectors that modulate the activity of these allosteric enzymes are of two types. Those activating and inhibiting effectors that bind at allosteric sites are called heterotropic effectors. (Thus there exist both positive and negative heterotropic effectors.)

These effectors can assume a vast diversity of chemical forms, ranging from simple inorganic molecules to complex nucleotides such as cyclic adenosine monophosphate (cAMP). Their single defining feature is that they are not identical to the substrate.

In many cases the substrate itself induces distant allosteric effects when it binds to the catalytic site. Substrates acting as effectors are said to be homotropic effectors.

When the substrate is the effector, it can act as such, either by binding to the substrate-binding site, or to an allosteric effector site. When the substrate binds to the catalytic site it transmits an activity-modulating effect to other subunits of the molecule. Often used as the model of a homotropic effector is hemoglobin, although it is not a branch-point enzyme and thus does not fit the definition on all counts.

There are two ways that enzymatic activity can be altered by effectors: the V_{max} can be increased or decreased, or the K_m can be raised or lowered. Enzymes whose K_m is altered by effectors are said to be K-type enzymes and the effector a K-type effector. If V_{max} is altered, the enzyme and effector are said to be V-type. Many allosteric enzymes respond to multiple effectors with V-type and K-type behaviour. Here again, hemoglobin is often used as a model to study allosteric interactions, although it is not strictly an enzyme.

In the preceding discussion we assumed that allosteric sites and catalytic sites were homogeneously present on every subunit of an allosteric enzyme. While this is often the case, there is another class of allosteric enzymes that are comprised of separate catalytic and regulatory subunits. The archetype of this class of enzymes is cAMP-dependent protein kinase (PKA), whose mechanism of activation is illustrated in the Figure below. The enzyme is tetrameric, containing two catalytic subunits and two regulatory subunits, and enzymatically inactive.

When intracellular cAMP levels rise, one molecule of cAMP binds to each regulatory subunit, causing the tetramer to dissociate into one regulatory dimer and two catalytic monomers. In the dissociated form, the catalytic subunits are fully active; they catalyze the phosphorylation of a number of other enzymes, such as those involved in regulating glycogen metabolism. The regulatory subunits have no catalytic activity.

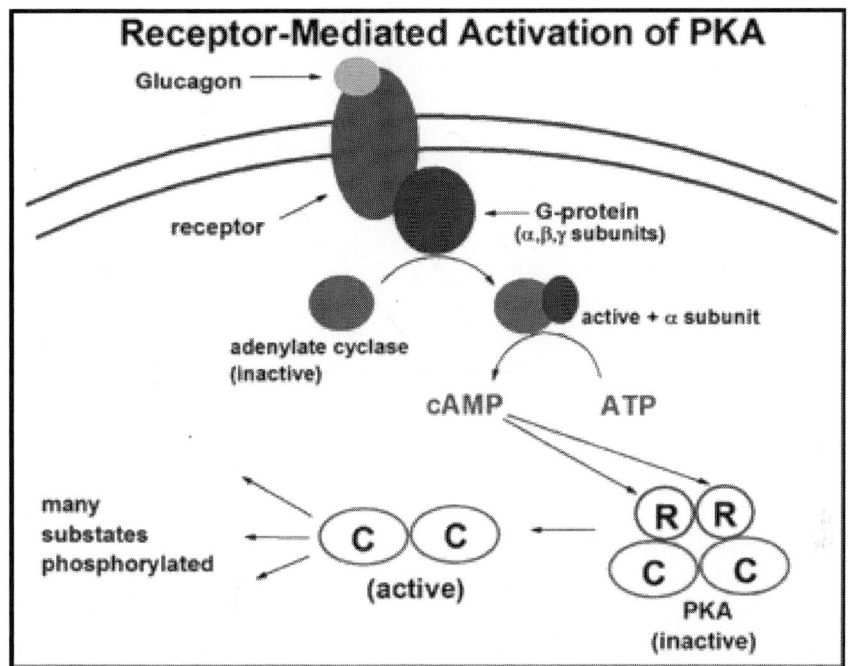

Representative pathway for the activation of cAMP-dependent protein kinase (PKA). In this example glucagon binds to its' cell-surface receptor, thereby activating the receptor. Activation of the receptor is coupled to the activation of a receptor-coupled G-protein (GTP-binding and hydrolyzing protein) composed of 3 subunits. Upon activation the á-subunit dissociates and binds to and activates adenylate cyclase. Adenylate cylcase then converts ATP to cyclic-AMP (cAMP). The cAMP thus produced then binds to the regulatory subunits of PKA leading to dissociation of the associated catalytic subunits. The catalytic subunits are inactive until dissociated from the regulatory subunits. Once released the catalytic subunits of PKA phosphorylate numerous substrate using ATP as the phosphate donor.

REGULATION OF ENZYME ACTIVITY

While it is clear that enzymes are responsible for the catalysis of almost all biochemical reactions, it is important to also recognize that rarely, if ever, do enzymatic reactions proceed in isolation. The most common scenario is that enzymes catalyze individual steps of multi-step metabolic pathways, as is the case with glycolysis, gluconeogenesis or the synthesis of fatty acids. As a consequence of these lock- step sequences of reactions, any given enzyme is dependent on the activity of preceding reaction steps for its substrate.

In humans, substrate concentration is dependent on food supply and is not usually a physiologically important mechanism for the routine regulation of enzyme activity. Enzyme concentration, by contrast, is continually modulated in response to physiological needs. Three principal mechanisms are known to regulate the concentration of active enzyme in tissues:

- Regulation of gene expression controls the quantity and rate of enzyme synthesis.
- Proteolytic enzyme activity determines the rate of enzyme degradation.
- Covalent modification of pre-existing pools of inactive proenzymes produces active enzymes.

Enzyme synthesis and proteolytic degradation are comparatively slow mechanisms for regulating enzyme concentration, with response times of hours, days or even weeks. Proenzyme activation is a more rapid method of increasing enzyme activity but, as a regulatory mechanism, it has the disadvantage of not being a reversible process. Proenzymes are generally synthesized in abundance, stored in secretory granules and covalently activated upon release from their storage sites. Examples of important proenzymes include pepsinogen, trypsinogen and chymotrypsinogen, which give rise to the proteolytic digestive enzymes. Likewise, many of the proteins involved in the cascade of chemical reactions responsible for blood clotting are synthesized as proenzymes. Other important proteins, such as peptide hormones and collagen, are also derived by covalent modification of precursors.

Another mechanism of regulating enzyme activity is to sequester enzymes in compartments where access to their substrates is limited. For example, the proteolysis of cell proteins and glycolipids by enzymes responsible for their degradation is controlled by sequestering these enzymes within the lysosome.

In contrast to regulatory mechanisms that alter enzyme concentration, there is an important group of regulatory mechanisms that do not affect enzyme concentration, are reversible and rapid in action, and actually carry out most of the moment to moment physiological regulation of enzyme activity. These mechanisms include allosteric regulation, regulation by reversible covalent modification and regulation by control proteins such as calmodulin. Reversible covalent modification is a major mechanism for the rapid and transient regulation of enzyme activity. The best examples, again, come from studies on the regulation of glycogen metabolism where phosphorylation of glycogen synthase and glycogen phosphorylase kinase results in the stimulation of glycogen degradation while glycogen synthesis is coordinately inhibited. Numerous other enzymes of intermediary metabolism are affected by phosphorylation, either positively or negatively. These covalent phosphorylations can be reversed by a separate sub-subclass of enzymes known as phosphatases. Recent research has indicated that the aberrant phosphorylation of growth factor and hormone receptors, as well as of proteins that regulate cell division, often leads to unregulated cell growth or cancer. The usual sites for phosphate addition to proteins are the serine, threonine and tyrosine R group hydroxyl residues.

INHIBITION OF ENZYME CATALYZED REACTIONS

To avoid dealing with curvilinear plots of enzyme catalyzed reactions, biochemists Lineweaver and Burk introduced an analysis of enzyme kinetics based on the following rearrangement of the Michaelis-Menten equation:

$$\frac{1}{V} = \left[\frac{K_m(1)}{V_{max}[S]} + \frac{1}{V_{max}} \right]$$

Plots of 1/v versus 1/[S] yield straight lines having a slope of K_m/V_{max} and an intercept on the ordinate at $1/V_{max}$.

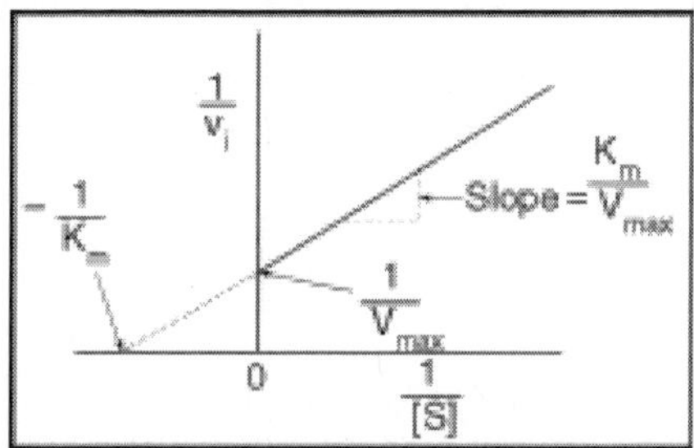

Fig. A Lineweaver-Burk Plot

An alternative linear transformation of the Michaelis-Menten equation is the Eadie-Hofstee transformation:

$$\frac{V}{[S]} = -V\left[\frac{1}{K_m}\right] + \left[\frac{V_{max}}{K_m}\right]$$

and when v/[S] is plotted on the y-axis versus v on the x-axis, the result is a linear plot with a slope of $-1/K_m$ and the value V_{max}/K_m as the intercept on the y-axis and V_{max} as the intercept on the x-axis.

Both the Lineweaver-Burk and Eadie-Hofstee transformation of the Michaelis-Menton equation are useful in the analysis of enzyme inhibition. Since most clinical drug therapy is based on inhibiting the activity of enzymes, analysis of enzyme reactions using the tools described above has been fundamental to the modern design of pharmaceuticals.

Well- known examples of such therapy include the use of methotrexate in cancer chemotherapy to semi-selectively inhibit DNA synthesis of malignant cells, the use of aspirin to inhibit the synthesis of prostaglandins which are at least partly responsible for the aches and pains of arthritis, and the use of sulfa drugs to inhibit the folic acid synthesis that is essential for the metabolism and growth of disease-causing bacteria. In addition, many poisons, such as cyanide, carbon monoxide and polychlorinated biphenols (PCBs) produce their life-threatening effects by means of enzyme inhibition.

Enzyme inhibitors fall into two broad classes: those causing irreversible inactivation of enzymes and those whose inhibitory effects can be reversed. Inhibitors of the first class usually cause an inactivating, covalent modification of enzyme structure. Cyanide is a classic example of an irreversible enzyme inhibitor: by covalently binding mitochondrial cytochrome oxidase, it inhibits all the reactions associated with electron transport. The kinetic effect of irreversible inhibitors is to decrease the concentration of active enzyme, thus decreasing the maximum possible concentration of ES complex. Since the limiting enzyme reaction rate is often k_2[ES], it is clear that under these circumstances the reduction of enzyme concentration will lead to decreased reaction rates.

Note that when enzymes in cells are only partially inhibited by irreversible inhibitors, the remaining unmodified enzyme molecules are not distinguishable from those in untreated cells; in particular, they have the same turnover number and the same K_m. Turnover number, related to V_{max}, is defined as the maximum number of moles of substrate that can be converted to product per mole of catalytic site per second. Irreversible inhibitors are usually considered to be poisons and are generally unsuitable for therapeutic purposes.

Reversible inhibitors can be divided into two main categories; competitive inhibitors and non-competitive inhibitors, with a third category, uncompetitive inhibitors, rarely encountered.

Inhibitor Type	Binding Site on Enzyme	Kinetic Effect
Competitive Inhibitor	Specifically at the catalytic site, where it competes with substrate for binding in a dynamic equilibrium- like process. Inhibition is reversible by substrate.	V_{max} is unchanged; K_m, as defined by [S] required for ½ maximal activity, is increased.
Noncompetitive Inhibitor	Binds E or ES complex other than at the catalytic site. Substrate binding unaltered, but ESI complex cannot form products. Inhibition cannot be reversed by substrate.	K_m appears unaltered; V_{max} is decreased proportionately to inhibitor concentration.
Uncompetitive Inhibitor	Binds only to ES complexes at locations other than the catalytic site. Substrate binding modifies enzyme structure, making inhibitor- binding site available. Inhibition cannot be reversed by substrate.	Apparent V_{max} decreased; K_m, as defined by [S] required for ½ maximal activity, is decreased.

The hallmark of all the reversible inhibitors is that when the inhibitor concentration drops, enzyme activity is regenerated. Usually these inhibitors bind to enzymes by non-covalent forces and the inhibitor maintains a reversible equilibrium with the enzyme. The equilibrium constant for the dissociation of enzyme inhibitor complexes is known as K_i:

$$K_i = \frac{[E][I]}{[EI]}$$

The importance of K_I is that in all enzyme reactions where substrate, inhibitor and enzyme interact, the normal $K_m$and or V_{max} for substrate enzyme interaction appear to be altered. These changes are a consequence of the influence of K_i on the overall rate equation for the reaction. The effects of K_i are best observed in Lineweaver-Burk plots.

Probably the best known reversible inhibitors are competitive inhibitors, which always bind at the catalytic or active site of the enzyme. Most drugs that alter enzyme activity are of this type. Competitive inhibitors are especially attractive as clinical modulators of enzyme activity because they offer two routes for the reversal of enzyme inhibition, while other reversible inhibitors offer only one. First, as with all kinds of reversible inhibitors, a decreasing concentration of the inhibitor reverses the equilibrium regenerating active free enzyme. Second, since substrate and competitive inhibitors both bind at the same site they compete with one another for binding.

Raising the concentration of substrate (S), while holding the concentration of inhibitor constant, provides the second route for reversal of competitive inhibition. The greater the proportion of substrate, the greater the proportion of enzyme present in competent ES complexes.

As noted earlier, high concentrations of substrate can displace virtually all competitive inhibitor bound to active sites. Thus, it is apparent that V_{max} should be unchanged by competitive inhibitors. This characteristic of competitive inhibitors is reflected in the identical vertical-axis intercepts of Lineweaver-Burk plots, with and without inhibitor.

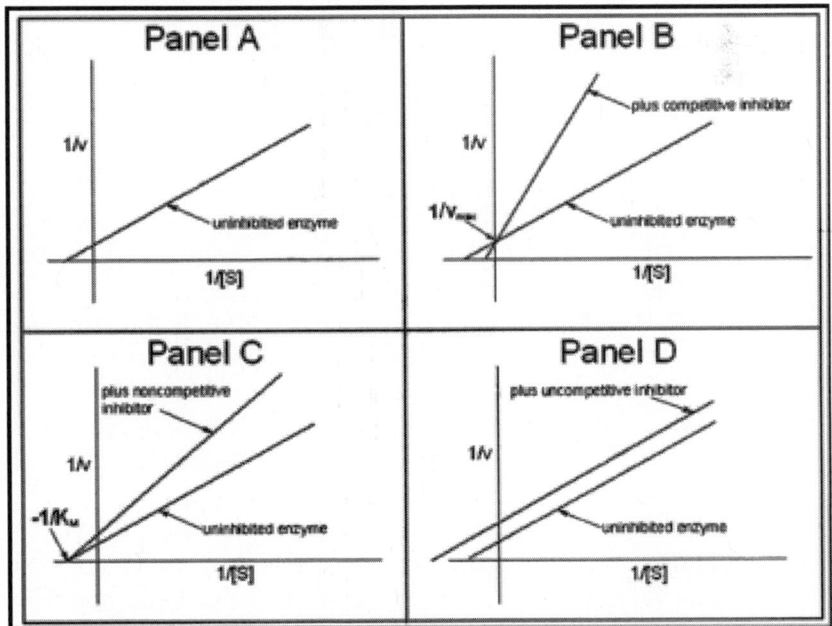

Fig. Lineweaver-Burk Plots of Inhibited Enzymes

Since attaining V_{max} requires appreciably higher substrate concentrations in the presence of competitive inhibitor, K_m (the substrate concentration at half maximal velocity) is also higher, as demonstrated by the differing negative intercepts on the horizontal axis in panel B.

Analogously, panel C illustrates that non-competitive inhibitors appear to have no effect on the intercept at the x-axis implying that non-competitive inhibitors have no effect on the K_m of the enzymes they inhibit. Since non-competitive inhibitors do not interfere in the equilibration of enzyme, substrate and ES complexes, the K_m's of Michaelis-Menten type enzymes are not expected to be affected by non-competitive inhibitors, as demonstrated by x-axis intercepts in panel C. However, because complexes that contain inhibitor (ESI) are incapable of progressing to reaction products, the effect of a noncompetitive inhibitor is to reduce the concentration of ES complexes that can advance to product. Since $V_{max} = k_2[E_{total}]$, and the concentration of competent E_{total} is diminished by the amount of ESI formed, non-competitive inhibitors are expected to decrease V_{max}, as illustrated by the y-axis intercepts in panel C.

A corresponding analysis of uncompetitive inhibition leads to the expectation that these inhibitors should change the apparent values of K_m as well as V_{max}. Changing both constants leads to double reciprocal plots, in which intercepts on the x and y axes are proportionately changed; this leads to the production of parallel lines in inhibited and uninhibited reactions.

TYPES OF MECHANISM FOR ENZYMIC CATALYSIS

The definitions given so far have been operational, *i.e.,* they provide a way of describing what may be observed independently of any interpretation that may been placed on it. Little has been said about *mechanisms,* i.e., the detailed descriptions of the chemical events that make up the catalytic process. nonetheless, the ultimate objective of most investigations in enzyme kinetics is to propose a mechanism. It is to be emphasized at the outset that a kinetic investigation can disprove a proposed mechanism but can never establish a mechanism beyond doubt. A mechanism may be consistent with all of the known facts, yet it is always possible to propose other mechanisms that are also consistent with the facts. The procedure that is adopted, and the only one that allows progress to be made, is to accept the simplest mechanism that is consistent with all the known facts. This is the principle of *Occam's razor.* A steady-state study of the effect of substrate concentrations on the rate, leading to an empirical rate equation, can often lead to a proposed mechanism. Such a mechanism can be tested by additional investigations, such as of the pre-steady-state kinetics, and of the effects of inhibitors, pH, temperature and solvent composition.It is not a practical proposition to institute a consistent system for naming mechanisms, although attempts to do this have been made, because except in the most trivial cases it is always simpler and clearer to specify a mechanism by reference to a scheme. non-etheless, certain terms occur frequently in the descriptions of mechanisms of enzymic catalysts and will be defined here.

The form of an enzyme that exists in solution in the absence of any substrate or other small molecule that can bind to it is called the *free enzyme.* An intermediate derived from the free enzyme by binding of a substrate molecule is called an *enzyme-substrate complex,* and terms such as *enzyme-product complex, enzyme-inhibitor complex, EA complex* may also be used by an obvious extension of this definition. A complex derived from the free enzyme and one other molecule is called a *binary complex*; one derived from the free enzyme and two other molecules is called a *ternary complex*; one derived from the free enzyme and three other molecules is called a *quaternary complex.* If the catalytic process proceeds through a second form of free enzyme that differs from the first by the presence of a covalently bound group that is transferred in the reaction this second form of free enzyme is called a *substituted enzyme.*

Complexes that do not undergo further reactions that are part of the catalytic pathway are called *dead-end complexes,* and the reactions producing them are called *dead-end reactions.* Enzyme-substrate complexes that do not lead to reaction, which are often but not necessarily dead-end complexes, are called *abortive* or *non-productive complexes.*

When a reaction proceeds through a series of steps that must occur in a specified order, *e.g.,* the substrates must bind in a particular order and the products are released in a particular order, it is said to obey a *compulsory-order mechanism.* (The alternative term *linear mechanism* is sometimes used, but it is discouraged because it can invite confusion with other uses of the term *linear* in enzyme kinetics.) When this is not the case the reaction is said to follow a *random-order mechanism* or a *branched mechanism.* These terms may also be applied to parts of mechanisms: for example, it may happen that substrates bind in random order but products are released in a compulsory order. The distinction between compulsory-order and random-order mechanisms is rarely absolute in practice. The term *preferred-order mechanism* may be used to emphasize that although more than one pathway exists most of the flux is through one of them. The term *random-order mechanism* does not, however, exclude this intermediate case unless explicitly stated to do so.

Elementary steps in which the enzyme forms complexes with small molecules are called *binding steps* and the reverse steps are called *release steps,* usually with a qualifier to indicate the type of species bound or released: *e.g., substrate binding step* or *product-release step, etc.* Elementary steps in which no binding or release occurs may be called *isomerizations.* An *allosteric* effector is one that acts by binding to the enzyme at a site different from the active site. There is no necessary connection between allosteric effects and co-operative effects, though they often occur together in real systems.

ENZYME ACTIVITY

The catalytic activity of an enzyme is the property measured by the increase in the rate of conversion of a specified chemical reaction that the enzyme produces in a specified assay system. This is an extensive quantity, and quantities derived from it include *specific catalytic activity* (used for following enzyme purification), which is the catalytic activity divided by the mass of protein, and *molar catalytic activity,* which is the catalytic activity divided by the amount of substance, either of enzyme catalytic centres or of multi-centre molecules. In the latter case it is important to specify whether the measurements refer to catalytic centres or to molecules. Molar catalytic activity has the dimensions of reciprocal time, like first-order rate constants; if it refers to catalytic centres it will approach k_0, the catalytic constant, if the conditions in the assay are such that v approaches V.

SUMMARY OF RECOMMENDED SYMBOLS AND UNITS

Although the table includes most alternatives mentioned in the text, it cannot cover all of the points of detailed explanation and should be taken in conjunction with the text. Although for brevity and convenience all volumes are expressed in dm^3, in all cases dm^3 may be replaced identically with L or l (litre), and amount-of-substance concentrations may be expressed in M (molar), which is identical with mol dm^{-3}. Both L and M may be used as primary units for forming multiples and submultiples such as mL, μM, *etc.*

Table. Recommended Symbols and their Units

Symbol	Meaning	Customary Unit
[A]	concentration of substrate A	mol dm^{-3}
$[A]_{0.5}$ or $[A]_{1/2}$	value of [A] at which $v = 0.5\ V$	mol dm^{-3}
[B]	concentration of substrate B	mol dm^{-3}
$[E]_0$ (or $[E]_t$ or $[E]_{stoich}$)	stoichiometric concentration of active centres	mol dm^{-3}
h (or n_H)	Hill coefficient	dimensionless
[I]	concentration of inhibitor I	mol dm^{-3}
k	rate constant of any order n	(mol $dm^{-3})^{1-n}$ s^{-1}
$\tilde{k}$	pH-independent value of k	as k
k^{app}	apparent value of k	as k
k_A, k_B	specificity constants for A, B	mol^{-1} dm^3 s^{-1}
k_A^Z(*etc.*)	reciprocal Dalziel coefficient	(various)
k_{cat} (or k_0)	catalytic constant	s^{-1}
k_i, k_{-i}	forward and reverse rate constants respectively for ith step	as k
K_i	inhibition constant (inhibition type unspecified)	mol dm^{-3}
K_{iA}	inhibition constant for A	mol dm^{-3}

K_{ic}	competitive inhibition constant	mol dm^{-3}
k_{ij}	rate constant for step from E_i to E_j	as k
K_{iu}	uncompetitive inhibition constant	mol dm^{-3}
K_m	Michaelis constant (or Michaelis concentration)	mol dm^{-3}
K_{mA}	Michaelis constant for A	mol dm^{-3}
K_s	substrate dissociation constant	mol dm^{-3}
K_{sA}	value of K_s for substrate A	mol dm^{-3}
k_0 (or k_{cat})	catalytic constant	s^{-1}
n_H (or h)	Hill coefficient	dimensionless
[Q]	concentration of activator Q	mol dm^{-3}
t	time	S
v	rate (or velocity) of reaction	mol dm^{-3} s^{-1}
V (or V_{max})	limiting rate (or maximum rate, or maximum velocity)	mol dm^{-3} s^{-1}
v_A	rate of consumption of A	mol dm^{-3} s^{-1}
v_i, v_{-i}, v_{ij}	chemical flux (or chemiflux) through step with rate constant k_i, k_{-i}, k_{ij}	mol dm^3 s^{-1}
v_{max}	true maximum value of v	mol dm^{-3} s^{-1}
v_Z	rate of formation of Z	mol dm^{-3} s^{-1}
v_0	initial rate of reaction	mol dm^{-3} s^{-1}
[Y]	concentration of product Y	mol dm^{-3}
[Z]	concentration of product Z	mol dm^{-3}
α_i	coefficient of $[A]^i$ in numerator of generalized rate expression	(mol dm^{-3})$^{1-i}$ s^{-1}
β_i	coefficient of $[A]_i$ in denominator of generalized rate expression	(mol^{-1} dm^3)^i
ε_a	degree of activation	dimensionless
ε_i	degree of inhibition	dimensionless
τ	relaxation time	S
ϕ_0*(etc.)*	Dalziel coefficient	(various)

The intermediate EA is said to be in a steady state. The use of this approximation to obtain an overall rate expression is known as the *steady-state treatment* or the *steady-state approximation.* At the very beginning of the reaction the concentration of EA in the above scheme is rising from zero to its steady-state value. The steady-state approximation is not valid during these early times and the kinetics are known as *pre-steady-state kinetics* or *transient-phase kinetics.* The transient phase of an enzyme-catalysed reaction usually occupies a very brief period of time (usually a small fraction of a second), and special techniques must be used for investigating this phase of the reaction.

The *rate of reaction* of an enzyme-catalysed reaction is not defined during the transient phase, because there is not a one-to-one stoichiometry between the reactants. In the steady state a one-to-one stoichiometry is established and the rate of reaction can be defined. This rate, extrapolated back to zero time, is called the *initial rate* and given the symbol v_0. The subscript 0 is normally omitted when no other kinds of rate are at issue, *i.e.,* when extended time courses are not being analysed.

The ratio [E] [A]/[EA] at equilibrium is called the *substrate dissociation constant* and given the symbol K_{sA} When only one substrate is in question the qualifier A may be omitted, and when it is included its location is a matter of typographical convenience. The substrate dissociation constant should not be confused with the Michaelis constant K_{mA}, to which it bears no necessary relationship in general. In the example given above K_{sA} is equal to the ratio k_{-1}/k_1.

MULTI-SUBSTRATE ENZYMES

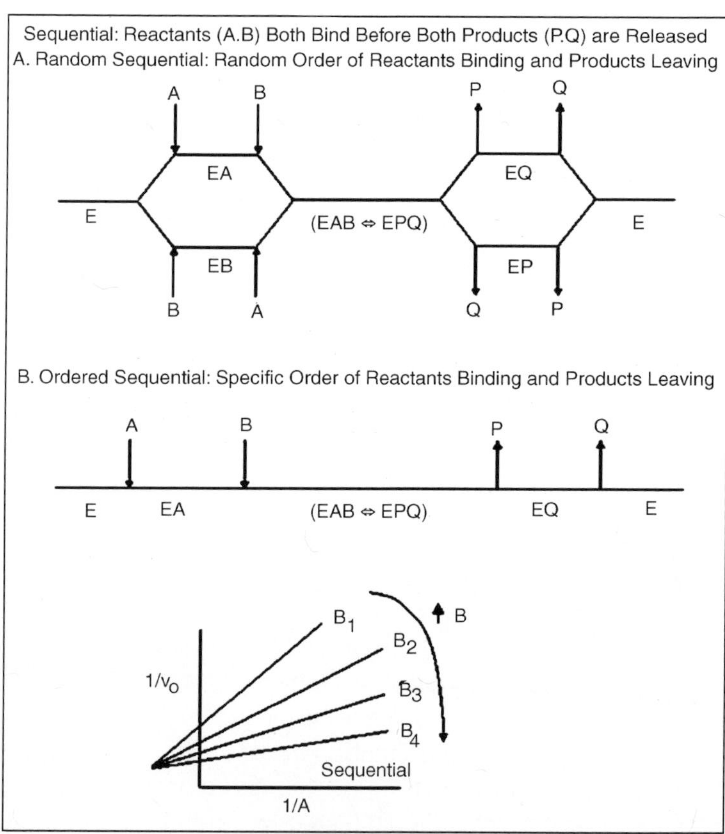

How do you resolve the enzymes kinetics of these more complicated systems? The answer is fairly straightforward. You keep one of the substrates (B, for example) fixed, and vary the other substrate (A) and obtain a series of hyperbolic plots of vo vs A at different fixed B concentrations. This would give a series of linear 1/v vs 1/A double-reciprocal plots (Lineweaver-Burk plots) as well. The pattern of Lineweaver-Burk plots depends on how the reactants and products interact with the enzyme.

Sequential Mechanism: In this mechanism, both substrates must bind to the enzyme before any products are made and released. The substrates might bind to the enzyme in a random fashion (A first then B or vice-versa) or in an ordered fashion (A first followed by B). An abbreviated notationscheme developed by W.W. Cleland is shown below for the sequential random and sequential ordered mechanisms. For both mechanisms, Lineweaver-Burk plots at varying A and different fixed values of B give a series of intersecting lines. Derivative curves can be solved to obtain appropriate kinetic constants.

Ping-Pong Mechanism: In this mechanism, one substrate bind first to the enzyme followed by product P release. Typically, product P is a fragment of the original substrate A. The rest of the substrate is covalently attached to the enzyme E, which is designated as E'. Now the second reactant, B, binds and reacts with the enzyme to form a covalent adduct with the A as it is covalentattached to the enzyme to form product Q. This is now released and the enzyme is restored to its initial form, E. This represents a ping-pong mechanism An abbreviated notationscheme is shown below for the ping-pong mechanisms. For this mechanism, Lineweaver-Burk plots at varying A and different fixed values of B give a series of parallel lines. An example of this type of reaction might be low molecular weight protein tyrosine phosphatase against the small substrate p-initrophenylphosphate (A) which binds to the enzyme covalently with the expulsion of the product P, the p-nitrophenol leaving group. Water (B) then comes in and covalently attacks the enzyme, forming an adduct with the covalently bound phosphate releasing it as inorganic phosphate. In this particular example, however, you can't vary the water concentration and it would be impossible to generate the parallel Lineweaver-Burk plots characteristic of ping-pong kinetics.

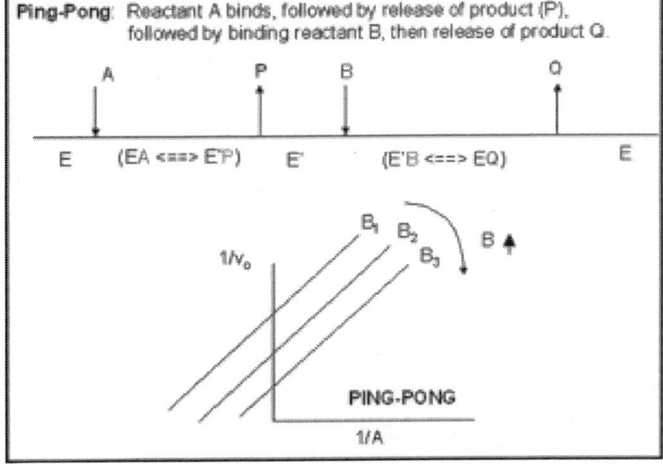

ALLOSTERIC ENZYMES

Many enzymes do not demonstrate hyperbolic saturation kinetics, or typical Michaelis-Menten kinetics. Graphs of initial velocity vs substrate demonstrate sigmoidal dependency of v on S, much as we discussed with hemoglobin binding of dioxygen. Enzymes that display this non Michaelis-Menten behaviour have common characteristics. They:

- Are multi-subunit
- Bind other ligands at sites other than the active site (allosteric sites)
- Can be either activated or inhibited by allosteric ligands
- Exist in two major conformational states, R and T
- Often control key reactions in major pathways, which must be regulated.

Examples of these enzymes include glycogen phosphorylase (the enzyme which breaks down intracellular glycogen reserves) and aspartate transcarbamyolase, which catalyses the first step in the synthesis of pyrimidine nucleotides. CTP is an allosteric inhibitor of this enzyme, which makes physiological sense since high levels of this pyrimidine nucleotide should inhibit the first enzyme in the synthesis of pyrimidines. ATP is a allosteric activator. This also makes sense since if high levels of the purine nucleotide ATP are present, one would also want to balance the level of pyrimidine nucleotides.

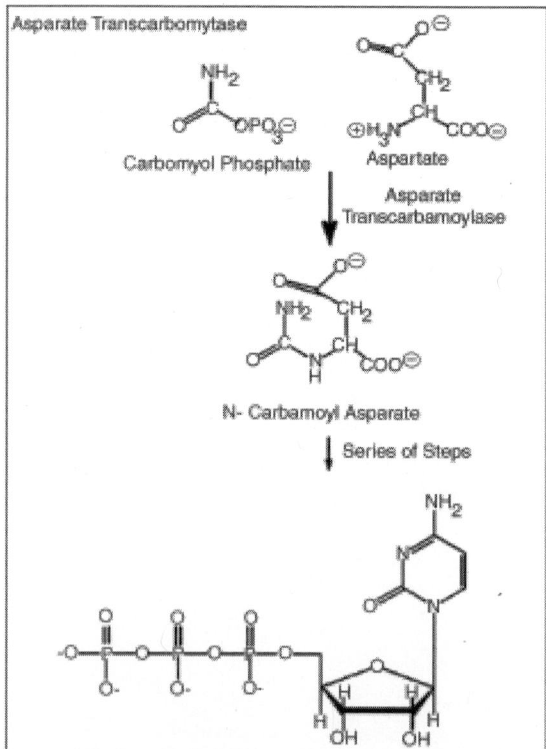

Fig. Aspartate Transcarbamoylase: Reactions

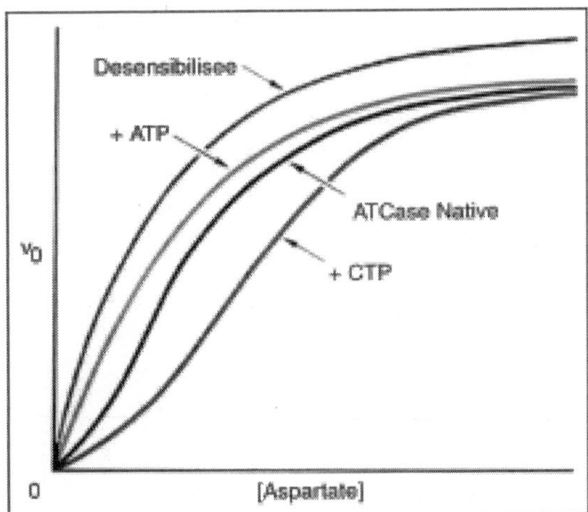

Fig. Aspartate Transcarbamoylase: Non Michaelis-Menten Kinetics

Earlierwe saw that cooperative binding equilibriumcould be modelled with the Hill Equation, which weintroduced through the equation Y = Ln/(Kd + Ln) = Ln/(P50n + Ln) where n is the cooperativity or Hill Coefficient. Likewise, for an enzyme which demonstrates cooperative (sigmoidal) initial rates plots,

$$v_o = V_m Sn/K0.50n + Ln)$$

When n=1, the equation reduces to the classical hyperbolic Michaelis Menten equation. For values of n>1, sigmoidal plots are observed. We found a more easily understandable molecular interpretation of cooperative binding of oxygen to hemoglobin using the MWC model (T and R states).

The MWC model has also been applied successfully to multi-subunit enzymes which display cooperative, sigmoidal kinetics. In this model, allosteric inhibitors (which often don't resemble the substrate) bind preferentially to the T state, leading tolower activity, while allosteric activators bind preferentially to the R state, leading to greater activity. Activators shift the vo vs S curve to the left while inhibitors shift it to the right (much like protons and carbon dioxide in hemoglobin binding). These allosteric ligands induce their effects by shifting the To ⇔ Ro equilibrium.

HOW DO ALLOSTERIC EFFECTORS CHANGE V_M AND K_M?

We have just studied how competitive, uncompetitive, and non-competitive (or mixed) inhibitors influence the apparent K_m and V_m values for enzymes that display Michaelis-Menten kinetics. How are Vm and Km influenced in allosteric enzymes? The example given above (ATCase), analogous to effects observed in hemoglobin:oxgyen binding, influence the apparent K_m, but not the V_m. Remember that in the case of hemoglobin binding curves, the allosteric activators and effectors we discussed shifted the sigmoidal binding curves to the left or right, but all reached a plateau at the same fractional saturation value of 1. Allosteric enzyme systems that behave like this are called K systems.

Enzymes in which allosteric regulators change V_m, called V systems, are also known. V systems display hyperbolicvo vs S curves in which activators display a greater apparent V_m and inhibitors display a lower Vm without affecting the apparent Km. In these systems, both the T and R forms have the same affinity for substrate (hence the same apparent Km). This would be analogous to a situation in the MWC model where KR/KT =1 which also gave hyperbolic, not sigmoidal Y vs a curves. This difference in V systems is that the R and T states have different catalytic rate constants, kcat, for turnover of the bound substrate (hence different apparent Vm values). In addition, the activator A and inhibitor I bind to the R and T forms with different affinity, which again shifts the To $\Leftrightarrow$ Ro equilibrium in the presence of the allosteric effectors.

Cooperative binding of dioxygen to hemoglobin, regulated by allosteric effectors (protons and carbon dioxide), was ideal for an oxygen transport system which must load and unload oxygen over a narrow range of oxygen concentrations and allosteric effectors. Allosteric enzymes are usually positioned at key metabolic steps which can be regulated to activate or inhibit whole pathways.

ENZYME REGULATION BY COVALENT MODIFICATION

Many enzymes are regulated not by allosteric ligands (activators and inhibitors), but by covalent modification. Often the covalent modification involves phosphorylation (by enzymes called kinases which transfer a phosphate from ATP to a Ser, Thr, or Tyr on the target enzyme) or phosphatases (which remove the phosphates from phospho-Ser, Thr, or Tyr in the target protein). In fact 1-2 per cent of all genes in the human genome code for kinases and phosphatases.

Integration of Binding, Facilitated Diffusion and Kinetics

By now you should realize the similarities among the chemical and mathematical equations used to describe binding, facilitated diffusion, and enzyme kinetics of macromolecule interactions with ligands and substrates. In simple, non-cooperative systems, it all starts with a macromolecule, usually a protein, binding a ligand, which can be a small molecule or another macromolecule.

$$M + L \Leftrightarrow ML$$

In some cases, the only function of M is to bind and release ligand (as in the case of myoglobin). In other cases, the ligand is transformed. In the simplest case, it is moved across a membrane down a concentration gradient (facilitated diffusion), a purely physical step.

$$R + A_{out} \Leftrightarrow RA \rightarrow R + A_{in}$$

In a yet more complicated case, the ligand can be transformed chemically (enzyme kinetics) into product.

$$E + S \Leftrightarrow ES \rightarrow E + P$$

In this chapter we added interactions (multiple substrates, inhibitors) and saw that the basic form of the binding, facilitated diffusion, and enzyme kinetic equations and graphs for non-cooperative system were extremely similar since the biological function always depended in some fashion on the concentration of the macromolecule complex (ML, RA, ES, EI, ESI *etc.*). The PowerPoint below reviews the similarity in the results of the mathematical analyses and resulting graphs showing the concentration dependencies of complex formation, facilitated diffusion, and enzyme-catalysed chemical reactions.

A FEEL FOR THE DISSOCIATION CONSTANT

In all of the curves shown in the above PowerPoint, a ligand concentration can be found at which the biological effect is half maximum (either activation or inhibition of a biological function). Remember, only in certain conditions is that number equal to the dissociation constant for ligand. This is clearly the case when the only interaction is a 1:1 binding of a macromolecule and ligand. If an inhibitor was unknowingly present during the direct or indirect measurement of a binding reaction, the ligand concentration at half-maximal binding would equal the apparent K_d, not the actual dissociation constant. In the case of facilitated diffusion and enzyme-catalysed chemical transformation of a single substrate, the ligand concentration at half-maximal binding is equal to the dissociation constant only in the rapid equilibrium assumption holds. It clearly doesn't in the steady state assumption, and clearly not in more complicated systems. Consider the case when a ligand binds a neurotransmitter receptor and alters intracellular the calcium ion concentration through a complicated signal transduction system.

If the step leading to the releases of stored intracellular calcium is several steps removed from the actual binding of ligand to a neurotransmitter receptor, the likelihood that the ligand concentration at half-maximal increase in intracellular calcium is equal to the K_d for ligand binding is small. If however, the actual ligand dissociation constant for the receptor can be determined (using radiolabeled ligand, for example), and it is equal to the ligand concentration for half-maximal calcium increase, then it might be argued the binding is "rate or effect determining". When inhibitors of biological function are used, IC_{50} values (the inhibitor concentration at which the response is reduced to 50 per cent) are usually reported.

FROM THE SIMPLE TO THE COMPLEX

Enzymes do not work in isolation. Most are part of complex synthetic/ degradative or signal transduction systems in which their biological response (product formation or degradation, pathway regulation, *etc.*) are functions of ligand concentration and are subject to signals further upstream or downstream in their resident pathways. How are the dose responses for biological activity modulated by other parts of the pathway? Although the equations we've derived don't directly predict the responses in lieu of knowing the concentrations, rate constants, and dissociation constants for all the steps in the pathway, an approach based on simple assumptions might illuminate the possibilities. A recent paper by Bashor et al. and an analysis of it by Pryciak describe the input and output responses of signalling enzymes attached "in series" to a scaffold protein and how different dose responses outputs can be engineered to meet the requirements of the pathway. Examples of different dose-response curves from Bashor analysis are shown below.

We've actually seen many of these response curves for individual protein/ enzymes or chemical reaction systems. Curve A represents a typical response curve which could be produced by a process governed by a simple M + L $\Leftrightarrow$ ML equilibrium. Curves B and C are sigmoidal and could be produced by a process governed by a multisubunit enzyme following the MWC model with different values for the parameters L and c. Curve D is similar to the output of consecutive irreversible reactions such as A $\rightarrow$ B $\rightarrow$ C, where the response would be similar to the rise and fall B with time (not concentration, however). One could envision a dose/response, however, that could produce a rise/fall as in curve D. Likewise Curve E would reflect the rise of C in the same chemical reaction.

To a first approximation, one could image that the responses shown below are produced by steps that are rate-limiting in the overall pathway such that the overall response is governed by those steps. Obviously a complex mathematical systems analysis involving the solving of matrices of differential equation for each step would be required for a more realistic understanding of actual biological responses, but I hope you can see that an understanding of real biological networks must start with a kinetic and thermodynamic understanding of the "simple" steps that constitute larger system pathways.

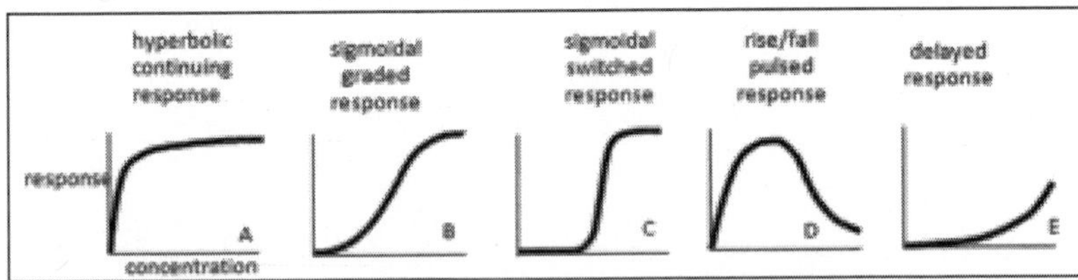

PASSIVE AND FACILITATED DIFFUSION

For some proteins, reversible binding is their sole function (consider the binding of dioxygen to myoglobin and hemoglobin). For many others, it is not. For those, what can happen next? The answer to that question clearly depends on the biological function of the macromolecule. We can simplify this process by adding one additional step as reflected in the equilibrium binding expression shown below:

$$M+L \leftrightarrow ML \leftrightarrow M+X$$

This expression indicates that the free ligand has changed in some fashion to x. We will consider two kinds of transformations:

1. L is a ligand on the outside of a biological membrane (Lout) which binds to a membrane protein receptor, R. This undergoes a conformational change (as we studied in the binding of dioxygen to hemoglobin) which leads to the expulsion of the bound ligand to the inside of the membrane (Lin). This can modelled with the simple equation: R+ Lout $\leftrightarrow$ RL $\leftrightarrow$ R+ Lin. This process is called facilitated diffusion and represents a physical as opposed to chemical process since no covalent bonds are made or broken. This process proceeds down a concentration or chemical potential gradient (Dm < 0) and hence is spontaneous (thermodynamically favoured). If the ligand concentration is higher inside the cell, net diffusion moves it to the outside of the cell. Passive (non-facilitated) diffusion is kinetically slow in the absence of a receptor since membranes present formidable barriers to the passage of polar molecules.
2. L is a ligand (or substrate S) which binds to a protein enzyme, E. The bound substrate is chemically altered to produce a new product, P, which dissociates from the enzyme. This can be expressed most simply as:

$$E+S \leftrightarrow ES \leftrightarrow E+ P.$$

This chapter will discuss diffusion processes. First, diffusion equations will be derived for cases not involving a binding receptor. The equation will show the rate of diffusion of a solute across a membrane from a region of high concentration to a region of low concentration (Dm$<$ 0) is a linear function of [DL] across the membrane. Next we will derive equations for receptor-mediated diffusion across a membrane - facilitated diffusion. We will deal with the situation when the solute must be transported up a concentration gradient (which requires ATP as an exogenous source of energy), a process called active transport, later in this web book.

PASSIVE DIFFUSION THROUGH A MEMBRANE - CASE 1: NO RECEPTOR

Before you study the derivation for simple passive diffusion, view the applet files below. Open up the following files with IE. On the top menu bar, select File, Open, Browse, My Computer, ADMain$ on Ad(S), Chem, Diffusion, then one of three folders:

1. Diffusion, and select the file diffusion1.html that will show diffusion of two particles through a membrane
2. Diffusion40, and select the file:diffusion40.html for 40 particles
3. Diffusion80, and select the file:diffusion80.html for 80 particles

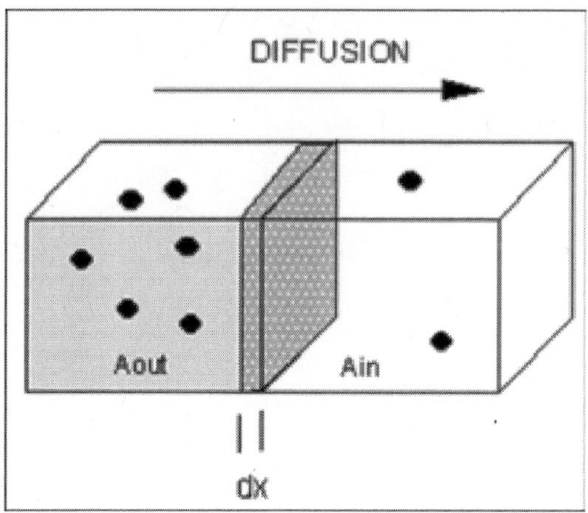

(Run each applet several times for 10 seconds and record the number of particles that pass through the membrane to the right hand side. Graph that number versus the initial number on the left hand side (20, 40, and 80, respectively, for the files above).

The flux of molecule A (J_A) across a membrane of thickness dx, is proportional to the concentration gradient across the membrane, dA/dx., which is expressed as Fick's First Law of Diffusion:

$J_A \propto \dfrac{dA}{dx}$ or $J_A= - D \dfrac{dA}{dx}$

where D is the diffusion coefficient. The negative sign is necessary since concentration increases in the opposite direction of net flux. For these derivations, we will assume the J_A is the initial flux. That is, the flux is measured for a short enough time that the relative concentrations of A on both sides of the membrane does not change significantly. It should be clear that eventually the net flux levels off to zero, when the concentrations of A on both sides of the membrane are equal (i.e. the chemical potential of A on both sides is identical).

This can also be expressed as JA = -L dmA/dx, which bridges kinetic and thermodynamic aspects of diffusion.

For J_A = - D dA/dx, dimensional analysis shows that

J = moles/area/sec = mol/cm^2.s = - (cm^2/s) mol/cm^3/cm.

Hence the units of D are cm^2/s.

Rearranging Fick's Law gives the following:

$$J_A= - D \dfrac{dA}{dx}$$

$$J_A \, dx = - D \, dA$$

Integrate both sides with x varying between 0 and x (the thickness of the membrane), and A from Aout to Ain.

JA x = -D [Ain - Aout] = D [Aout - Ain] or

JA = (D/x) [Aout - Ain] or

$$JA = P\ [Aout - Ain]$$

where P is the permeability coefficient, which has units of $cm^2/s/cm$ or cm/s.

A plot of J_A vs $(A_{out} - A_{in})$ is linear, with a slope of D/x

FACILITATED DIFFUSION THROUGH MEMBRANE - CASE 2: RECEPTOR-MEDIATED

Consider the following mechanism:

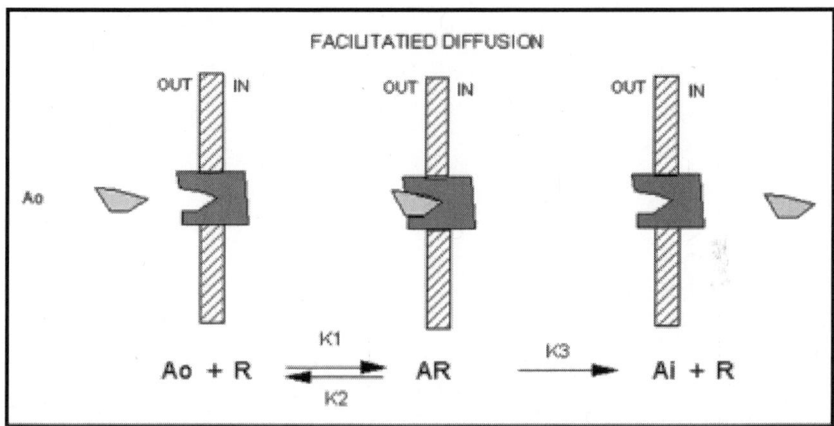

Let's assume that for this system the initial flux will be measured. We would like to derive equations which show J as a function of Aout (assuming that Ain is negligible over the time course of measuring the initial flux. Also assume that the J facilitated is much greater than J passive. In contrast to passive diffusion, JA is not proportional to Aout but rather to [Abound].

Consider this example to help you understand that proportionality. Pretend that the receptor is a truck which can carry one particle across the membrane at a time (i.e. 1/1 stoichiometry). Also assume that the particle can't get across without being carried by the truck. If there are no trucks in the membrane, no load can be delivered.

If there are trucks in the membrane but no particles in them, no load will be delivered. As the number of particles available to be loaded into the truck increase, the truck will have an increased chance to be loaded (depending of course on the affinity of the particle for the truck). If the number of loaded trucks is doubled, the number of particles dumped to the other side will double. Therefore, by analogy,

JA is proportional to [RA], or,

Equation 1: JA = const [RA] = k3 [RA]

How can we calculate RA when we know A and R? Let us assume that A total (A0) is much greater than R0, as is the likely biological case, and Ain = 0. We can calculate RA using the following equations, and the same procedure we used for the derivation of the binding equation

ML = (MoL)/(Kd + L)

Equation 2 - Dissociation constant: Kd = ([A]eq[R]eq)/[AR]eq = ([A][R])/[RA] (units of molarity)

Equation 3 - Mass Balance of R: Ro = R + RA so R = Ro -RA

Since, we will assume that A0 is much greater than R0, we will not need the mass balance for A (which is Ao = A + RA) Substitute 3 into 2:

(RA)Kd = (Ro)A - (RRA)A

(RA)Kd + (RA)A = (Ro)A

Equation 4: RA = (Ro)A/(Kd + A)

Substitute 4 into 1 gives the final equation,

Equation 5:JA = const [RA] = k3 [RA] = k3 (Ro)A/(Kd + A).

It should be clear to you from this equations that:

- A plot of JA vs A is hyperbolic
- JA = 0 when A = 0.
- JA = Jmax when A is much greater than Kd
- A = Kd when JA = Jmax/2.

These are the same conditions we detailed for our understanding of the binding equation

$$ML = (M_oL)/(K_d + L)$$

RAPID EQUILIBRIUM ASSUMPTION

This derivation is based on the assumption that the relative concentrations of A, R, and AR can be determined by the K_d for the interactions and the concentrations of each species during the early part of diffusion (i.e. under initial rate conditions).

Remember under these conditions, Aout does not change much with time. Is this a valid assumption? Aout binds to R with a second order rate constant k_1. RA has two fates.

It can dissociate with a first order rate constant k_2 to Aout + R (to give the original species), or dissociate with a first order rate constant of k_3 to give Ain + R (as A moves across the membrane). If we assume that $k_2 >> k_3$ (i.e. that the complex falls apart much more quickly than A is carried in), then the relative ratios of A, R, and RA can be described by Kd. Alternatively, you can think about it this way.

If A binds to R, most of A will dissociate, and a small amount will be carried across the membrane. If this happened, then R is now free, and will quickly bind Aout and Re-equilibrate. This occurs since the most likely fate of bound A is to dissociate, not to be carried across the membrane, since $k_3 << k_2$.

DIFFUSION OF IONS AND MOLECULES ACROSS MEMBRANES

As we saw previously, the permeability coefficients of synthetic membranes (liposomes) to solutes is related to the size and polarity of the solute.

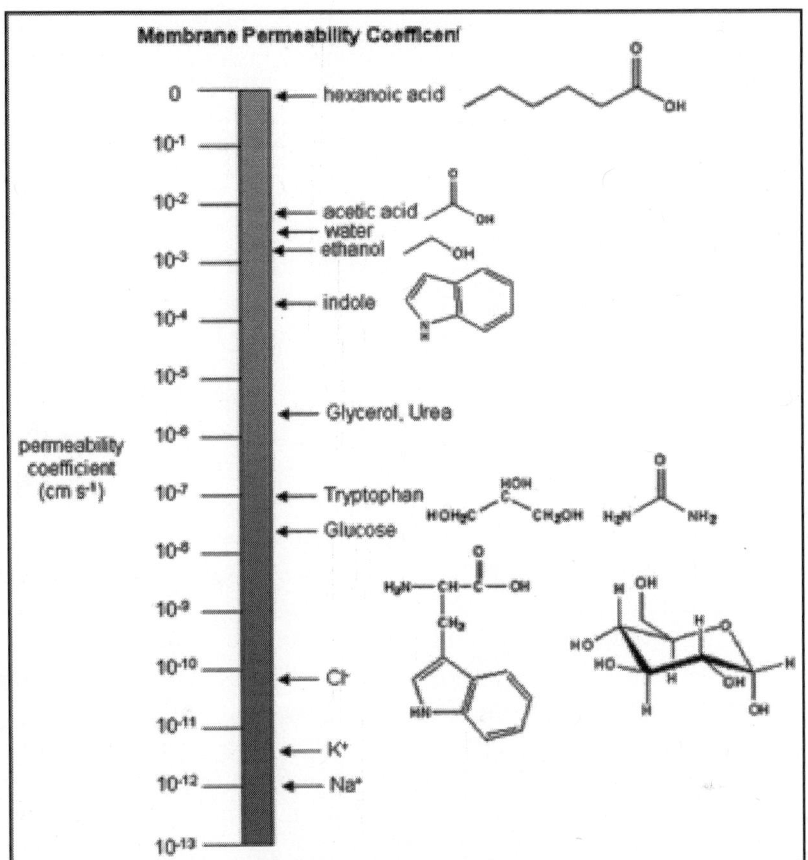

Fig. Permeability Coefficients

Smaller, higher charge density ions (like Na^+) have a lower permeability coefficient than do larger, lower charge density ions (like K^+). What about natural membranes? Look at the chart below.

Permeability coefficient (cm/s) of natural and synthetic membranes to D-Glucose and D-mannitol at 25°C.

Membrane Preparation	D-Glucose	D-Mannitol
Synthetic Lipid Bilayer	2.4 x 10-10	4.4 x 10-11
Calculated Passive Diffusion	4 x 10-9	3 x 10-9
Intact Human Erythrocyte (red blood cell)	2.0 x 10-4	5 x 10-9

Why do the permeability coefficients differ for D-Glc and D-Mannitol across the red blood cell membrane?

- Animation of Facilitated Diffusion
- Animation of Diffusion and Active Transport from the HHMI

"Receptors" in Facilitated Diffusion

Two types of proteins are involved in facilitated diffusion. Carrier proteins (also called permeases or transporters) such as the glucose transporter (GLUT1) move solute molecules across a membrane, Channels facilitate diffusion of ions down a concentration gradient. In the former cases, the ligand binds the receptor (permease, transport protein) which induces a conformational change in the receptor as illustrated in the animation above

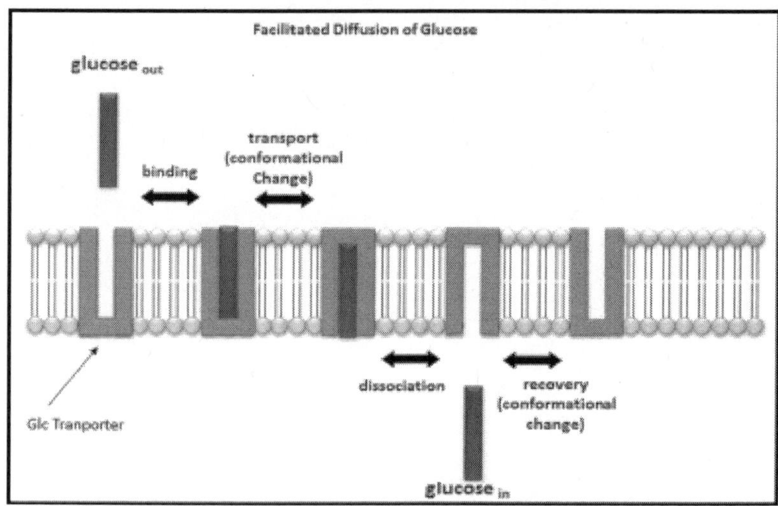

Fig. Models for Facilitated Diffusion of Glucose.

In the latter case, a ligand can bind to the receptor (channel protein) which induces a conformational change in the receptor, a "ligand-gated" channel through the membrane. This process would lead to the diffusion of many ions across the membrane (down a concentration gradient) until the channel closes (which can be induced by ligand dissociation or other events). Clearly, the mathematics we derived for the carrier proteins does not apply to the channel proteins. In addition, there are other ways to "gate" open a channel protein, which we will discuss later. Also some transporters can move solute molecules across a membrane against a concentration gradient. These proteins require an external energy source (like ATP or coupling to the favourable collapse of a second transmembrane gradient) to drive this thermodynamically unfavored process.

- Animation ofribitol diffusion through a glycerol channel using interactive molecular dynamics from
- Animation of lactose diffusion through the LacY receptor

Both links above are from the Theoretical and Computational Biophysics group at the Beckman Institute, University of Illinois at Urbana-Champaign. These molecular dynamic simulations were made with VMD/NAMD/BioCoRE/JMV/other software support developed by the Group with NIH support.

Recently, the x-ray structures of two transporters that are powered by the collapse of a second gradient, were reported.

- Although it is clear that conformational changes in membrane proteins must occur for activity to be expressed, the exact structural rearrangements of these proteins has been difficult to determine given the lack of structural data for membrane proteins, especially as contrasted to water-soluble enzymes. Transporters allow slow (102 to 10-5 s-1) movement of ligands through pores which must be alternately opened and closed to allow extracellular and intracellular access/egress. They need at a minimum an "open-to-out" and "open-to-in" conformational states. The pores must be transient and not continuous. Structural studies have been performed on E. Coli Lac Y Permease, a member of the largest transporter family, Major Facilitator Superfamily (MFS). These proteins have two halves with symmetrical halves each with 6 transmembrane domains.

Structural work by Singh et al. on the Leu Transporter (LeuT), a member of the solute carrier 6 or sodium coupled transporters, which is an active transporter requiring movement of Na ion into the cells to power the uptake of Leu, show an "open-to-out" and occluded binding state for ligand (Leu). Tryptophan, a competitive non-transportable inhibitors binds to the open-to-out state, but is too large for the obligate occluded state so it is not transported.

Carriers of ligands need not be proteins. Ionophores are small organic molecules which can bind metal ions and move them down a concentration gradient across a bilayer. Most ionophores are not resident in the cell membrane. Rather they are mobile carriers of ions. They bind ions in solution, ferry them through the membrane, and then release them on the other side. As with receptor carrier proteins, they work in both directions but move ions in a net fashion down the concentration gradient of the ion. One example is valinomycin, a natural product of Streptomyces fulvissimus. It is a cyclic peptide consisting of L and D-Val along with L-lactate and D-hydroxyisovalerate, connected through both ester and amide bonds.

The structure of the valinomycin in its K+ form give clues to its function. The six Val carbonyl oxygens bind the K+ ion. The hydrophilic groups are pointed towards the centre, while the hydrophobic groups point to the outside of the structure, allowing the K+ ion to be sequesters in a polar environment as the non-polar exterior of the complex passes through the membrane. This ionophore is specific for K+ and binds the smaller Na+ ion weakly. This can be accounted for by two factors. The smaller sodium ion doesn't bind as tightly to the chelating carbonyl oxygens. Also, the sodium ion has a higher charge density, so the Na+/water interactions must be more stable and more difficult to break than those to K+. The ion must be desolvated before it binds to the complex. Other ionophores are specific for other ions.

Some ionophores, like gramicidin, from Bacillus brevis, forms a pore in the membrane through which different types of Group I ions may flow. It is a 15-residue peptide consisting of alternating D and L amino acids of significant hydrophobicity. The peptide appears to form a dimer helix (not an alpha helix, however), in a bilayer membrane.

AQUAPORINS

Aquaporins are tetrameric complexes that facilitate water diffusion through bilayers. They have amazing selectivity as they don't pass other solute or even protons (H3O+ or naked H+ ions).

Gram negative bacteria (as well as mitochondria in eukaryotic cells) have protein complexes called porins (different from the aquaporin discussed above). The monomer porin forms a trimer in the membrane which forms a pore allowing small solute molecules necessary for bacterial cell growth to pass. The porin proteins share a 16 stranded anti-parallel beta barrel as a common motif. Solute molecules can pass through the pore created by the beta barrel. An example, maltoporin, is shown below:

An Even Bigger Pore: The Nuclear Pore Complex Channels have pores which can be gated open and allow selective flow of ions. Porins have larger entrances which allow larger molecules to pass the bilayer. The biggest pore structure known is the nuclear pore complex, which has a combined molecular mass around 125,000,000! Its job is to shuttle small molecules through passive diffusion down a concentration gradient through the pore. In addition it moves large molecules and molecular structures (proteins, RNA, and perhaps ribosomes) across the nuclear membrane in a process which requires energy. The proteins that comprise this complex are called nucleoporins, of which there appears to be around 30 in yeast. Large proteins that pass through the pore must first be bound to a cargo receptor, which can move the "cargo" across the pore with concomitant GTP hydrolysis.

DRUGS AND DIFFUSION

One of the biggest challenges in medical drug development is the synthesis of drugs that can diffuse across the cell membrane. This requires that the drug be sufficiently non-polar while at the same time it must be sufficiently polar to have reasonable aqueous solubility, allowing blood transport. One way around this problem is to develop a water soluble drug and a protein "receptor" which would allow drug passage across the membrane. A novel mechanism not based on facilitated diffusion has been developed that allows certain protein sequences to transduce the drug across the membrane (a process called protein transduction or molecular transportation). Researchers have looked to nature to find proteins which can move across the membrane and adapted them for this process.

Several viral proteins (including TAT from the HIV virus) possess such properties, which require the presence of a short "transporter" amino acid sequence in the protein.

Drugs are covalently attached to the sequence, and carried through the membrane by the short protein sequence. This mechanism does not fit the criteria of facilitated diffusion since the required protein fragment is not a classical receptor. It is amphiphilic and can pass through the membrane even if synthesized in the lab using D-amino acids or if the sequence is scrambled.

The fusion domain of the TAT protein is ionic (containing 5-15 Arg residues) and probably interacts initially with negatively charged glycosoaminoglycans on the cell surface. Many different types of drugs can be delivered in this fashion (large to small, proteins, nucleic acids, and large liposomes, *etc.*). The company Cyclacel is developing a proprietary drug system using another protein transducer, Penetratin, a fragment of the antennapedia protein from Drosophila.

Another approach is to design artificial receptors. For example, a ligand that might ordinarily bind to a protein could be covalently modified with a hydrophobic group (often a cholesterol derivative) which would allow it to partition into the cell membrane, exposing the ligand on the cell surface. The surface ligand can then bind its target protein.

If the protein is multivalent (can bind more than one ligand per protein, such as an antibody), lateral diffusion and clustering of protein-artificial receptor complexes in the membrane can occur, as well as the formation of lipid rafts. Similar to other ligand-receptor interactions that display such properties, these membrane changes can lead to endocytosis of the protein-artificial ligand complex.

9

Chemical Potential Energy

CHEMICAL POTENTIAL

Let us recall the experimental setup. Allowing two systems to exchange energy leads to an equilibration of their temperatures T. The energies E_1 , E_2 of the two subsystems will then only fluctuate - usually just slightly - around their average values.

Let us now assume that the systems can also exchange particles. In such a situation we will again find some initial equilibration after which the particle numbers N_1 , N_2 will only slightly fluctuate around their mean values. The flow of particles from one subsystem to the other will come to an end as soon as the free energy $A = A_1 + A_2$ of the combined system tends to a constant value. This happens when. The quantity that determines the equilibrium is therefore $\mu \equiv (\partial A / \partial N)_V$; the subsystems will trade particles until $\mu_1 = \mu_2$. (Compare this definition of the chemical potential with that of temperature,.)

Example: The chemical potential of Argon ($m = 39.95\ m_H$) at normal conditions is

Simulation in the Canonical Ensemble: Monte Carlo Calculation

We may once more find an appropriate numerical rule for the correct "browsing" of all states with given (N, V, T) (in place of (N, V, E)). Since it is now the Boltzmann factor that gives the statistical weight of a microstate $\vec{\Gamma} \equiv \{\vec{r}_i, \vec{v}_i\}$, the average value of some quantity $\alpha(\vec{\Gamma})$ is given by

$$\langle a \rangle_{N,V,T} = \int a(\vec{\Gamma}) \exp[-E(\vec{\Gamma})/kT] d\vec{\Gamma} \Big/ \int \exp[-E(\vec{\Gamma})/kT] d\vec{\Gamma}$$

If depends only on the particle positions and not on velocities - and this is true for important quantities such potential energy, virial, *etc.* - then we may even confine the weighted integral to the -dimensional configurational subspace $\vec{\Gamma}_c \equiv \{\vec{r}_i\}$ of full phase space:

$$\langle a \rangle_{N,V,T} = \int a(\vec{\Gamma}) \exp[-E_{pot}(\vec{\Gamma}_e)/kT] d\vec{\Gamma}_c \Big/ \int \exp[-E_{pot}(\vec{\Gamma}_e)/kT] d\vec{\Gamma}_e$$

In order to find an estimate for $\langle a \rangle$ we formally replace the integrals by sums and construct a long sequence of randomly sampled states $\vec{\Gamma}_e(m); m=1,...M$ with the requirement that the relative frequency of microstates in the sequence be proportional to their Boltzmann factors. In other words, configurations with high potential energy should occur less frequently than states with small E_{pot}.

The customary method to produce such a biased random sequence (a so-called "Markov chain") is called "Metropolis technique", after its inventor Nicholas Metropolis.

Now, since the Boltzmann factor is already contained in the frequency of states in the sequence we can compute the Boltzmann-weighted average simply according to $\langle a \rangle = \frac{1}{M} \sum_{m=1}^{M} a(\vec{\Gamma}_e(m))$.

There is also a special version of the molecular dynamics simulation method that may be used to perambulate the canonical distribution. We recall that in MD simulations normally the total system energy $E_{tot} = E_{kin} + E_{pot}$ is held constant, which is roughly equivalent to sampling the microcanonical ensemble. However, by introducing a kind of numerical thermostat we may at each time step adjust all particle velocities so as to keep E_{kin} either constant (isokinetic MD simulation) oder near a mean value such that $\langle E_{kin} \rangle \propto T$ (isothermal MD).

POTENTIAL ENERGY

Consider a body moving in a conservative force- field *f*(*r*). Let us arbitrarily pick some point *O* in this field. We can define a function *U*(*r*) which possesses a unique value at every point in the field. The value of this function associated with some general point *R* is simply,

$$U(R) = -\int_O^R f \cdot dr.$$

In other words, *U*(*R*) is just the energy transferred to the field (*i.e.,* minus the work done by the field) when the body moves from point O to point *R*. Of course, the value of *U* at point *O* is zero: *i.e.,* $U(O) = 0$. Note that the above definition uniquely specifies *U*(*R*), since the work done when a body moves between two points in a conservative force- field is independent of the path taken between these points.

Furthermore, the above definition would make no sense in a non-conservative field, since the work done when a body moves between two points in such a field is dependent on the chosen path: hence, *U*(*R*) would have an infinite number of different values corresponding to the infinite number of different paths the body could take between points *O* and *R*.

According to the work-energy theorem,

$$\text{DK} = \int_O^R f \cdot dr.$$

In other words, the net change in the kinetic energy of the body, as it moves from point O to point R, is equal to the work done on the body by the force-field during this process. However, comparing with Eq., we can see that

$$\text{DK} = U(O) - U(R) = -\text{D}U.$$

In other words, the increase in the kinetic energy of the body, as it moves from point O to point R, is equal to the decrease in the function U evaluated between these same two points. Another way of putting this is

$$E = K + U = \text{constant}$$

i.e., the sum of the kinetic energy and the function U remains constant as the body moves around in the force-field. It should be clear, by now, that the function U represents some form of potential energy. The above discussion leads to the following important conclusions. Firstly, it should be possible to associate a potential energy (*i.e.,* an energy a body pos-sesses by virtue of its position) with any conservative force- field. Secondly, any force-field for which we can define a potential energy must necessarily be con-servative.

For instance, the existence of gravitational potential energy is proof that gravitational fields are conservative. Thirdly, the concept of potential energy is meaningless in a non-conservative force- field (since the potential energy at a given point cannot be uniquely defined). Fourthly, potential energy is only defined to within an arbitrary additive constant. In other words, the point in space at which we set the potential energy to zero can be chosen at will. This im-plies that only differences in potential energies between different points in space have any physical significance.

For instance, we have seen that the definition of gravitational potential energy is $U = mgz$, where z represents height above the ground. However, we could just as well write $U = mg\,(z - z_0)$, where z_0 is the height of some arbitrarily chosen reference point (*e.g.,* the top of Mount Ever-est, or 0the bottom of the Dead Sea). Fifthly, the difference in potential energy between two points represents the net energy transferred to the associated force-field when a body moves between these two points. In other words, potential energy is not, strictly speaking, a property of the body—instead, it is a property of the force- field within which the body moves.

KINETIC ENERGY OF SYSTEMS

For a single point, or a rigid body that is not rotating, the kinetic energy goes to zero when the body stops.

However, for systems containing multiple independently moving bodies, which may exert forces between themselves, and may (or may not) be rotating; this is no longer true.

This energy is called 'internal energy'.

The kinetic energy of the system at any instant in time is simply the sum of the kinetic energies of the masses- including the kinetic energy due to the rotations. An example would be the solar system. In the center of mass frame of the solar system, the Sun is (almost) stationary, but the planets and planetoids are in motion about it. Thus even in a stationary center of mass frame, there is still kinetic energy present.

However, recalculating the energy from different frames would be tedious, but there is a trick. The kinetic energy of the system from a different inertial frame can be calculated simply from the sum of the kinetic energy in the center of mass frame and adding on the energy that the total mass of bodies in the center of mass frame would have if it were moving at the relative speed between the two frames.

This may be simply shown: let V be the relative speed of the frame *k* from the center of mass frame *i* :

$$E_k = \int \frac{v_k^2 dm}{2} = \int \frac{(v_i + V)^2 dm}{2} = \int \frac{(v_i^2 + 2v_i V + V^2)dm}{2}$$

$$= \int \frac{v_i^2 dm}{2} + v \int v_i dm + \frac{V^2}{2} \int dm$$

However, let

$$\int \frac{v_i^2 dm}{2} = E_i$$

the kinetic energy in the center of mass frame, $\int v_i dm$ would be simply the total momentum which is by definition zero in the center of mass frame, and let the total mass:

$$\int dm = M.$$

Substituting, we get:

$$E_k = E_i + \frac{MV^2}{2}$$

The kinetic energy of a system thus depends on the inertial frame of reference and it is lowest with respect to the center of mass reference frame, i.e., in a frame of reference in which the center of mass is stationary. In any other frame of reference there is an additional kinetic energy corresponding to the total mass moving at the speed of the center of mass.

Thus it sometimes is convenient to split the total kinetic energy of this body into the sum of the body's center-of-mass translational kinetic energy and the energy of rotation around the center of mass rotational energy:

$$E_k = E_t + E_r$$

where:

E_k is the total kinetic energy

E_t is the translational kinetic energy

E_r is the *rotational energy* or *angular kinetic energy* in the rest frame

Thus the kinetic energy of a tennis ball in flight has is the kinetic energy due to its rotation, plus the kinetic energy due to its translation.

Rotational Systems

If a rigid body is rotating, its *rotational kinetic energy* or angular kinetic energy about any line through the center of mass, and r is the distance of any mass dm from that line is simply sum of kinetic energies of its moving parts, and thus it is equal to:

$$E_r = \int \frac{v^2 dm}{2} = \int \frac{(rw)^2 dm}{2} = \frac{w^2}{2}\int r^2 dm = \frac{w^2}{2} I = \frac{1}{2} I w^2$$

where:

I is the body's moment of inertia

ω is the body's angular velocity.

The moment of inertia must be taken about an axis through the center of mass and the rotation measured by ω must be around that axis.

KINETIC ENERGY CORRECTION

Before claiming that this formula explains the fine structure of the hydrogen atom, however, one needs to be careful. The correction is of the order a^4, which means it of the order v^4, where v is the electron speed. The kinetic energy used in the Hamiltonian when solving the Schrφdinger equation was just $p^2/2m$, which contributed to order a^2. However, the next term in the expansion of the true relativistic kinetic energy is of order $p^4$and hence will contribute to order a^4. So if one wishes to quote the energy splittings of the hydrogen atom accurate to order a^4, one had better include the contribution from this further correction. The relativistic kinetic energy of the electron can be expanded in terms of momentum as,

$$T = \frac{p^2}{2m} - \frac{p^4}{8m^3c^2} + \ldots$$

Therefore, the correction to the Hamiltonian is,

$$\Delta H_{rel} = -\frac{1}{8m^3c^2} p^4.$$

At first sight, this looks quite complicated, since it involves the operator $p^4 = \hbar^4 \nabla^4$. However, one can make use of the fact that,

$$\frac{p^2}{2m} = En - V$$

to get,

$$\Delta H_{rel} = -\frac{1}{2mc^2}\left(E_n^2 - 2E_n V + V^2\right).$$

With V = $-ze^2/r$, applying first-order perturbation theory to this Hamiltonian reduces to the problem of finding the expectation values of r^{-1} and r^{-2}. This can be done with some effort, and the result is,

$$\Delta E_{rel} = -(Za)^4 mc^2 \frac{1}{2n}\left[\frac{1}{\left(I+\frac{1}{2}\right)} - \frac{3}{4}\right].$$

Combining equations 57 and 52 and using the fact that j=l–1/2,l+1/2, the complete energy correction to order $(Za)^4$may be written

$$\Delta E_{fs} = \Delta E_{rel} + \Delta E_{so} = -(Za)^4 mc^2 \frac{1}{2n}\left[\frac{1}{\left(j+\frac{1}{2}\right)} - \frac{3}{4n}\right].$$

This energy correction depends only on j and is called the fine structure of the hydrogen atom, since it is of order $a^2 \sim 10^{-4}$times smaller than the principle energy splittings. This is why a is known as the fine-structure constant.

The fine structure of the hydrogen atom. Note that all levels are shifted down from the Bohr energies, and that for every n and l there are two states corresponding to j=l–1/2 and j=l+1/2, except for s states.

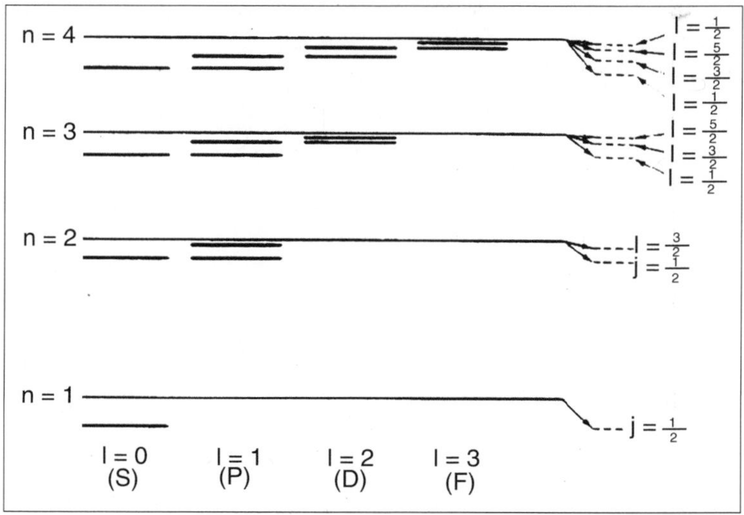

Fig. The Fine Structure of the Hydrogen Atom.
The Diagram is Not to Scale.

Also note that states with the same n and j but different l have the same energies, though this will be shown later not to be true due an effect know as the Lamb shift. As an aside, these fine structure splittings were derived by

Sommerfeld by modifying the Bohr theory to allow elliptical orbits and then calculating the energy differences between the different states due to differences in the average velocity of the electron.

By using the wrong method he got exactly the right answer, a coincidence which caused much confusion at the time. Strictly speaking, equation 58 has only been shown to be correct for $I \neq 0$ states, although it turns out to be correct for all l. To do the calculation correctly for l=0, one needs to include the effect of an additional term in the Hamiltonian known as the Darwin term, which is purely an effect of relativistic quantum mechanics and can only be understood in the context of the Dirac theory.

It is therefore appropriate to discuss the Dirac theory to achieve a more complete understanding of the fine structure of the hydrogen atom.

PRINCIPLE OF WORK AND ENERGY FOR A RIGID BODY

If a particle or body is acted upon only by conservative forces energy is conserved. This means that the total kinetic and potential energy in the system remains constant, and does not change. Such a system has no friction forces acting on it, and as such is an idealized simplification for solving problems using energy calculations.

Common examples of conservative forces acting on a particle or body are gravitational forces and (elastic) spring forces.

The arbitrary path traveled by the particle may be due to the presence of other forces also acting on the particle, but we do not need to consider them, since the work done by gravity is unaffected by them and can therefore be treated independently.

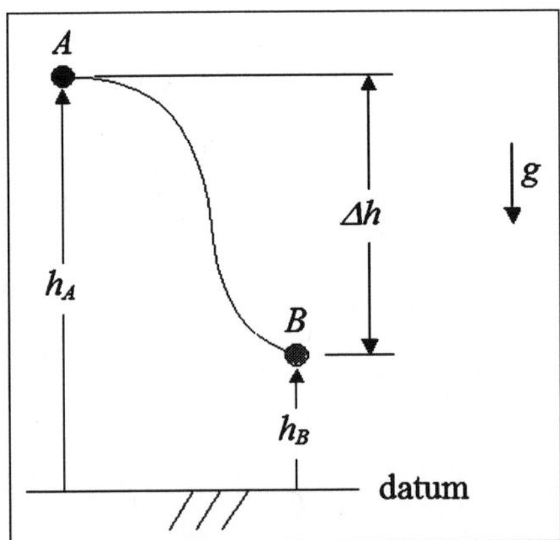

If Δh is equal to the final vertical position (at point B) minus the initial vertical position (at point A) then the work (W_g) done by gravity on the particle is given by the following scalar equation:

$$W_g = -mg(\Delta h)\backslash = -mg(h_B - h_A)$$

where m is the mass of the particle, and g is the acceleration due to gravity.

A negative sign is present in the above equation because vertical displacement (Δh) acting in the same direction as gravity ($\Delta h < 0$) must yield positive work, and vertical displacement acting in the opposite direction of gravity ($\Delta h > 0$) must yield negative work. So, if B is lower than A we have $\Delta h < 0$, and the negative sign accounts for this, making the work positive. On the other hand, if B is higher than A we have $\Delta h > 0$, and the negative sign accounts for this, making the work negative.

The expression for W_g is useful because no matter how complicated the path from A to B is, we only need to know the change in vertical height W_g and we can find the work done on the particle by gravity.

Work Done By An Elastic Spring

Consider a spring force acting on a particle, as shown below. The spring is attached to a wall at point O, where it can pivot. At position A of the particle, the spring is stretched or compressed by an amount s_A from its equilibrium (unstretched) position. At position B of the particle, the spring is stretched or compressed by an amount s_B from its equilibrium position. The arbitrary path traveled by the particle from A to B is represented by the blue curve.

Note that the dashed line represents the equilibrium position of the spring, where the spring is unstretched. Once again, the possible presence of other forces acting on the particle is irrelevant to this discussion, since we are only focusing on the work done by the spring.

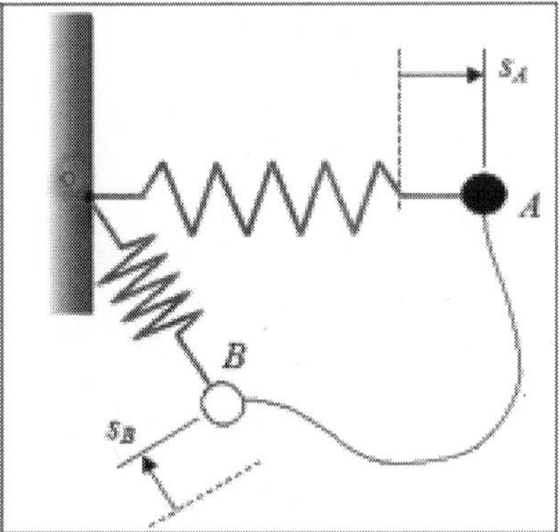

The work done by the spring (W_s) on the particle as it moves from A to $B$is given by the following scalar equation:

$$W_s = -\left(\frac{1}{2}ks_B^2 - \frac{1}{2}ks_A^2\right)$$

where k is the spring constant.

For the above equation we are assuming we have a spring that obeys Hooke's.

Therefore, the work done by the spring depends only on the position of *A*and *B* (relative to the position of point *O*), since this is what determines the amount of stretch or compression in the spring (s_A and s_B).

RELATIVISTIC KINETIC ENERGY OF RIGID BODIES

In special relativity, we must change the expression for linear momentum. Integrating by parts, we get:

$$E_r = \int v \cdot dp =$$

$$\int v \cdot d(m\gamma v) = m\gamma v \cdot v - \int m\gamma v \cdot dv = m\gamma v^2 - \frac{m}{2} = \int \gamma d(v^2)$$

Remembering that $\gamma = \left(1 - v^2 / c^2\right)^{-1/2}$, we get:

$$E_k = m\gamma v^2 - \frac{-mc^2}{2} \int \gamma d\left(1 - v^2 / c^2\right)$$
$$= m\gamma v^2 + mc^2\left(1 - v^2 / c^2\right)^{1/2} + C$$

And thus:

$$E_k = m\gamma\left(v^2 + c^2\left(1 - v^2 / c^2\right)\right) + C$$
$$= m\gamma\left(v^2 + c^2 - v^2\right) + C = m\gamma c^2 + C$$

The constant of integration is found by observing that $\gamma = 1$ when v = 0, so we get the usual formula:

$$E_k = m\gamma c^2 - mc^2 = \frac{mc^2}{\sqrt{1 - v^2 / c^2}} - mc^2$$

If a body's speed is a significant fraction of the speed of light, it is necessary to use relativistic mechanics (the theory of relativity as expounded by Albert Einstein) to calculate its kinetic energy.

For a relativistic object the momentum p is equal to:

$$p = \frac{mc^2}{\sqrt{1 - (v / c)^2}},$$

where *m* is the rest mass, *v* is the object's speed, and *c* is the speed of light in vacuum. Thus the work expended accelerating an object from rest to a relativistic speed is:

$$Ek = \frac{mc^2}{\sqrt{1 - (v / c)^2}} - mc^2$$

The equation shows that the energy of an object approaches infinity as the velocity *v* approaches the speed of light *c*, thus it is impossible to accelerate an object across this boundary. The mathematical by-product of this calculation is the mass-energy equivalence formula—the body at rest must have energy content equal to:

$$E_{rest} = mc^2$$

At a low speed (v<<c), the relativistic kinetic energy may be approximated well by the classical kinetic energy. This is done by binomial approximation. Indeed, taking Taylor expansion for square root and keeping first two terms we get:

$$E_k \approx mc^2\left(1+\frac{1}{2}v^2/c^2\right) - mc^2 = \frac{1}{2}mv^2.$$

So, the total energy E can be partitioned into the energy of the rest mass plus the traditional Newtonian kinetic energy at low speed.

When objects move at a speed much slower than light (*e.g.* in everyday phenomena on Earth), the first two terms of the series predominate. The next term in the approximation is small for low speeds, and can be found by extending the expansion into a Taylor series by one more term:

$$E \approx mc^2\left(1+\frac{1}{2}v^2/c^2+\frac{3}{8}v^4/c^4\right) = mc^2+\frac{1}{2}mv^2+\frac{3}{8}mv^4/c^2.$$

For example, for a speed of 10 km/s the correction to the Newtonian kinetic energy is 0.07 J/kg (on a Newtonian kinetic energy of 50 MJ/kg) and for a speed of 100 km/s it is 710 J/kg (on a Newtonian kinetic energy of 5 GJ/kg), *etc.*

For higher speeds, the formula for the relativistic kinetic energy is derived by simply subtracting the rest mass energy from the total energy:

$$E = m\gamma c^2 - mc^2 = mc^2\left(\frac{1}{\sqrt{1-(v/c)}}-1\right)$$

The relation between kinetic energy and momentum is more complicated in this case, and is given by the equation:

$$E_k = \sqrt{p^2c^2+m^2c^4} - mc^2$$

This can also be expanded as a Taylor series, the first term of which is the simple expression from Newtonian mechanics.

What this suggests is that the formulas for energy and momentum are not special and axiomatic, but rather concepts which emerge from the equation of mass with energy and the principles of relativity.

Quantum Mechanical Kinetic Energy of Rigid Bodies

In the realm of quantum mechanics, the expectation value of the electron kinetic energy, $\langle\hat{T}\rangle$, for a system of electrons described by the wavefunction $|\psi\rangle$ is a sum of 1-electron operator expectation values:

$$\langle \hat{T} \rangle = -\frac{\hbar^2}{2m_e} \left\langle \psi \left| \sum_{i=1}^{N} \nabla_i^2 \right| \psi \right\rangle$$

where m_e is the mass of the electron and ∇_i^2 is the Laplacian operator acting upon the coordinates of the ith electron and the summation runs over all electrons. Notice that this is the quantized version of the non-relativistic expression for kinetic energy in terms of momentum:

$$E_k = \frac{p^2}{2m}$$

The density functional formalism of quantum mechanics requires knowledge of the electron density *only*, i.e., it formally does not require knowledge of the wavefunction. Given an electron density $\rho(r)$, the exact N-electron kinetic energy functional is unknown; however, for the specific case of a 1-electron system, the kinetic energy can be written as

$$T(\rho) = \frac{1}{8} \int \frac{\nabla\rho(r) \cdot \nabla\rho(r)}{\rho(r)} d^3r$$

where $T[\rho]$ is known as the Von Weizsacker kinetic energy functional.

DERIVATION AND DEFINITION

The work done accelerating a particle during the infinitesimal time interval *dt* is given by the dot product of *force* and *displacement*:

$$F \cdot dx = F \cdot vdt = \frac{dp}{dt} \cdot vdt = v \cdot dp = v \cdot d(mv)$$

Applying the product rule we see that:

$$d(v \cdot v) = (dv) \cdot v + v \cdot (dv) = 2(v \cdot dv)$$

Therefore (assuming constant mass), the following can be seen:

$$v \cdot d(mv) = \frac{m}{2} d(v \cdot v) = \frac{m}{2} dv^2 = d\left(\frac{mv^2}{2}\right)$$

Since this is a total differential (that is, it only depends on the final state, not how the particle got there), we can integrate it and call the result kinetic energy:

$$E_k = \int F \cdot dx = \int v \cdot dp = \frac{mv^2}{2}$$

This equation states that the kinetic energy (E_k) is equal to the integral of the dot product of the velocity (v) of a body and the infinitesimal change of the body's momentum (p). It is assumed that the body starts with no kinetic energy when it is at rest (motionless).

EQUATION OF MOTION FOR A RIGID BODY

The explicit form of the laws of mechanics depend on the frame used to reference the motions. The reference frame is frequently the background of the event, the earth being the most common reference frame. For example, Newton's second law is written as,

$$\sum \underline{F} = m\underline{a}$$

when the acceleration $\underline{a}$ is calculated relative to an inertial reference frame. The points of an inertial frame have no acceleration (purely translating with constant or zero velocity, and no rotation). On the other hand, this same law will be written as,

$$\sum \underline{F} = m\underline{a} + m\underline{a}_0 .$$

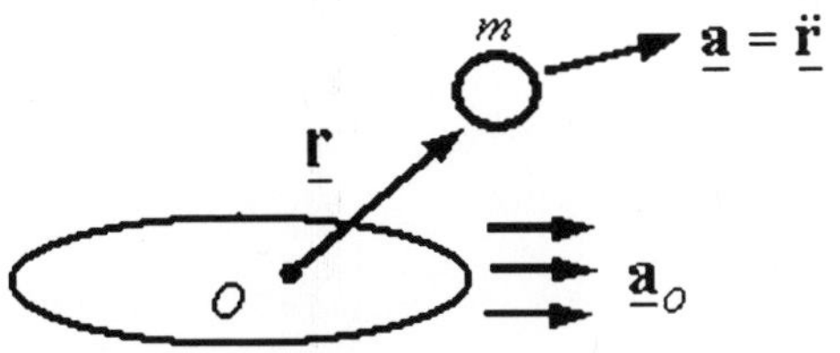

Reference frame in pure transiation with acceleration $\underline{\mathbf{a}}_o$

if the acceleration $\underline{a}$ is calculated with respect to a frame which is purely translating with an acceleration of $\underline{a}_0$. Therefore, it is important to consider the reference frame carefully before using any particular form of a given law. Most of the time, the laws of motion are written relative to an inertial frame, then from this form one can derive the form of the law relative to other frames. The laws of motion for a rigid body are known as Euler's laws. Euler gave two laws for the motion of a rigid body. The two laws written relative to an inertial reference frame are,

$$\sum \underline{F} = \frac{d}{dt}(\underline{G})$$

$$\sum \underline{M}_0 = \frac{d}{dt}(\underline{H}_0)$$

where O is a *fixed point* on the inertial reference frame. The first of Euler's two laws describes how the forces control the "translational" motion of the rigid body (*i.e.,* the change of the velocity of the center of mass, $\underline{V}_{CM}$). The second of Euler's two laws describes how the change of angular momentum of the rigid body is controlled by the moment of forces and couples applied on the body. The laws of Euler are written for a body of fixed matter (*i.e.,* matter can not be added to the body, matter can not be subtracted from the body, nor can matter be replaced by other matter). An alternate form of these laws which replaces the need to have a fixed point O on an inertial frame is obtained by using the center of mass. This law is derived below and given by,

$$\sum \underline{F} = \frac{d}{dt}(\underline{G})$$

$$\sum \underline{M}_{CM} = \frac{d}{dt}(\underline{H}_{CM})$$

where CM stands for the center of mass.

A third alternate form of Euler's laws can be derived from the second alternate form. This form replaces the center of mass by an arbitrary point P which may be moving in any way (it can be accelerating). This form of Euler's laws are written as,

$$\sum \underline{F} = \frac{d}{dt}(\underline{G})$$

$$\sum \underline{M}_p = \frac{d}{dt}(\underline{H}_{CM}) + \underline{r}_{CM/P}$$

where $\underline{r}_{CM/P}$ is the position of the center of mass relative to the selected point P. Details of how to obtain this form is also given below. A fourth form of Euler's laws is given as

$$\sum \underline{F} = \frac{d}{dt}(\underline{G})$$

$$\sum \underline{M}_p + \underline{r}_{P/CM} \times \sum \underline{F} = \frac{d}{dt}(\underline{H}_{CM})$$

where $\underline{r}_{P/CM}$ is the position of point P with respect to the center of mass, and as above P may be selected arbitrarily.

It will also be shown that the linear momentum of a rigid body is given by the mass times the velocity of the center of mass of the body to give the alternate form of Euler's first law as,

$$\sum \underline{F} = m\underline{a}_{CM}$$

where m is the total mass of the rigid body and $\underline{a}_{CM}$ is the acceleration of the center of mass of the body.

10

Organic Reaction and Organic Chemistry

FUNCTIONAL GROUPS AND ORGANIC REACTIONS

The hydrocarbons that are essential components of every organic compound. Hydrocarbons molecules are chemically inert (fairly unreactive). Most organic reactions involve substituents on the hydrocarbon chain. These substituents determine the major chemical properties of the individual organic compounds. The substituent, which is the focus of the present section of organic chemistry, is called a functional group. The term functional group is used because it is the substituent that determines how the compound functions chemically (how the compound reacts).

The type of chemical reactions will be determined by the functional group because it is the chemically active part (site) on the molecule. Organic chemists classify organic compounds according to their functional group. The chemist uses the symbol R- to represent a hydrocarbon chain (alkyl) or ring (cyclic).

Halocarbon

Halocarbons are a class of organic compounds containing a hydrocarbon parent covalently bonded with fluorine, chlorine, bromine or iodine.

The IUPAC rules for naming halocarbons uses the same rules as is used for Alkanes if the parent chain is saturated, or for Alkenes or Alkynes if the parent chain is unsaturated. The halogen(s) is treated as a branched group and is located on the continuous chain of carbons as you would locate and name any alkyl branch.

For example, the structure below would be called

```
      Br
      |
C – C – C – C – C – C – C  2,3-dibromoheptane
              |
              Br
```

The longest continuous chain of carbons with the halogen atoms attached is seven. The chain is numbered from the end that will give the lowest carbon number designators for the halogen atoms and any existing alkyl branch.

Physical Properties

Boiling Points

Halocarbons are more polar than the hydrocarbon families. Therefore the boiling points will be larger for the same sized molecules. The London Dispersion and the dipole-dipole interactions are the intermolecular forces holding the halocarbon molecules together.

Water Solubility

The halocarbons are insoluble in water because there is very little attraction that the halide molecules have for the solvent water. Hydrogen bonding is not present and for that reason most halocarbons are insoluble in water but tend to be soluble in hydrocarbon solvents and other alkyl halide solvents.

Density of Alkyl Halides

The densities of halocarbons tend to be greater than that of water because of the relatively massive halogen atoms.

Production of Halocarbons

Halocarbons can be produced using two methods:

- Substitution Reaction (inorganic chemistry calls this single replacement)

A halogen is mixed with a saturated hydrocarbon; the halogen atom substitutes hydrogen from the hydrocarbon chain. This reaction can be shown by the general equation:

$$X_2 + R - H \rightarrow R - X + HX$$

Example:

$$Cl_2(g) + CH_3 - CH_3(g) \rightarrow CH_3\text{-}CH_2Cl(g) + HCl(g)$$

chloroethane (only one possible isomer)

- Addition Reaction (inorganic chemistry calls this combination or composition)

A halogen is mixed with an unsaturated hydrocarbon; a halogen atom adds on to each of the two carbons from the carbon-carbon double bond in the hydrocarbon chain. This reaction can be shown by the general equation:

$$X_2 + R - C = C - R' \rightarrow R - \underset{}{\overset{X}{\overset{|}{C}}} - \overset{X}{\overset{|}{C}} - R'$$

R and R' represent alkyl groups

Example:

$$Cl_2(g) + CH_3 - CH_3(g) \rightarrow CH_2Cl - CH_2Cl(g)$$

1,2-dichloroethane (more than one isomer possible but only one can be produced by this reaction)

Alcohol

Alcohols are a family of organic compounds that share a common chemically active group called the hydroxyl group (– O-H). Do not confuse the hydroxyl group with the hydroxide ion (OH^-) (a negatively charged ion). The hydroxyl group is electrically neutral and forms a covalent bond.

The hydroxyl group (– OH) represents the position in an alcohol molecule where the chemical change takes place. The general symbol to represent any alcohol with only one hydroxyl group attached to one of its carbons is R –OH. Alcohols can have two or more hydroxyl groups attached to the molecule and are called "diols", "triols".

Naming

Simple alcohols can be named using the common name approach of identifying and naming the alkyl portion and follow it with the word alcohol. For example, CH_3 – OH could be called methyl alcohol since the CH_3 group is the methyl group.

This method works as long as you have simple alcohols. When the alcohols become more complex and more branched, then IUPAC has come up with a set of rules that are used to name any alcohol regardless of its complexity. These rules are as follows:

- Identify the longest continuous (parent) chain of carbons with the hydroxyl group attached to one of the carbons in the chain.
- Number the chain so that the hydroxyl group is attached to the lowest numbered carbons.
- Identify and locate the other branches on the chain so that they are named alphabetically and their carbon number is hyphenated onto the front of the name. If more than one of the same group is present use the Greek prefix attached to the branch name. (di=2, tri=3, etc)
- After all the branches have been named and located then attach the carbon number that is attached to the hydroxyl group onto the alkane name associated with the number of carbons found in the parent chain in step 1. Drop the "e" on the alkane name and attach the IUPAC ending ol to the rest of the alkane name.
- For alcohols with more than one hydroxyl group you would have to locate two or more carbons which the hydroxyl groups were attached and hyphenate those carbon numbers to the front of the alkane name and attach "diol" if two are involved, "triol" if three hydroxyl groups were involved.

Examples:

CH3
OH

| |

CH_3–CH–CH_2–CH_2–CH_2–CH_2–OH   C–C–C–C–OH
5-methyl-1-hexanol   1,2-butandiol

Physical Properties

Boiling Points

The boiling points of alcohols are in general much higher than comparably sized hydrocarbons. There are extra intermolecular forces between the alcohol molecules due to the hydrogen bonding.

There is no such hydrogen bonding between the hydrocarbon molecules because all the hydrogen atoms are bonded to carbon and not oxygen or nitrogen. Because there is no hydrogen bonding, the forces between the hydrocarbon molecules are much weaker and the molecules can be much more easily vapourized. As an example, the boiling point of ethanol, CH_3–CH_2–OH, is 78°C; ethane has a boiling point of –88.5°C.

Water Solubility

The alcohols are soluble in water because there is very strong attraction between alcohol molecules and the solvent water molecules. Hydrogen bonding is present in both and for that reason most alcohols are soluble in water. Alcohols also have a non-polar part of the molecule that tends to make them soluble in hydrocarbon solvents.

SUBSTITUTION REACTIONS

These are reactions in which one atom in a molecule is replaced by another atom or group of atoms. Free radical substitution involves breaking a carbon-hydrogen bond in alkanes such as:

- Methane (CH_4)
- Ethane(CH_3CH_3)
- Propane ($CH_3CH_2CH_3$)

A new bond is then formed to something else.

It also happens in alkyl groups like methyl, ethyl (and so on) wherever these appear in more complicated molecules.

- Methyl (CH_3)
- Ethyl (CH_3CH_2)

For example, ethanoic acid is CH_3COOH and contains a methyl group. The carbon-hydrogen bonds in the methyl group behave just like those in methane, and can be broken and replaced by something else in the same way. A simple example of substitution is the reaction between methane and chlorine in the presence of UV light (or sunlight).

$$CH_4 + Cl_2 \rightarrow CH_3Cl + HCl$$

Notice that one of the hydrogen atoms in the methane has been replaced by a chlorine atom. That's substitution.

Free Radical Reactions

Free radicals are atoms or groups of atoms which have a single unpaired electron. A free radical substitution reaction is one involving these radicals. Free radicals are formed if a bond splits evenly - each atom getting one of the two electrons. The name given to this is *homolytic fission*.

To show that a species (either an atom or a group of atoms) is a free radical, the symbol is written with a dot attached to show the unpaired electron.

For example:

- A chlorine radical (Cl)
- A methyl radical (CH_3)

Reaction Between Methane and Chlorine

Here we are going to discuss the mechanism for the free radical substitution reaction between methane and chlorine. If a mixture of methane and chlorine is exposed to a flame, it explodes - producing carbon and hydrogen chloride. This isn't a very useful reaction! The reaction we are going to explore is a more gentle one between methane and chlorine in the presence of ultraviolet light - typically sunlight. This is a good example of a *photochemical reaction* - a reaction brought about by light.

$$CH_4 + Cl_2 \rightarrow CH_3Cl + HCl$$

The organic product is chloromethane.

One of the hydrogen atoms in the methane has been replaced by a chlorine atom, so this is a substitution reaction. However, the reaction doesn't stop there, and all the hydrogens in the methane can in turn be replaced by chlorine atoms.

Chain Initiation

The chain is initiated (started) by UV light breaking a chlorine molecule into free radicals.

$$Cl_2 \rightarrow 2Cl$$

Chain Propagation Reactions

These are the reactions which keep the chain going.

$$CH_4 + Cl \rightarrow CH_3 + HCl$$

$$CH_3 + Cl_2 \rightarrow CH_3Cl + Cl$$

Chain Termination Reactions

These are reactions which remove free radicals from the system without replacing them by.new ones.

$$2Cl \rightarrow Cl_2$$
$$CH_3 + Cl \rightarrow CH_3Cl$$
$$CH_3 + CH_3 \rightarrow CH_3CH_3$$

REACTION BETWEEN METHANE AND BROMINE

This reaction between methane and bromine happens in the presence of ultraviolet light - typically sunlight. This is a good example of a *photochemical reaction* - a reaction brought about by light.

$$CH_4 + Br_2 \rightarrow CH_3Br + HBr$$

The organic product is bromomethane.

One of the hydrogen atoms in the methane has been replaced by a bromine atom, so this is a substitution reaction. However, the reaction doesn't stop there, and all the hydrogens in the methane can in turn be replaced by bromine atoms.

Mechanism

The mechanism involves a *chain reaction.* During a chain reaction, for every reactive species you start off with, a new one is generated at the end - and this keeps the process going.

The over-all process is known as free radical substitution, or as a free radical chain reaction.

Chain Initiation

The chain is initiated (started) by UV light breaking a bromine molecule into free radicals.

$$Br_2 \rightarrow 2Br$$

Chain Propagation Reactions

These are the reactions which keep the chain going.

$$CH_4 + Br \rightarrow CH_3 + HBr$$
$$CH_3 + Br_2 \rightarrow CH_3Br + Br$$

Chain Termination Reactions

These are reactions which remove free radicals from the system without replacing them by new ones.

$$2Br \rightarrow Br_2$$
$$CH_3 + Br \rightarrow CH_3Br$$
$$CH_3 + CH_3 \rightarrow CH_3CH_3$$

ELECTROPHILIC ADDITION

Electrophiles

An electrophile is something which is attracted to electron-rich regions in other molecules or ions. Because it is attracted to a negative region, an electrophile must be something which carries either a full positive charge, or has a slight positive charge on it somewhere.

Ethene and the other alkenes are attacked by electrophiles. The electrophile is normally the slightly positive (δ+) end of a molecule like hydrogen bromide, HBr. Electrophiles are strongly attracted to the exposed electrons in the pi bond and reactions happen because of that initial attraction - as you will see shortly.

$$\overset{\delta+}{H} — \overset{\delta-}{Br}$$

One might wonder why fully positive ions like sodium, Na^+, don't react with ethene. Although these ions may well be attracted to the pi bond, there is no possibility of the process going any further to form bonds between sodium and carbon, because sodium forms ionic bonds, whereas carbon normally forms covalent ones.

Addition Reactions

In a sense, the pi bond is an unnecessary bond. The structure would hold together perfectly well with a single bond rather than a double bond. The pi bond often breaks and the electrons in it are used to join other atoms (or groups of atoms) onto the ethene molecule. In other words, ethene undergoes addition reactions.

For example, using a general molecule X-Y

$$H_2C{=}CH_2 + X{-}Y \longrightarrow H{-}CH(X){-}CH(Y){-}H$$

Electrophilic Addition Reactions

An addition reaction is a reaction in which two molecules join together to make a bigger one. Nothing is lost in the process. All the atoms in the original molecules are found in the bigger one. An electrophilic addition reaction is an addition reaction which happens because what we think of as the "important" molecule is attacked by an electrophile.

The "important" molecule has a region of high electron density which is attacked by something carrying some degree of positive charge. Electrophilic addition reaction occurs in many of the reactions of compounds containing carbon-carbon double bonds - the alkenes.

The Structure of Ethane

Lets start by looking at ethene, because it is the simplest molecule containing a carbon-carbon double bond. What is true of C=C in ethene will be equally true of C=C in more complicated alkenes.

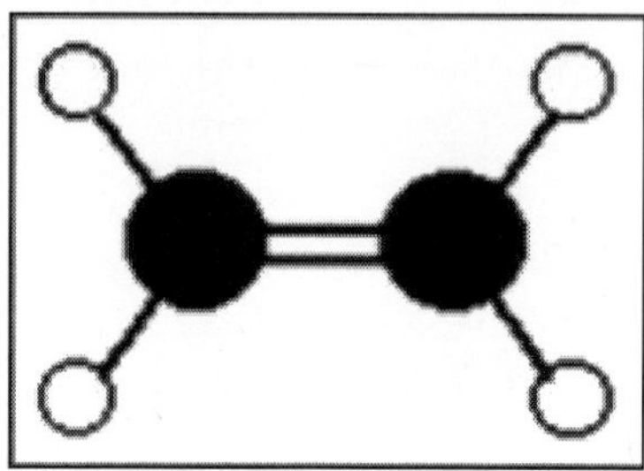

.. Ethene, C_2H_4, is often modeled as shown above. The double bond between the carbon atoms is, of course, two pairs of shared electrons. What the diagram doesn't show is that the two pairs aren't the same as each other. One of the pairs of electrons is held on the line between the two carbon nuclei as one would expect, but the other is held in a molecular orbital above and below the plane of the molecule. A molecular orbital is a region of space within the molecule where there is a high probability of finding a particular pair of electrons.

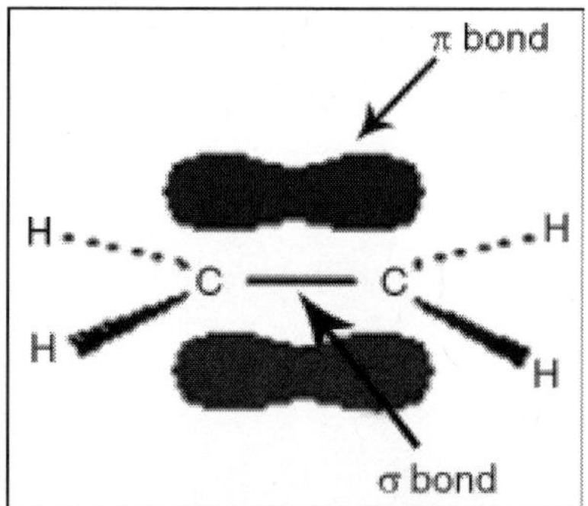

In this diagram, the line between the two carbon atoms represents a normal bond - the pair of shared electrons lies in a molecular orbital on the line between the two nuclei where you would expect them to be. This sort of bond is called a sigma bond.

The other pair of electrons is found somewhere in the shaded part above and below the plane of the molecule. This bond is called a pi bond. The electrons in the pi bond are free to move around *anywhere* in this shaded region and can move freely from one half to the other.

The pi electrons are not as fully under the control of the carbon nuclei as the electrons in the sigma bond and, because they lie exposed above and below the rest of the molecule, they are relatively open to attack by other things.

Electrophilic Addition Mechanism

The Mechanism for.the Reaction Between Ethene and a Molecule X-Y

It is very unlikely that any two different atoms joined together will have the same electronegativity. We are going to assume that Y is more electronegative than X, so that the pair of electrons is pulled slightly towards the Y end of the bond. That means that the X atom carries a slight positive charge.

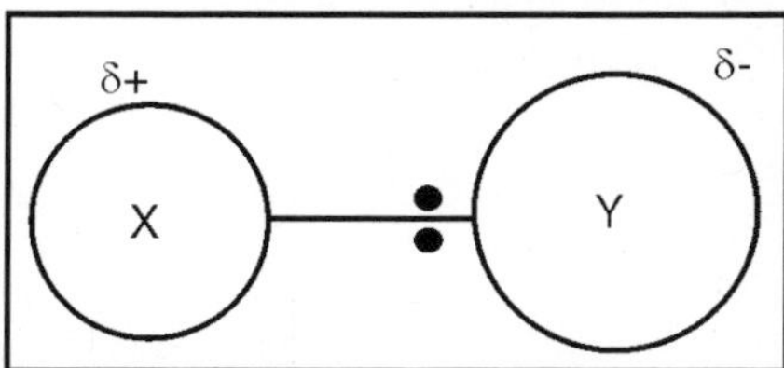

The slightly positive X atom is an electrophile and is attracted to the exposed pi bond in the ethene. Now imagine what happens as they approach each other.

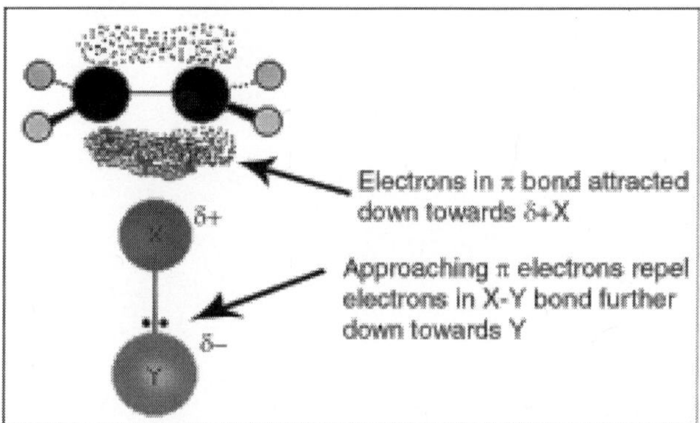

One are now much more likely to find the electrons in the half of the pi bond nearest the XY. As the process continues, the two electrons in the pi bond move even further towards the X until a covalent bond is made. The electrons in the X-Y bond are pushed entirely onto the Y to give a negative Y^- ion.

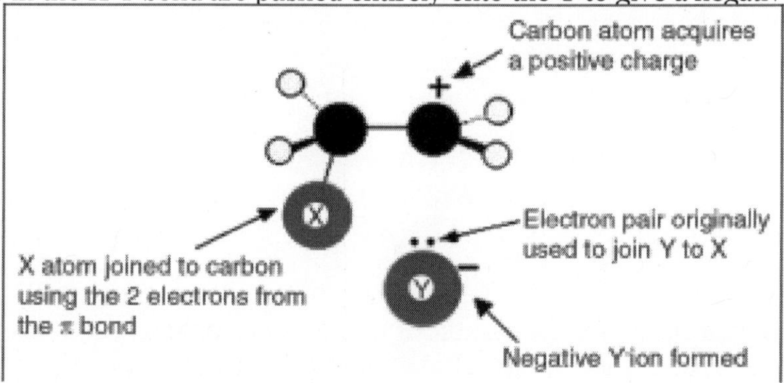

An ion in which the positive charge is carried on a carbon atom is called a *carbocation* or a *carbonium ion* (an older term). In the final stage of the reaction the electrons in the lone pair on the Y^- ion are strongly attracted towards the positive carbon atom.

They move towards it and form a co-ordinate (dative covalent) bond between the Y and the carbon.

REACTION BETWEEN SYMMETRICAL ALKENES AND SULPHURIC ACID

The Electrophilic Addition Reaction between Ethene and Sulphuric Acid.

Alkenes react with concentrated sulphuric acid in the cold to produce alkyl hydrogensulphates. Ethene reacts to give ethyl hydrogensulphate.

$$CH_2 = CH_2 + H_2SO_4 \rightarrow CH_3CH_2OSO_2OH$$

The structure of the product molecule is sometimes written as $CH_3CH_2HSO_4$, but the version in the equation is better because it shows how all the atoms are linked up.

Sulphuric Acid as an Electrophile

The hydrogen atoms are attached to very electronegative oxygen atoms which means that the hydrogens will have a slight positive charge while the oxygens will be slightly negative.

In the mechanism, we just focus on one of the hydrogen to oxygen bonds, because the other one is too far from the carbon-carbon double bond to be involved in any way.

The Mechanism for the Reaction between Cyclohexene and Sulphuric Acid

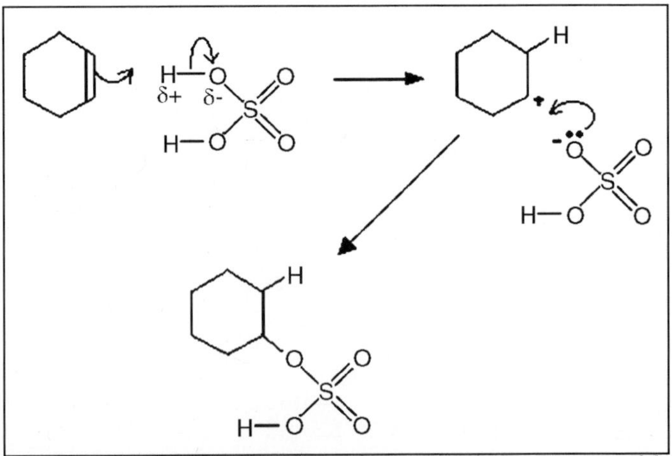

Having worked out the structure of the product, one could now write a simple equation for the reaction if one wants to.

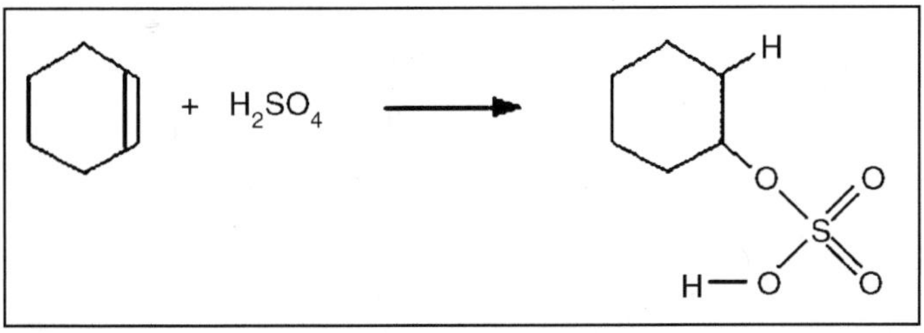

The Mechanism

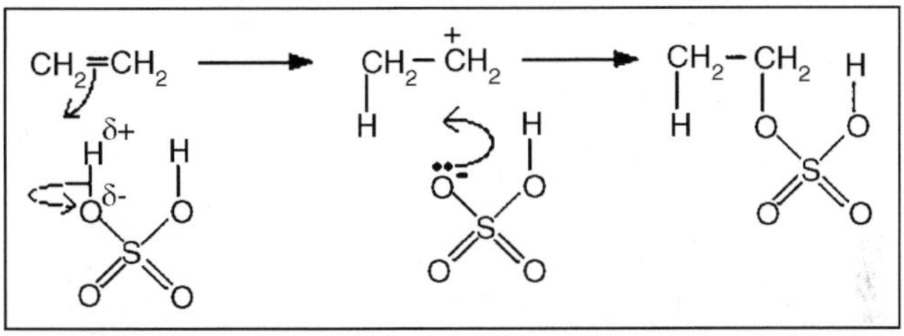

Reaction between Symmetrical Alkenes and the Hydrogen Halides

Hydrogen halides include hydrogen chloride and hydrogen bromide.

Electrophilic Addition Reactions Involving Hydrogen Bromide.

Alkenes react with hydrogen bromide in the cold. The double bond breaks and a hydrogen atom ends up attached to one of the carbons and a bromine atom to the other.

$$CH_2{=}CH_2 + HBr \rightarrow CH_3CH_2Br$$

With Cyclohexene we get Bromocyclohexane

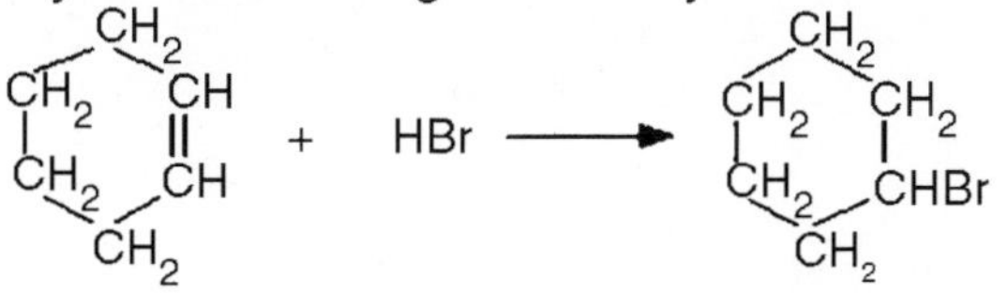

The structures of the cyclohexene and the bromocyclohexane are often

In the case of ethene, bromoethane is formed. With cyclohexene we get bromocyclohexane.The structures of the cyclohexene and the bromocyclohexane are often simplified.

Mechanism

The reactions are examples of *electrophilic addition.* With ethene and HBr: and with cyclohexene:

$CH_2=CH_2$ + $H^{\delta+}-Br^{\delta-}$ → $CH_2(H)-\overset{+}{C}H_2$ + Br^- → $CH_2(H)-CH_2(Br)$

And with cyclohexene:

cyclohexene + $\underset{\delta+}{H}-\underset{\delta-}{Br}$ → carbocation (H, +) + $:Br^-$ → product (H, Br)

Electrophilic Addition Reactions Involving the other Hydrogen Halides

Hydrogen chloride and the other hydrogen halides add on in exactly the same way. For example, hydrogen chloride adds to ethene to make chloroethane:

$$CH_2 = CH_2 + HCl \rightarrow CH_3CH_2Cl$$

The only difference is in how fast the reactions happen with the different hydrogen halides. The rate of reaction increases as you go from HF to HCl to HBr to HI.

HF	slowest reaction
HCl	
HBr	
HI	fastest reaction

The reason for this is that as the halogen atoms get bigger, the strength of the hydrogen-halogen bond falls. Bond strengths (measured in kilojoules per mole) are:

H-F	568
H-Cl	432
H-Br	366
H-I	298

As you have seen in the HBr case, in the first step of the mechanism the hydrogen-halogen bond gets broken. If the bond is weaker, it will break more readily and so the reaction is more likely to happen.

Mechanism

The reactions are still examples of *electrophilic addition*.

With ethene and HCl, for example:

$$CH_2=CH_2 \; (H^{\delta+}-Cl^{\delta-}) \longrightarrow CH_2(H)-\overset{+}{C}H_2 \; (Cl^-) \longrightarrow CH_2(H)-CH_2(Cl)$$

This is exactly the same as the mechanism for the reaction between ethene and HBr, except that we've replaced Br by Cl. All the other mechanisms for symmetrical alkenes and the hydrogen halides would be done in the same way.

ORGANIC CHEMISTRY

Organic chemistry is a discipline within chemistry which involves the scientific study of the structure, properties, composition, reactions, and preparation (by synthesis or by other means) of hydrocarbons and their derivatives. These compounds may contain any number of other elements, including hydrogen, nitrogen, oxygen, the halogens as well as phosphorus, silicon and sulfur..

Because of their unique properties, multi-carbon compounds exhibit extremely large variety and the range of application of organic compounds is enormous. They form the basis of, or are important constituents of many products (paints, plastics, food, explosives, drugs, petrochemicals, to name but a few) and (apart from a very few exceptions) they form the basis of all earthly life processes. Organic chemistry, like all areas of science, evolves with particular waves of innovation. These innovations are motivated by practical considerations as well as theoretical innovations.

The area is however underpinned financially by the very large applications in polymer science, pharmaceutical chemistry, and agrichemicals.

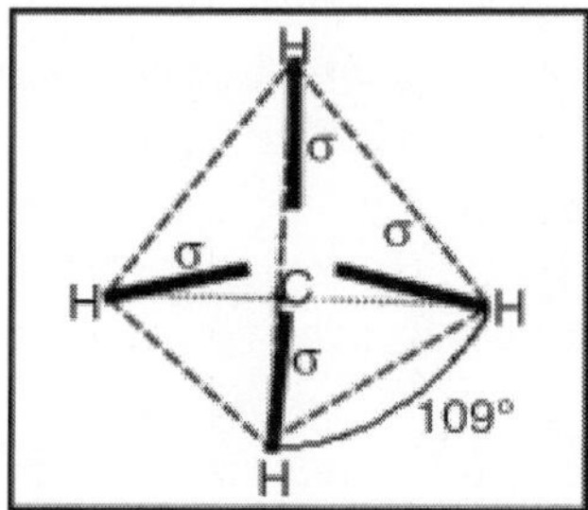

Fig. Structure of the Methane Molecule: the Simplest Hydrocarbon Compound.

PROPERTIES

Physical properties of organic compounds typically of interest include both quantitative and qualitative features. Quantitative information include melting point, boiling point, and index of refraction. Qualitative properties include odour, solubility, and colour.

Melting and Boiling Properties

In contrast to many inorganic materials, typical organic compounds typically melt and boil. In earlier times, the melting point (m.p.) and boiling point (b.p.) provided crucial information on the purity and identity of organic compounds.

The melting and boiling points correlate with the polarity of the molecules and their molecular weight.

Some organic compounds, especially symmetrical ones, sublime, that is they evaporate without melting.

A well known example of a sublimable organic compound is para-dichlorobenzene, the odiferous constituent of mothballs. Organic compounds are usually not very stable at temperatures above 300 °C, although some exceptions exist.

Colour

Organic compounds are typically colorless or white. The situation is quite different for organic compounds that contain several adjacent multiple bonds. These compounds, where the double bonds are "conjugated" can be deeply colored.

The biological pigments carotene and heme illustrate the relationship between "conjugation" and colour. Impure organic compounds, as well as many biological materials, often are yellow or brownish owing to the presence of trace amounts of intensely colored impurities.

Solubility

Neutral organic compounds tend to be hydrophobic, that is they are less soluble in water than in organic solvents. Exceptions include organic compounds that contain ionizable groups as well as low molecular weight alcohols, amines, and carboxylic acids where hydrogen bonding occurs.

Organic compounds tend to dissolve in organic solvents. Solvents can be either pure substances like ether or ethyl alcohol, or mixtures, such as the paraffinic solvents such as the various petroleum ethers and white spirits, or the range of pure or mixed aromatic solvents obtained from petroleum or tar fractions by physical separation or by chemical conversion. Solubility in the different solvents depends upon the solvent type and on the functional groups if present.

Solid State Properties

Various specialized properties are of interest depending on applications, *e.g.* thermo-mechanical and electro-mechanical such as piezoelectricity, electrical conductivity and electro-optical (*e.g.* non-linear optics) properties. For historical reasons, such properties are mainly the subjects of the areas of polymer science and materials science.

ORGANIC MOLECULES

To understand life as we know it, we must first understand a little bit of organic chemistry. Organic molecules contain both carbon and hydrogen. Though many organic chemicals also contain other elements, it is the carbon-hydrogen bond that defines them as organic.

Organic chemistry defines life. Just as there are millions of different types of living organisms on this planet, there are millions of different organic molecules, each with different chemical and physical properties.

There are organic chemicals that make up your hair, your skin, your fingernails, and so on. The diversity of organic chemicals is due to the versatility of the carbon atom. Why is carbon such a special element? Let's look at its chemistry in a little more detail.

Carbon (C) appears in the second row of the periodic table and has four bonding electrons in its valence shell. Similar to other non-metals, carbon needs eight electrons to satisfy its valence shell. Carbon therefore forms four bonds with other atoms (each bond consisting of one of carbon's electrons and one of the bonding atom's electrons).

Every valence electron participates in bonding, thus a carbon atom's bonds will be distributed evenly over the atom's surface. These bonds form a tetrahedron (a pyramid with a spike at the top), as illustrated below:

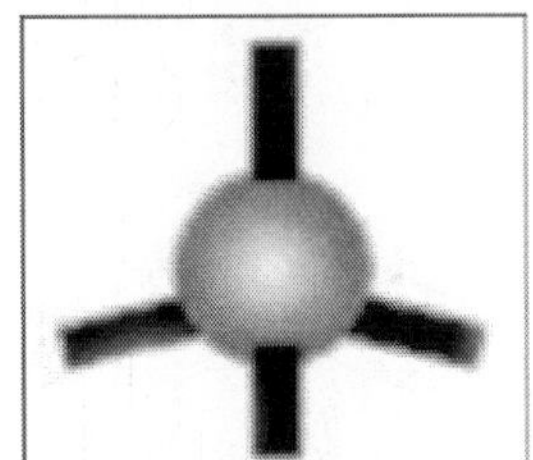

Fig. Carbon Forms 4 Bonds.

Organic chemicals get their diversity from the many different ways carbon can bond to other atoms. The simplest organic chemicals, called hydrocarbons, contain only carbon and hydrogen atoms;The simplest hydrocarbon (called methane) contains a single carbon atom bonded to four hydrogen atoms:

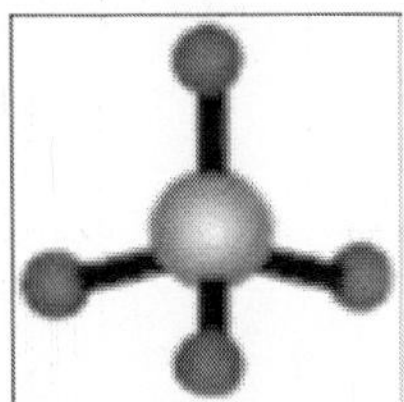

Fig. Methane - A Carbon Atom Bonded to 4 Hydrogen Atoms.

But carbon can bond to other carbon atoms in addition to hydrogen, as illustrated in the molecule ethane below:

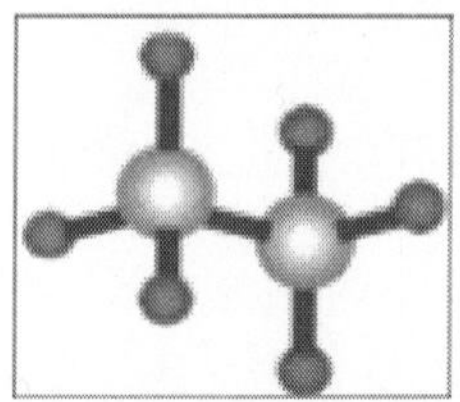

Fig. Ethane - A Carbon-Carbon Bond.

In fact, the uniqueness of carbon comes from the fact that it can bond to itself in many different ways. Carbon atoms can form long chains:

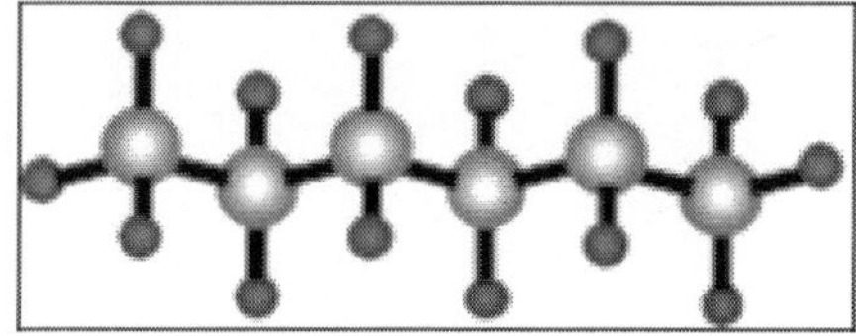

Fig. Hexane - a 6-Carbon Chain.

Branched Chains:

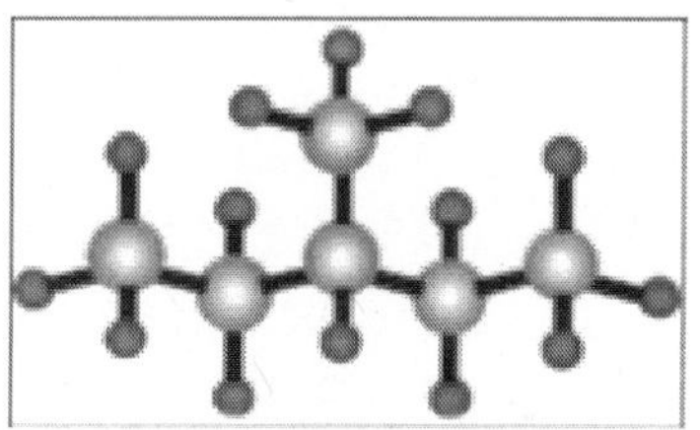

Fig. Isohexane - a Branched-Carbon Chain.

Rings:

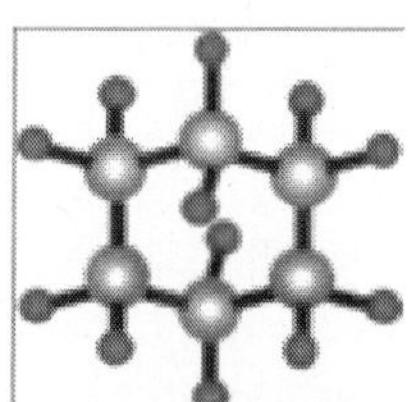

Fig. Cyclohexane - a Ringed Hydrocarbon.

There appears to be almost no limit to the number of different structures that carbon can form. To add to the complexity of organic chemistry, neighboring carbon atoms can form double and triple bonds in addition to single carbon-carbon bonds:

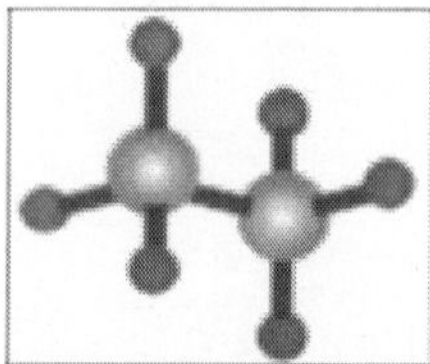

Fig. Single Bonding .

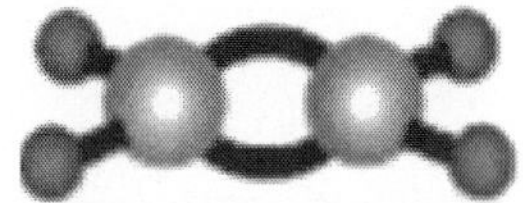

Fig. Double Bonding.

Fig. Triple Bonding

Keep in mind that each carbon atom forms four bonds. As the number of bonds between any two carbon atoms increases, the number of hydrogen atoms in the molecule decreases.

SIMPLE HYDROCARBONS

The simplest hydrocarbons are those that contain only carbon and hydrogen. These simple hydrocarbons come in three varieties depending on the type of carbon-carbon bonds that occur in the molecule. Alkanes are the first class of simple hydrocarbons and contain only carbon-carbon single bonds.

The alkanes are named by combining a prefix that describes the number of carbon atoms in the molecule with the root ending "ane". The names and prefixes for the first ten alkanes are given in the following table.

CarbonAtoms	Prefix	Alkane Name	Chemical Formula	Structural Formula
1	**Meth**	**Methane**	CH_4	CH_4
2	**Eth**	**Ethane**	C_2H_6	CH_3CH
3	**Prop**	**Propane**	C_3H_8	$CH_3CH_2CH_3$
4	**But**	**Butane**	C_4H_{10}	$CH_3CH_2CH_2CH_3$
5	**Pent**	**Pentane**	C_5H_{12}	$CH_3CH_2CH_2CH_2CH_3$
6	**Hex**	**Hexane**	C_6H_{14}	...
7	**Hept**	**Heptane**	C_7H_{16}	
8	**Oct**	**Octane**	C_8H_{18}	
9	**Non**	**Nonane**	C_9H_{20}	
10	**Dec**	**Decane**	$C_{10}H_{22}$	

The chemical formula for any alkane is given by the expression C_nH_{2n+2}. The structural formula, shown for the first five alkanes in the table, shows each carbon atom and the elements that are attached to it. This structural formula is important when we begin to discuss more complex hydrocarbons. The simple alkanes share many properties in common.

All enter into combustion reactions with oxygen to produce carbon dioxide and water vapour. In other words, many alkanes are flammable. This makes them good fuels. For example, methane is the principle component of natural gas, and butane is common lighter fluid.

$$CH_4 + 2O_2 \rightarrow CO_2 + 2H_2O$$

The Combustion of Methane

The second class of simple hydrocarbons, the alkenes, consists of molecules that contain at least one double-bonded carbon pair. Alkenes follow the same naming convention used for alkanes. A prefix (to describe the number of carbon atoms) is combined with the ending "ene" to denote an alkene. Ethene, for example is the two- carbon molecule that contains one double bond. The chemical formula for the simple alkenes follows the expression C_nH_{2n}. Because one of the carbon pairs is double bonded, simple alkenes have two fewer hydrogen atoms than alkanes.

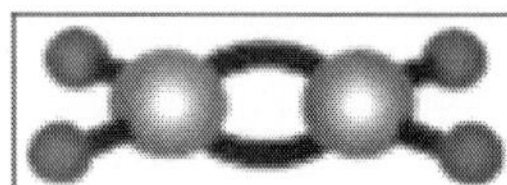

Fig. Ethene

Alkynes are the third class of simple hydrocarbons and are molecules that contain at least one triple-bonded carbon pair. Like the alkanes and alkenes, alkynes are named by combining a prefix with the ending "yne" to denote the triple bond. The chemical formula for the simple alkynes follows the expression C_nH_{2n-2}.

Fig. Ethyne

ISOMERS

Because carbon can bond in so many different ways, a single molecule can have different bonding configurations. Consider the two molecules illustrated in figure.

Isomers of C_6H_{14} are:

- $CH_3CH_2CH_2CH_2CH_2CH_3$

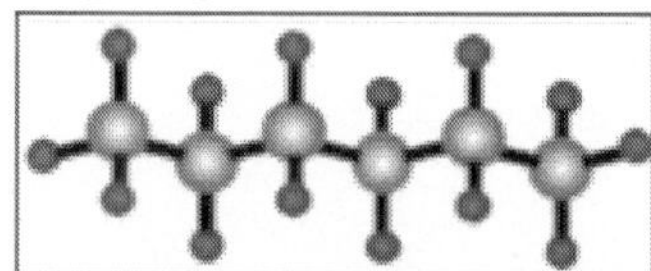

Fig. Molecules Have Identical Chemical Formulas but Different Structural Formulas.

- $CH_3CH_2CH(CH_3)CH_2CH_3$

$$\begin{array}{c} \quad\; CH_3 \\ \quad\; | \\ CH_3CH_2CHCH_2CH_3 \end{array}$$

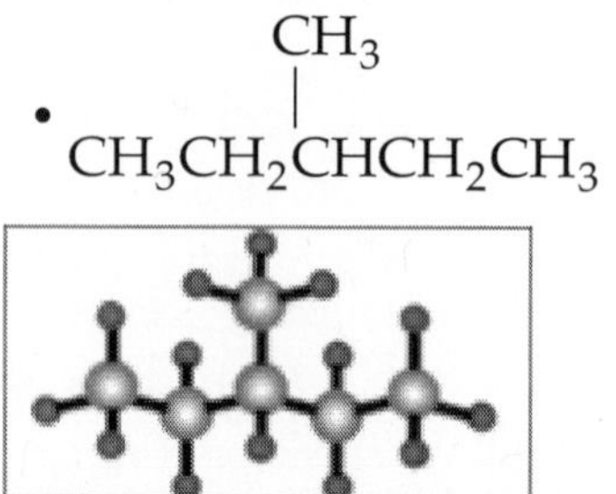

Fig. Molecules Have Identical Chemical Formulas but Different Structural Formulas.

Both molecules have identical chemical formulas; however, their structural formulas (and thus some chemical properties) are different. These two molecules are called isomers. Isomers are molecules that have the same chemical formula but different structural formulas.

FUNCTIONAL GROUPS

In addition to carbon and hydrogen, hydrocarbons can also contain other elements. In fact, many common groups of atoms can occur within organic molecules, these groups of atoms are called functional groups. One good example is the hydroxyl functional group.

The hydroxyl group consists of a single oxygen atom bound to a single hydrogen atom (-OH). The group of hydrocarbons that contain a hydroxyl functional group is called alcohols. The alcohols are named in a similar fashion to the simple hydrocarbons, a prefix is attached to a root ending (in this case "anol") that designates the alcohol.

The existence of the functional group completely changes the chemical properties of the molecule. Ethane, the two-carbon alkane, is a gas at room temperature; ethanol, the two-carbon alcohol, is a liquid. Ethanol, common drinking alcohol, is the active ingredient in "alcoholic" beverages such as beer and wine.

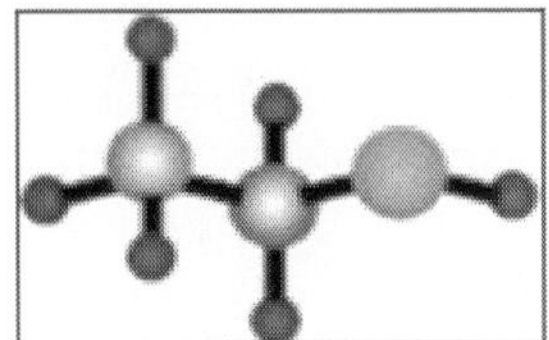

Fig. Ethanol

TOXICITY

Living matter including food, drugs are collections organic compounds, so the potential beneficial and harmful aspects of organic compounds span the entire range required for life to some of the most dangerous materials known.

Indicative of the power of organic chemistry, the biomolecule called botulism toxin ("Botox") is lethal at the level of less than a microgram.

THE NATURE OF ORGANIC CHEMISTRY

There are over six million organic compounds characterized, including the foods we eat, (made of carbohydrates, lipids, proteins, and vitamins), furs and feathers, hides and skins, and the organisms they came from. not to mention plastics, synthetic and natural fibres, dyes and drugs, insecticides and herbicides, ingredients in perfumes and flavoring agents, and petroleum products.

The name *organic* chemistry came from the word organism. Prior to 1828, all organic compounds had been obtained from organisms or their remains. The belief then was that the synthesis of organic compounds from inorganic compounds in the laboratory was impossible. All efforts had failed, and scientists became convinced that some "vital force" that living organisms had was necessary to make an organic compound. The synthesis of urea from inorganic substances in 1828 led to the disappearance of this *vital force theory*.

THE UNIQUENESS OF CARBON

The great number of carbon compounds is possible because of the ability of carbon to form stron covalent bonds to eath other *while also holding the atoms of other nonmetals strongly*. Chains of carbon atoms can be thousands of atoms long, as in polyethylene.

Polyethylene chain:

```
     H  H  H  H  H  H  H  H  H  H  H
     |  |  |  |  |  |  |  |  |  |  |
H -  C- C- C- C- C- C- C- C- C- C- C-  etc.
     |  |  |  |  |  |  |  |  |  |  |
     H  H  H  H  H  H  H  H  H  H  H
```

Isomers

Isomerism is another reason why there are so many organic compounds. Isomers are compounds with identical molecular composition but their structures are arranged differently. Depending on how they are arranged, they may have similar or different properties. Because the number of carbons per molecule increases as the compound gets more complex, the number of possible isomers for any given formula becomes very, very large.

Formula	Number of Isomers
C_8H_{18}	**18**
$C_{10}H_{22}$	**75**
$C_{20}H_{42}$	**366,319**
$C_{40}H_{82}$	**6.25 x 10^{13} (approx.)**

Examples of isomers:

```
     H  H  H  H
     |  |  |  |
H -  C- C- C- C- H
     |  |  |  |
     H  H  H  H
```

Butane

```
          H
          |
  H  H -  C - H  H
  |       |      |
H-C-------C------C-H
  |       |      |
  H       H      H
```

2-Methyl Propane

Butane and 2-methyl propane both have the molecular formul.a C_4H_{10}.

```
        H   H
        |   |
    H - C - C - O - H
        |   |
        H   H
```

Ethanol

```
        H       H
        |       |
    H - C - O - C - H
        |       |
        H       H
```

Dimethyl Ether

Ethanol and dimethyl ether both have the molecular formula C_2H_6O.

ORGANIC FAMILIES AND THEIR FUNCTIONAL GROUPS

Such a huge number of organic compounds requires organization. They are sorted into *organic families* defined by *functional groups*. Functional groups are small structural units within molecules at which most of the compound's chemical reactions occur.

For example, two of the most important families are the *alcohols* and the *carboxylic acids*. Their functional groups, the alcohol group and the carboxyl group, respectively, distinguishes them from the rest of the other types of organic compounds.

```
                      H
                      |
                  H — C — O — H
                      |
— O — H               H
Alcohol Group     Methanol (methyl alcohol)

                      H   O
                      |   ||
    O             H — C — C — O — H
    ||                |
— C — O — H           H
Carboxyl Group    Acetic Acid (ethanoic acid )
```

IMPORTANT FAMILIES OF ORGANIC COMPOUNDS

- Alkanes:.Only single bonds
- Alkenes: Double bond(s) between two carbons
- Alkynes: Triple bond(s) between two carbons
- Alcohols:

R — O — H

- Ethers:

R — O — R'

- Aldehydes:

$$R-\overset{\displaystyle O}{\overset{\|}{C}}-H$$

- Ketone:

$$R-\overset{\displaystyle O}{\overset{\|}{C}}-\overset{\displaystyle R''(H)}{\overset{|}{N}}-R'(H)$$

- Carboxylic Acids:

$$R-\overset{\displaystyle O}{\overset{\|}{C}}-O-H$$

- Esters:

$$R-\overset{\displaystyle O}{\overset{\|}{C}}-O-R'$$

- Amines:

$$R-\overset{\displaystyle R''(H)}{\overset{\|}{N}}-O-R'(H)$$

- Amides:

$$R-\overset{\displaystyle O}{\overset{\|}{C}}-\overset{\displaystyle R''(H)}{\overset{|}{N}}-R'(H)$$

R, R', and R" represent hydrocarbon groups (*alkyl groups*)

The type of properties associated with a compound is different depending on the family it is in. For example, the alkanes have just C–C and C–H single bonds.

Since C and H are so alike in electronegativity, they are least able to attract ions or polar molecules, and least able to interact with them. The significance of this will be explained later.

CONDENSED STRUCTURES

To save space and time, condensed structures may be used to simplify the drawing and writing of structural formulas of organic compounds. C-H bonds are understood, and CH_3 means that three hydrogen atoms are bonded to a carbon atom.

For example: CH_3 is understood to be

```
    H
    |
  H-C -
    |
    H
```

CH_3CH_2OH is understood to be:

```
     H    H
     |    |
H-  C-   C-  O- H
     |    |
     H    H
```

CH_3OCH_2 is understood to be:

```
     H    H
     |    |
H-  C-   O-  C- H
     |    |
     H    H
```

Functional Groups and Polar Reactants

When a polar group of atoms, like the OH group or NH group are attached to a carbon, the molecule has a polar site. It may undergo chemical reactions when it attracts polar and ionic reactants, but it will usually be near this functional group. This is why compounds with the same functional group have similar chemical properties, or kinds of reactions. For example, in the amine group, both methylamine and ethylamine give the same kinds of reactions.

$$CH_3NH_2 + HCl \rightarrow CH_3NH_3^+ + Cl^-$$

methylamine + hydrogen chloride → methylammonium ion + chloride ion

$$CH_3CH_2NH_2 + HCl \rightarrow CH_3CH_2NH_3^+ + Cl^-$$

ethylamine + hydrogen chloride → ethylammonium ion + chloride ion

Only the NH_2 group of the amines change in the previous two reactions, and it changes the same way in both.

The Symbol R in Structural Formulas

The symbol R, stands for the word *radical*, and in organic chemistry, represents alkane-like groups. It is sometimes used to simplify equations or to summarize them. For example, we may write:

$$RNH_2 + HCl \rightarrow RHN_3 + Cl^-$$

The R takes care of any alkane group that you may substitute in. It takes care of all the equations, including methylamine, ethylamine, propylamine, butylamine, *etc.* It gives a simpler view of how one group, not just the individual compound, changes into another and how each group affects its properties.

ALDEHYDES AND KETONES

NAMING ALDEHYDES

The parent chain is the longest chain that includes the aldehyde group. It's name is made by replacing the ending *-e* of parent alkane name with *-al*. The numbering of the chain always starts with the carbon of the aldehyde group being 1. For example: (parent chain will be bold for visual clarity)

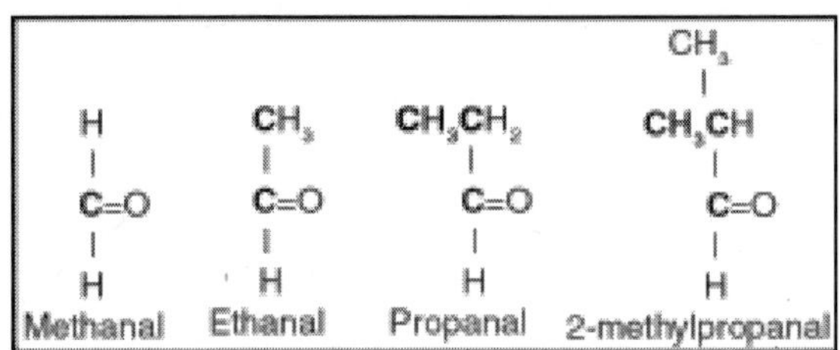

NAMING KETONES

In naming ketones, the parent chain includes the carbonyl group and is numbered so that the carbonyl location is the lowest number. The number of the location must be part of the name whenever there would be an uncertainty. For example: (parent chain will be bold for visual clarity

```
                                                  CH3
                                                   |
CH3                        CH3CH2              CH3CH
 |                            |                    |
C=O                          C=O                  C=O
 |                            |                    |
CH3                          CH3CH2CH3            CH2CH2CH3
Propanone (acetone)  3-Hexanone              2-Methyl-3-Hexanone
```

HYDROGENATION OF ALDEHYDES AND KETONES

Just like the double bond in alkenes, aldehydes and ketones also add hydrogen by breaking the double bond under roughly the same conditions—metal catalyst, heat, and pressure.

For example:

```
CH3          CH3                    CH3            CH3          OH
 |            |                      |              |            |
C=O + H-H → H-C-O-H or CH3CH2OH     C=O + H-H ==> H-C-O-H or CH3CHCH3
 |            |                      |              |
 H            H                     CH3            CH3
Ethanal + Hydrogen → Ethanol        propanone + hydrogen → 2-propanol
```

OXIDATION OF ALDEHYDES

Aldehydes are the most easily oxidized of all organic compounds, though ketones resist oxidation strongly. They can also be slowly oxidized by oxygen in the air. In this example, potassium dichromate, together with hydrogen sulfate, easily oxidizes ethanal to ethanoic *acid*:

$$\underset{\text{ethanal}}{3CH_3CHO} + K_2Cr_2O_7 + 4H_2SO_4 \rightarrow \underset{\text{ethanoic acid}}{3CH_3COOH} + Cr(SO_4)_3 + K_2SO_4 4H_2O$$

CARBOXYLIC ACIDS

NAMING CARBOXYLIC ACIDS

The parent chain must include the carboxyl carbon, which is given position number 1. The name of the alkane attached is changed by replacing the *-e* with *-oic acid*.

For example:

- $HCOOH \rightarrow$ Methanoic acid
- $CH_3COOH \rightarrow$ Ethanoic acid (acetic acid)
- $CH_3CH_2CH_2COOH \rightarrow$ Butanoic acid

Because carboxylic acids have both a lone oxygen and an OH group, they are strongly hydrogen-bonded to each other, therefore having high boiling points. The carboxyl group is weakly acidic and all carboxylic acids neutralize OH^-.

$$RCOOH + NaOH \rightarrow RCOONa + H_2O$$

Esters of Carboxylic Acids

Carboxylic acids are used to synthesize two important derivatives, *esters* and *amides*. In esters, the OH of the carboxyl group is replaced by OR. To name esters, first have the name of the akyl group attached to the O, followed by the name of the parent carboxylic acid after its name is altered by changing *-ic acid* to *-ate*.

For exmample: $CH_3COO(CH_2)_7CH_3$

Octyl ethanoate: Esters are responsible for or mimic many pleasant fragrances in nature.

Ester	Similar Aroma
$HCOOCH_2CH_3$	**Rum**
$HCOOCH_2CH(CH_3)_2$	**Raspberries**
$CH_3COO(CH_2)_4CH_3$	**Bananas**
$CH_3COO(CH_2)_2CH(CH_3)_2$	**Pears**
$CH_3COO(CH_2)_7CH_3$	**Oranges**
$CH_3(CH_2)_2COOCH_2CH_3$	**Pineapples**
$CH_3(CH_2)_2COO(CH_2)_4CH_3$	**Apricots**

In an mixture with carboxylic alcohol, an equilibrium reaction forms ester and water in the presence of heat and a strong acid catalyst.

$$RCOOH + HOR' \leftrightarrow RCOOR' + H_2O$$

For example:

$$CH_3CH_2CH_2COOH + HOCH_2CH_3 \ll CH_3CH_2CH_2COOCH_2CH_3 + H_2O$$

butanoic acid + ethanol ↔ ethyl butanoate + water

Hydrolysis and Saponification of Esters

An ester is hydrolyzed to its parent acid and alcohol when it is heated together with a stoichiometric excess of water (with acid catalyst)—this is the reverse of the reaction above.

Saponification, is the reaction of an ester with a strong base, like sodium hydroxide.

For example:

$$CH_3COOCH_2CH_3 + NaOH \rightarrow CH_3COO^- + Na^+ + HOCH_2CH_3$$

Ethyl Ethanoate + Sodium Hydroxide → Ethanoate Ion + Sodium Ion + Ethanol

Sodium Ethanoate Salt

The mixtures of salt of long-chain carboxylic acids are the ingredients of ordinary soap.

Amides of Carboxylic Acids

In amides, the OH of the carboxyl group is replaced by nitrogen holding any combination of H atoms or hydrocarbon groups.

The names of simple amides are devised by writing the name of the parent carboxylic acid and replacing the ending, *-oic acid*, with *-amide For example*:

$$CH_3CONH_2$$

ethanamide (acetamide)

The preparation of simple amides are similar to those of esters, but they are in an excess of ammonia.

For example:

$$RCOOH + NH_3 \rightarrow RCONH_2 + H_2O$$

carboxylic acid + ammonia → simple amide + water

When amides are in water, they are changed back to the carboxylic acid and ammonia. Unlike amines, the amides are non-basic even though they have the NH_2 group in simple amides.

This is because of the O atom in the carbonyl group, which is very electronegative. This tightens the electrons on N so that it is unable to accept a proton.

ALDEHYDES, KETONES, THEIR DERIVATIVES AND ANALOGUES

ALDEHYDES, THIOALDEHYDES AND THEIR ANALOGUES

The generic term "aldehyde" refers to compounds containing a –CHO group attached to a carbon atom. Aldehydes corresponding to carboxylic acids with trivial names are named by changing the "-ic acid" or "–ioc acid" ending of the acid name to "-aldehyde". Acyclic mono- and dialdehydes are named by adding the suffixes "–al" or "c–dial" to the name of the acyclic hydrocarbon with the same number of carbon atoms, eliding the final "e" of the hydrocarbon name before "a".

Other aldehydes are named by adding the suffix "–carbaldehyde" to the name of a parent hydride. When a group having priority for citation as a principal characteristic group is present, an aldehyde group is described by the prefix "formyl–". In names of natural products, conversion of an implied CH_3 group to an aldehyde is indicated by the prefix "oxo–".

Examples:

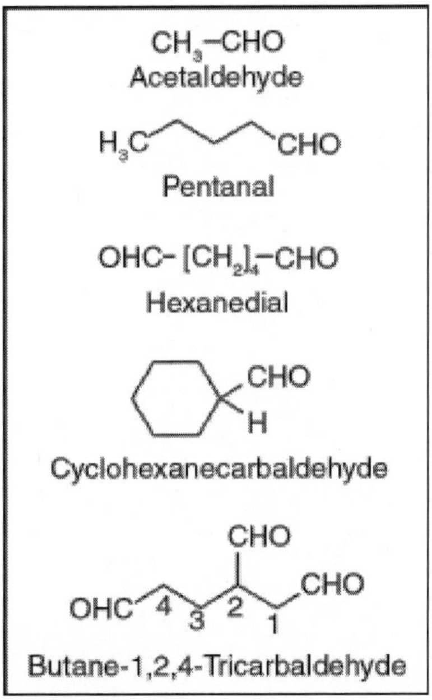

CH_3–CHO
Acetaldehyde

H_3C–CH_2–CH_2–CH_2–CHO
Pentanal

OHC–$[CH_2]_4$–CHO
Hexanedial

CHO, H
Cyclohexanecarbaldehyde

CHO, CHO, OHC, 4, 3, 2, 1
Butane-1,2,4-Tricarbaldehyde

Chalcogen analogues of aldehydes are named using suffixes such as "-thial", "-selenal", "-carbothialdehyde", and "-carboselenaldehyde" and prefixes such as "thioformyl-" and "thioxo-". The use of the prefixes "thio-", "seleno-", and "telluro-" with trivial names of aldehydes such as acetaldehyde to indicate replacement of an aldehydic oxygen atom is not included in these recommendations.

Examples:

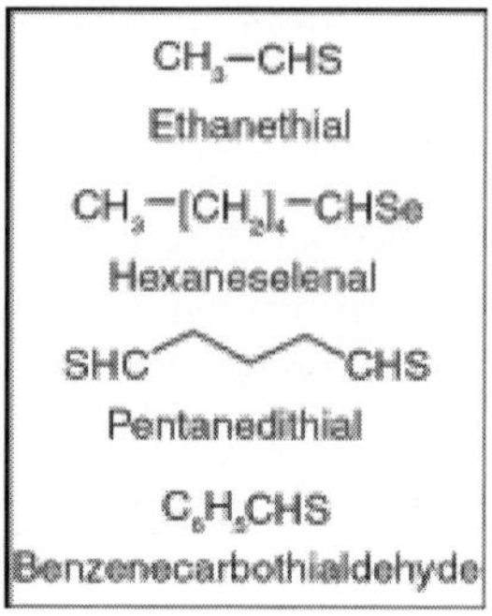

$CH_3{-}CHS$
Ethanethial

$CH_3{-}[CH_2]_4{-}CHSe$
Hexaneselenal

SHC~~~CHS
Pentanedithial

C_6H_5CHS
Benzenecarbothialdehyde

Ketones, Thioketones, and Their Analogues

Ketones

The generic term "ketone" refers to compounds containing a carbonyl group, >C=O, joined to two carbon atoms. Ketones are named substitutively by adding a suffix such as "-one", and "-dione" to the name of a parent hydride with elision of the final "e" of the parent hydride, if any, before "o".

When a group having priority for citation as principal characteristic group is present, a ketone is described by the prefix "oxo-". Functional class names for monoketones and vicinal diketones, *etc.*, are formed by citing the prefix names for the two groups attached to the carbonyl group(s) in alphabetical order followed by the class name "ketone", "diketone", *etc.*, as a separate word.

Examples:

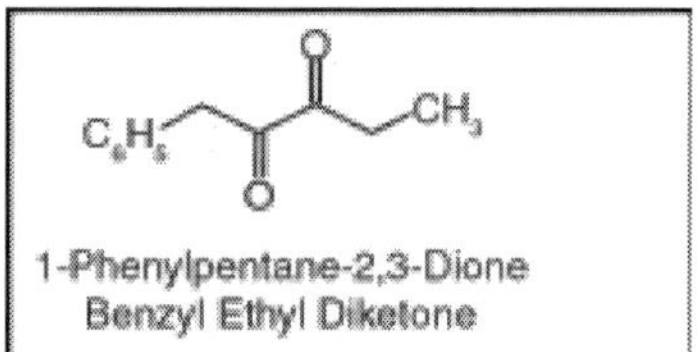

1-Phenylpentane-2,3-Dione
Benzyl Ethyl Diketone

Hexane-2,4-Dione

4-Oxocyxlohexane-1-Carboxylic Acid

Diketones derived from cyclic parent hydrides having the maximum number of non-cumulative double bonds by conversion of two –CH= groups into >CO groups with rearrangement of double bonds to a quinonoid structure may be named alternatively by adding the suffix "-quinone" to the name of the aromatic parent hydride.

Example:

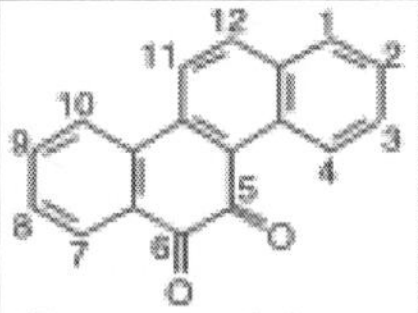

Chrysene-5,6-Quinone
Chrysene-5,6,dion

Acyl derivatives of benzene or naphthalene have been named by changing the "-ic acid" or "-oic acid" ending of a trivial name of the acid corresponding to the acyl group to "-ophenone" or "-onaphthone".

Only the names acetophenone, propiophenone, and benzophenone are retained in these recommendations.

Acyl derivatives of cyclic parent hydrides are named by prefixing the substituent name derived from the cyclic parent hydride to the name of the acyclic ketone.

Example:

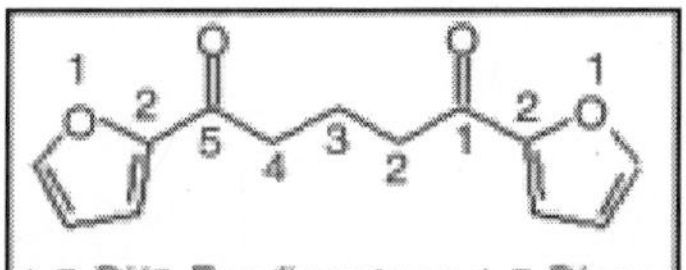

1,5-Di(2-Furyl)pentane-1,5-Dione

Chalcogen analogues of ketones

Chalcogen analogues of ketones are named by using suffixes such as "-thione" and "-selone", and prefix names such as "thioxo-" and "selenoxo-". The use of prefixes such as "thio-" and "seleno-" with the trivial names of ketones, such as acetone, to indicate replacement of the ketonic oxygen atom with a chalcogen atom is not recommended.

Examples:

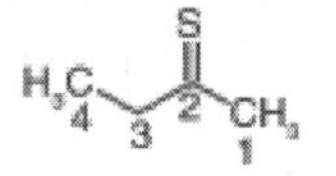

Butane-2-Thione

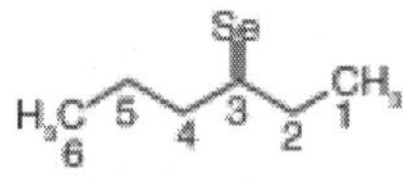

Hexane-3-Selone

KETENES

The compound $CH_2=C=O$ is a functional parent compound with the name "ketene".

Derivatives can be named by:

- Citing all substituents as prefixes
- Using the principles for naming ketones

Examples:

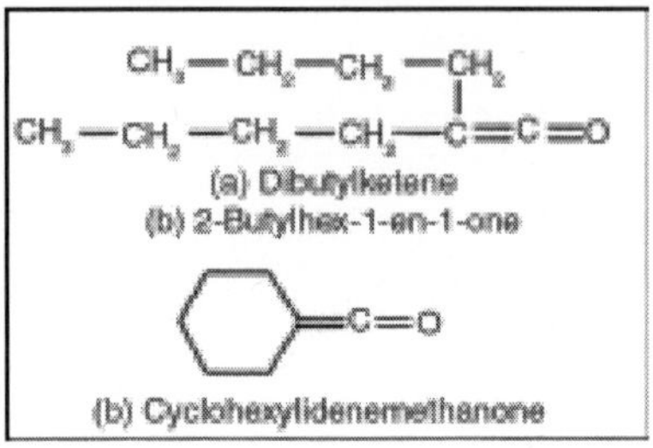

$CH_3-CH_2-CH_2-CH_2$
$CH_3-CH_2-CH_2-CH_2-C{=}C{=}O$

(a) Dibutylketene
(b) 2-Butylhex-1-en-1-one

(b) Cyclohexylidenemethanone

ACETALS, HEMIACETALS, ACYLALS AND THEIR ANALOGUES

Acetals

Compounds with the general structure RR'C(O—R") (O—R'"), where R and/or R' may be, but need not be, hydrogen, but R" and R'" cannot be hydrigen, are termed generically "acetals". "Ketals" constitute a subclass of acetals wherein neither R nor R' may be hydrogen. Acetals are named substitutively as "alkoxy-", "aryloxy-", *etc.*, derivatives of an appropriate parent hydride or functional parent compound. Alternatively, according to functional class nomenclature, the name of the appropriate aldehyde or ketone is followed by that of the *O*-substituent(s), which is in turn followed, after a space, by the word "acetal" or "ketal".

Example:

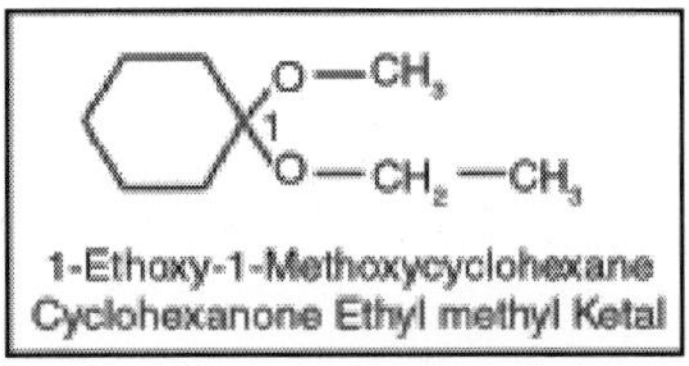

1-Ethoxy-1-Methoxycyclohexane
Cyclohexanone Ethyl methyl Ketal

Sulfur analogues of acetals and ketals with the general structures RR' C (S—R") (S—R'") or RR'C (O—R") (S—R'"), are termed generically "dithioacetals" or "monothioacetals", respectively. They are named substitutively as "alkylsulfanyl-", "arylsulfanyl-", "alkyloxy-" (or "alkoxy-"), or "aryloxy-", derivatives, as appropriate, of a parent hydride.

They may also be named by functional class nomenclature in the same way as acetals. Capital italic latter locants are used to provide structural specificity. Selenium and tellurium and mixed analogues are treated in the same way as their sulfur analogues; generically, they are "monoselenoacetals", "ditelluroacetals", "selenothioacetals", *etc.*, and are named substitutively using prefixes such as "alkylselanyl-", "aryltellanyl", *etc.*

Example:

CH_3-CH_2-S
$CH_3-CH_2-CH_2-CH_2-CH_2-S-CH_2-CH_3$
5 4 3 2 1
1,1-Bis(ethylsulfanyl)pentane
Pentanal Diethyl Dithioacetal

Hemiacetals

Compounds with the general structure are termed generically "hemiacetals". Hemiacetals are named substitutively as alkoxy-, aryloxy-, *etc.*, derivatives of an appropriate hydroxy parent compound, such as an alcohol, and by functional class nomenclature in the same way as acetals using the class name "hemiacetal".

Example:

CH_3-CH_2-O
$CH_3-CH_2-CH_2-CH-OH$
4 3 2 1
1-Ethoxybutan-1-ol
Butanal Ethyl Hemiacetal

Sulfur analogues of hemiacetals with general structures RR'C(SH)(S–R") and RR'C(OH)(S–R")or RR'C(SH)(O–R")are generically "dithiohemiacetals" or "monothiohemiacetals", respectively. They are named substitutively as "alkylsulfanyl-", "arylsulfanyl-", "alkyloxy-" (or "alkoxy-"), or "aryloxy-", derivatives, as appropriate, of a hydroxy parent compound, such as an alcohol, or of a thiol parent compound.

They may also be named by functional class nomenclature in the same way as hemiacetals. Selenium and tellurium and mixed analogues are treated in the same way as their sulfur analogues; generically, they are "monoselenohemiacetals", "ditellurohemiacetals", "selenothiohemia-cetals", *etc.*, and are named substitutively as derivatives of a hydroxy parent compound or a chalcogen analogue, such as a "thiol" or "selenol", using prefixes such as "alkylselanyl-" and "aryltellanyl-", *etc.*

Example:

CH_3-CH_2-S
$CH_3-CH_2-CH-SH$
3 2 1
1-(Ethylsulfanyl)Propane-1-Thiol
Propanal Ethyl Dithiohemiacetal

Acylals.3

Compounds with the general structure R–CH(OCO–R')$_2$, RR'C(OCO–R")$_2$, *etc.*, are generically called "acylas". Specific compounds are named as esters.

Example:

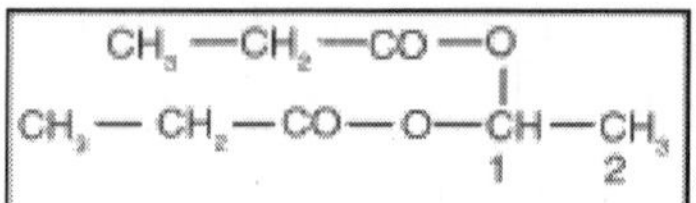

$CH_3—CH_2—CO—O$
$CH_3—CH_2—CO—O—CH—CH_3$ (1, 2)

Ethane-1,1-Diyl Dipropionate

ACYLOINS

α-Hydroxy ketones, RCH(OH)—CO—R, in which R is an alkyl, aryl, or heterocyclic group, are generically called "acyloins" and are named by substitutive nomenclature. Names ending in "-oin" are not included in these recommendations.

Examples:

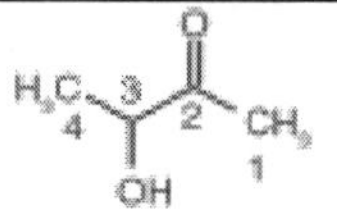

3-Hydroxybutan-2-One

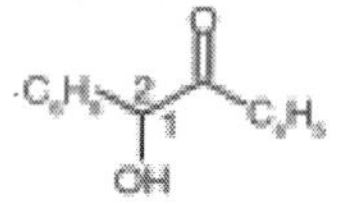

2-Hydroxy-1,2-Diphenylethan-1-One

Nitrogenous Derivatives of Carbonyl Compounds

Oximes

Compounds having the general structure R–CH=N–OH or RR'C=N–OH are called generically "oximes" and have been further classified as "aldoximes" and "ketoximes", respectively. They are named according to the principles of functional class nomenclature by placing the class name "oxime" as a separate word after the name of the aldehyde RCHO or the ketone RR'C=O, respectively; or substitutively by use of the prefix "hydroxyimino-" attached to the name of a parent hydride or parent substituent prefix. Compounds containing the group =N–OR may be named as *O*-substituted oximes or as alkoxy-substituted imines.

Examples:

Cinnamaldehyde Oxime

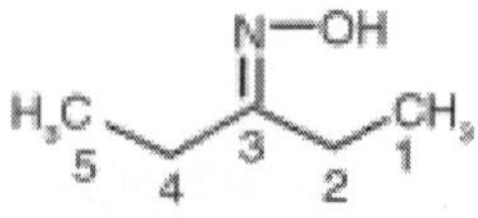

Pentan-3-One Oxime

Benzaldehyde Oxime

Hydrazones

Compounds having the general structure RCH=N–NH_2 or RR'C=N–NH_2 are called "hydrazones" generically and are named according to functional class nomenclature by placing the class name "hydrazone" as a separate word following the name of the corresponding aldehyde or ketone, or substitutively by means of the prefix "hydrazono-" attached to the name of an appropriate parent hydride or parent substitutive prefix.

Example:

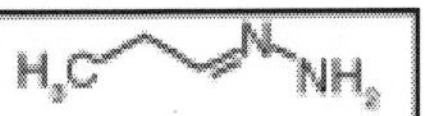

Propanal Hydrazone

Azines

Compounds having the general structure RCH=N–N=CHR or RR'C=N–N=CRR' are called "azines" generically and are named substitutively as derivatives of diazane or as an assembly of identical units using the prefix "azino-". According to functional class nomenclature, they may be named by adding the class name "azine" as a separate word after the name of the corresponding aldehyde or ketone.

Examples:

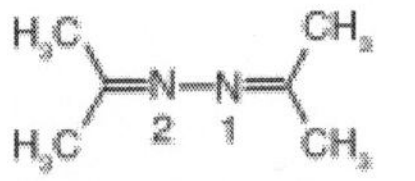

1,2-Diisopropylidenediazane
Acetone Azine

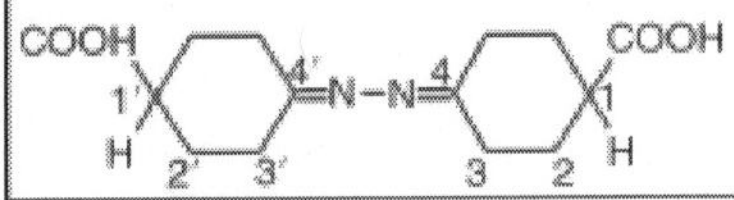

Compounds having the general structure R=N–R=R' (RR') are named as "ylidenehydrazones" of the preferred carbonyl compound or by using the prefix "hydrazono-".

Examples:

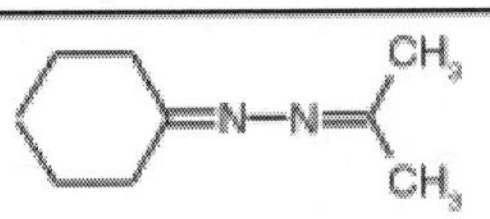

Cyclohexanone isopropylidene hydrazone

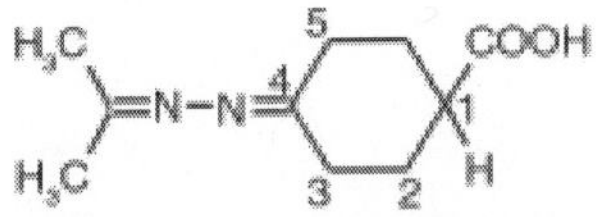

4-(Isopropylidenehydrazono)cyclohexa-2,5-diene-1-carboxylic Acid

Other Nitrogen Derivatives of Carbonyl Compounds

Derivatives of semicarbazideof the types NH_2—CO—NH—N=CH—R and NH_2—CO—NH—N=CRR'may be named substitutively by using the functional

$$\underset{4}{NH_2} - \underset{3}{CO} - \underset{2}{NH} - \underset{1}{NH_2}$$

parent compound name "semicarbazide", or the prefix name "semicarbazono-", or according to principles of functional class nomenclature, by adding the class name "semicarbazone" as a separate word after the name of the aldehyde RCHO or ketone RR'C=O.

Chalcogen analogues are named in the same way on the basis of functional parent compound names such as "selenosemicarbazide" and prefix names such as "thiosemicarbazono-", or functional class names such as "thiosemicarbazone".

11

Chemical Dynamics

Some Basic Ideas. Until now, our major concern has been whether or not a chemical reaction of interest may or may not occur under specified conditions of temperature, pressure, and concentration. The answer to this question is in the realm of thermodynamics. If thermodynamics tells us that a reaction may occur, then we must be concerned with how rapidly and by what pathway it will occur, and how we may influence this. The speed (rate) at which a reaction takes place is a matter of chemical dynamics, or alternately, chemical kinetics. Chemical dynamics (kinetics) is the study of the rates (speeds) and mechanisms (pathways) of chemical reactions.

We have briefly encountered the rate concept before, when we discussed the manner in which vapour pressure changes with time, and when we presented dynamic equilibrium as a situation of balanced, opposing rates. Although chemical kinetics is a well-developed field experimentally, we are still a long way from the theoretical grasp of rate that we have of equilibrium. Thus our predictive ability in the realm of kinetics is very limited.

Thermodynamics may predict that a reaction should proceed essentially completely. Whether it does so depends on the rate with which it occurs. For example, consider the reaction below at 298 K:

$C(graph) + O_2(g) \rightarrow CO_2(g)$ [$DG^o_R = -394.4$ kJ]

Thermodynamics predicts that the reaction should go to completion. But carbon (coal) deposits exist in the atmosphere without noticeably reacting; and the "lead" in our pencils does not burst spontaneously into flame. The reaction is favored, but is very slow at 298 K. We say that C(graph) is thermodynamically unstable in the presence of O_2(g) at 298 K, but is kinetically inert. As a second case, consider the acid-base neutralization reaction:

$H_3O^+(aq) + OH^-(aq) \rightarrow 2H_2O(l)$ [$DG^o_R = -58.28$ kJ]

Thermodynamics predicts that this reaction, too, should go to completion. In fact, it occurs almost instantaneously; equilibrium is attained as rapidly as the reactants are mixed. Thus we say that H_3O^+ is both thermodynamically unstable and kinetically labile in the presence of OH^-. Two sets of descriptive words are necessary to distinguish thermodynamics and kinetics:

Thermodynamics: stable DG > 0; unstable DG < 0

Kinetics: inert (reaction is slow); labile (reaction is fast).

As we have seen above, there is no necessary correlation between "stable" and "inert", or "unstable" and "labile". This lack of correlation reflects the fact that thermodynamics deals with values of state functions in various equilibrium states of a system; kinetics deals with the rate at which a path between states is traversed.

Reaction rates are extremely important in many ways in our lives. Sustained muscle action depends on the continuous supply of energy produced by glucose combustion in the mitochondria of cells. Energy production must be fairly rapid, otherwise sustained action is not possible. The extremely rapid (explosive) burning of gasoline in the internal combustion engine provides the motive power for automobiles. Food cooks (or, as the chemist would say, oxidizes) extremely slowly at room temperature, but in a relatively short time at more elevated temperature. Our first concern must be to discuss the manner in which reaction rates are expressed, and to introduce some of the terminology of chemical dynamics.

What is a rate? In general, a rate is a number expressing how much a quantity changes in a convenient interval of time. Thus an interest rate is the amount by which a sum of money increases in 1 year; travel rate in an automobile or airplane is the distance covered in 1 hour; heart rate is the number of heartbeats per minute; thus a rate is the change in a quantity per unit time:

$$\text{Rate} = \text{Dquantity/Dt}$$

Note that the units of rate must be "quantity/time, or "quantity-time^{-1}". Thus interest rate has units of dollars/year; travel rate, miles/hour; and heart rate; beats/minute.

How are we to express the rate of a chemical reaction? To develop the answer to this question, we will begin with the simple chemical process in equation. The equation describes quantitatively the conversion of ozone to oxygen that occurs in the earth's upper atmosphere.

$$2O_3(g) \rightarrow 3O_2(g)$$

When this (or any other) reaction takes place, it is the amounts of reactants and products that change. As we know, amount can be expressed in either mass, moles, or molecules. The coefficients in the chemical equation relate both molecules and moles of reactants and products but do not directly relate masses. So we will agree that our measure of amount should be either molecules or moles. Further, since work in the laboratory is carried out on the macroscopic scale, moles is the more suitable unit of amount.

So we might describe the rate of process either in terms of the number of moles of O_3 used or of O_2 produced per unit time. However, there is a major difficulty with this approach. The number of moles of ozone used will depend upon the volume of atmosphere that we choose to examine! We cannot specify a rate without also specifying the volume examined. Rather than use moles/

time, then, we choose to divide by the appropriate volume to give moles/litre-time, or concentration/time. Rates expressed in these units are independent of the actual amount of reaction carried out. At this point then we can express the rate of reaction in either of two ways:

Rate = Moles O_3 used per litre per time = –D(conc O_3)/Dtime

Rate = Moles O_2 produced per litre per time = D(conc O_2)/Dtime

The units of the rate are then moles per litre per time, where the time unit should be the one most convenient for the reaction being measured. Chemists agree that the rate of a reaction should be expressed as a positive number.

As reaction proceeds, the concentration of ozone becomes smaller while that of oxygen becomes larger. Thus the change in ozone concentration, D(conc O_3), is a negative quantity. To produce a positive rate, it is necessary to place the minus sign preceding the D(conc) term, as has been done. In general, whenever the rate of a reaction is expressed in terms of the change in concentration of a reactant, D(conc reactant) is negative and the minus sign is necessary. The minus sign is unnecessary because D(conc O_2) is positive.

There is an additional difficulty, though. We would like to specify a single number representing the rate of a reaction under given conditions. However, the two expressions above give different numbers for the rate, because dioxygen appears 1.5 times faster than ozone disappears (i.e., in the same time that 2 moles of ozone react, 3 moles of dioxygen appear). This can be fixed by normalizing the two expressions in with the stoichiometric coefficients. Dividing D(conc O_3) by the coefficient 2 gives the same number as dividing D(conc O_2) by the coefficient 3:

Rate = –1/2 * D(conc O_3)/Dtime = 1/3 * D(conc O_2)/Dtime

Replacing "conc O_3" with the brackets to represent molar concentratration:

Rate = –1/2 * D[O_3]/Dt = 1/3 * D[O_2]/Dt

The symbol, D, signifies a finite (i.e., measurable) change in a quantity.

In the limit as the length of the time interval, Dt, approaches zero, we replace D(conc) and Dt with the differential (infinitesimal) quantities D(conc) and dt to obtain the differential expression of rate.

Rate = –1/2 D[O_3]/dt = 1/3 D[O_2]/dt

The appropriate formulation for finite difference methods of analysis.

For the completely general chemical reaction, proceeding from left to right from reactants, *A* and *B*, to products, D and *F*, the rate of reaction can be represented using any one of the expressions:

$$aA + bB \rightarrow dD + fF$$

Rate = –1/*a* D[*A*]/*dt* = –1/*b* D[*B*]/*dt* = 1/D D[D]/*dt* = 1/*f* D[*F*]/*dt*

As discussed earlier, negative signs are necessary when rate is expressed in terms of disappearance of a reactant; and division by the stoichiometric coefficient guarantees that the same numerical value of rate is obtained from all four expressions.

Example: In the reaction of ozone to produce oxygen, it is found that under certain conditions of temperature and concentration, 0.0360 moles of ozone per litre react in a 2-hour period. How much dioxygen is produced in this time period? What is the average rate of reaction over this time period?

Solution: The chemical equation, tells us that 3 moles of O_2 are produced for each 2 moles of ozone reacted. Therefore

0.0360 moles ozone * (3 moles O_2/2 moles O_3) = 0.0540 moles dioxygen are produced.

The average rate of reaction can be expressed in either of two ways:

Rate = –1/2 D[O_3]/Dt = –1/2 (-0.0360 moles/L)/2 hours = 0.009 moles/L-hour

Rate = –1/3 D[O_2]/Dt = 1/3 (0.0540 moles/L)/2 hours = 0.009 moles/L-hour

Both expressions give the same positive rate of reaction.

Example: The rate was studied at 55°C by measuring the concentration of t-butyl bromide (*t*-BuBr) as a function of time. The data acquired are given below. Use the data to estimate the rate of reaction 20 seconds after reaction was begun.

t-BuBr + H_2O → *t*-BuOH + HBr

Time, s	**[*t*-BuBr]**
0	0.100
10	0.0876
20	0.0768
30	0.0672
40	0.0590
50	0.0517
60	0.0453
80	0.0348
100	0.0267
120	0.0205
180	0.0093
240	0.0042

We begin by plotting the data to obtain a picture of the manner in which the concentration of t-BuBr varies with time. A plot of [*t*-BuBr] versus time is given. The shape of the plot is typical of a rate process: the concentration of reactant changes very rapidly at first, but as time goes on, the change in concentration per unit time becomes less and less as the concentration of *t*-BuBr approaches its final (equilibrium) value. We have seen this characteristic time plot before in our discussions of vapour pressure and approach to chemical equilibrium. Most concentration-time plots for a reaction in progress have this same general appearance.

Two approaches are possible for obtaining the rate at the 20-second mark of the reaction. The finite difference expression of rate can be used to calculate the average rate during the time interval preceding or following the 20-second mark. Both average rates are calculated below:

Average rate between 10 and 20 seconds
$= -(\text{conc at } 20 \text{ s} - \text{conc at } 10 \text{ s})/(20 - 10)\ \text{Ms}^{-1}$
$= -(0.0768 - 0.0876)/(20 - 10)$
$= 1.08 \times 10^{-3}$ moles L^{-1} s^{-1}

Average rate between 20 and 30 seconds
$= -(0.0672 - 0.0768)/(30 - 20)$
$= 0.96 \times 10^{-3}$ moles L^{-1} s^{-1}

The average rate over the 20 – 30 second time interval is less than that over the preceding interval because the reaction slows down as reactants are used up. This slowing down is readily apparent in the shape of the plot. However, the values are not too much different, so either may be used as an estimate of the rate of reaction at the 20 second mark.

The second approach is to use the differential expression of rate. Just as dy/dx at $x = x_0$ is the slope of a plot of y versus x at $x = x_0$, D[*t*-BuBr]/dt at a particular time, t_0, is the slope of a plot of [*t*-BuBr] versus time at the particular time, t_0. This slope is the same as the slope of the tangent to the curve at t_0. The tangent at $t = 20$ s is shown in the figure. It's slope is readily calculated to be 1.01×10^{-3} moles L^{-1} s^{-1}. As we might have expected, this value gives a rate intermediate between the values obtained above, based on finite time intervals.

Because the slope of the tangent at a particular time represents the rate of reaction at that time, the rate obtained in this manner is called the instantaneous rate at the particular time. It is analogous to the speed that you measure when you glance at your automobile speedometer at a particular instant. As we have shown in this example, the instantaneous rate can be obtained by constructing tangents; or it can be estimated using the average rate over either the preceding or succeeding interval of time.

We can summarize our introductory discussion of reaction rate as follows:

- From a plot of [reactant] or [product] versus time, we can measure the rate at any particular time during the reaction as the slope of the tangent to the curve at that time; the initial rate is the slope of the tangent at t = 0.
- Rate decreases as the reaction proceeds towards equilibrium; this is evident from the shape of the concentration-time plot.
- The units of rate are [ ]/time, or moles/L-time.

COMPLEX REACTIONS: PARALLEL PROCESSES, CONSECUTIVE PROCESSES, AND REVERSIBLE PROCESSES

Thus far we have discussed relatively simple reactions, which occur by mechanisms that involve one step that is substantially slower than the others, and which go to completion. In this situation, rate laws may be readily written from the mechanism and compared with the laboratory-determined rate law.

As you might expect, most reactions are not so simple. In this section, we briefly describe three frequently encountered complications: parallel reactions (two or more simultaneous processes leading to the same or different products); consecutive reactions having comparable rates; reversible reactions (those in which the reverse process plays an important role in determining the observed rate).

Parallel Reactions

Parallel reaction paths may be represented schematically as

$$A + B \rightarrow P\ [k_1]$$
$$A + C \rightarrow Q\ [k_2]$$

To simplify discussion, we assume that each reaction occurs by a single bimolecular elementary step. If so, then we may write the rate laws directly. The rate of disappearance of A is the sum of the rates at which it is used

$$\text{Rate} = -\text{D}[A]/dt = k_1[A][B] + k_2[A][C] = \{k_1[B] + k_2[C]\}[A]$$

If pseudo-order conditions are imposed, this simplifies to the pseudo first-order rate law.

$$\text{Rate} = -\text{D}[A]/dt = (k_1' + k_2')[A]$$

where $k_1' = k_1[B]$ and $k_2' = k_2[C]$. A plot of Rate versus $[A]$ is linear, with slope $k_1' + k_2'$. Thus if one follows the reaction rate by monitoring $[A]$, only the sum of the two rate constants is obtained. If it is possible to monitor $[P]$ or $[Q]$, k_1' or k_2' may be separately determined, and the other one calculated by difference.

Example:. The overall reaction below is thought to occur by two different pathways operating simultaneously.

$$Pt(NH_3)_2Cl_2 + H_2O \rightarrow Pt(NH_3)_2Cl(H_2O)^+ + Cl^-$$

The first pathway involves attack by H_2O at the platinum atom, with simultaneous loss of Cl^-. Product is thus formed in a single elementary step with the same equation as the overall reaction, with rate constant k_1. In the second pathway, Cl^- leaves the platinum compound in a slow first step; this is followed by a rapid second step in which H_2O forms a bond with platinum:

$$Pt(NH_3)Cl_2 \rightarrow Pt(NH_3)Cl^+ + Cl^-\ [k_1, \text{slow}]$$
$$Pt(NH_3)_2Cl^+ + H_2O \rightarrow Pt(NH_3)_2Cl(H_2O)^+\ [k_2, \text{rapid}]$$

What rate law will be exhibited if this hypothesis is correct?

Solution: The overall rate law is the sum of the rate laws for the two parallel processes:

$$\text{Rate} = \text{Rate mechanism 1} + \text{Rate mechanism 2}$$
$$= k_1[Pt(NH_3)_2Cl_2][H_2O] + k_2[Pt(NH_3)_2Cl_2]$$

(Be sure that you understand why the rate laws for the two mechanisms have these forms.)

Consecutive Reactions

We have previously examined systems consisting of two consecutive reactions under what might be called limiting conditions: either the first reaction was much more rapid than the second and could be treated as an equilibrium;

or the first reaction was much slower than the second and completely determined the overall reaction rate. It is instructive to examine how the concentrations of *A, B,* and *C* vary with time in these situations, as well as the one in which the two steps have comparable rate constants.

Consider the following two step mechanism:

$$A \rightarrow B\ [k_1]$$
$$B \rightarrow C\ [k_2]$$

We present three situations involving different relative sizes of k_1 and k_2. We assume for simplicity that only *A* is present initially; $[B] = [C] = 0$ at $t = 0$. When $k_1 >> k_2$, reaction goes to completion before gets under way. When $k_1 = 20k_2$, the concentrations of *A, B,* and *C* vary. A rapidly disappears and is replaced by *B*; *B* then very slowly converts to *C*. If the reaction rate is monitored by measuring the time dependence of [*C*], only the rate constant k_2 is obtained.

On the other hand, if [*A*] is monitored, its rapid disappearance is governed by k_1, which is obtainable from either a plot of ln [*A*] versus time, or a plot of –D[*A*]/*dt* vs [*A*]. When $k_1 << k_2$, *B* is formed very slowly via, but is then immediately consumed.

The concentration of *B* thus never rises much above zero, and *C* appears at the same rate that *A* disappears. This corresponds to the second set of limiting conditions mentioned above. When $k_2 = 20k_1$, the concentrations of the 3 species vary. The rate constant determined experimentally is k_1, the smaller of the two. In this situation, it is very difficult to measure k_2 as [*B*] is never appreciable. Only if *B* can be prepared independently under conditions where it does not convert to *C* can the rate constant, k_2, be measured.

There are many situations in which k_1 and k_2 are approximately equal. We illustrate these by looking at the case in which $k_1 = k_2$. The time dependence of the concentrations of *A, B,* and *C*. At the very beginning of reaction, $t = 0$, the rate of the second reaction is zero because $[B] = 0$. The initial rate is the rate of disappearance of *A*, from which the rate constant k_1 can be determined. Before reaction has progressed very far, however, [*B*] has risen sufficiently that the second process occurs simultaneously with the first.

This tends to drain the concentration of *B* as it is converted to *C*. In the Figure, we see [*B*] rise to a maximum, then fall off because its rate of production from *A* slows as *A* is consumed. At any time, it must be true that $[A] + [B] + [C] = \text{constant} = [A]_o$. If [*C*] is monitored and [*A*] is calculated from the value of k_1, $[A]_o$, and *t*, [*B*] can be calculated as a function of time. It is then possible to determine k_2 from appropriate plots.

Finally, the mechanism and b can be modified slightly to allow reversibility in the first (but not the second) step:

$$A \Leftrightarrow B\ [k_1, k_{-1}]$$
$$B \rightarrow C\ [k_2]$$

(When $k_{-1} = 0$ this mechanism reduces to the). The time dependence of the concentrations of *A, B,* and *C* assuming $k_1, k_{-1} >> k_2$. This corresponds to the first limiting situation mentioned at the beginning of this section. The concentration behaviour for a short time early in the reaction. It is very clear from this figure that A and B rapidly come to equilibrium and maintain a constant concentration ratio thereafter. The total of their concentrations is then very slowly reduced as B converts to C.

Using finite difference methods to calculate the concentrations of the three species, A, B, and C, at many times during the course of reaction. These repetitive calculations were carried out using an electronic spreadsheet. Table contains the data used. It would be instructive for you to set up your own spreadsheet according to the information to see that you can produce plots similar.

Reversible Reactions

Our last example of complexity in chemical dynamics will focus on reversible reactions: those that approach an equilibrium position in which appreciable amounts of both reactants and products are present. The simplest such process is shown below.

$$A \Leftrightarrow B\ [k_1, k_{-1}]$$

k_1 and k_{-1} are the rate constants for the forward and reverse processes, respectively, which are both assumed to be first order. The rate law follows:

$$\text{Rate} = -\text{D}[A]/dt = k_1[A] - k_{-1}[B]$$

This equation states that the rate of change in the concentration of A is the difference between its rate of consumption via the forward process and its rate of production via the reverse process. If the initial concentration of B is 0, then at any point during the process, $[B] = [A]_0 - [A]$. Equation becomes

$$-\text{D}[A]/dt = (k_1 + k_{-1})[A] - k_{-1}[A]_0$$

This can be integrated, but the resulting expression is not readily plottable. Thus, even though the reaction is first-order, a plot of ln[A] versus t will not be linear. To eliminate the troublesome $k_{-1}[A]_0$ term, we recognize that applies at all points during reaction, including equilibrium. Letting $[A]_e$ represent the equilibrium concentration of A, reaction holds when the reaction reaches equilibrium:

$$-\text{D}[A]_e/dt = (k_1 + k_{-1})[A]_e - k_{-1}[A]_0$$

Subtracting equation eliminates the $k_{-1}[A]_0$ term, giving

$$-\text{D}([A] - [A]_e)/dt = (k_1 + k_{-1})([A] - [A]_e) = k_{tot}([A]-[A]_e)$$

This equation is readily integrated in the variable $[A] - [A]_e$, which is the deviation of the concentration of A from the equilibrium value. Thus equation says that in a reversible first-order process, the approach to equilibrium is governed by first-order kinetics. A plot of $\ln([A] - [A]_e)$ against t will be linear, with slope $-k_{tot}$, the sum of the forward and reverse rate constants. We make several observations about reversible first-order processes.

- The rate constant for approach to equilibrium is the sum of forward and reverse rate constants because equilibrium is approached by simultaneous reaction in both directions.

- When reaction is irreversible, $k_{-1} = 0$ and $[A]_e = 0$; thus reduces (with $n = 1$).
- When $k_1 >> k_{-1}$, the measured rate constant, k_{tot}, is essentially equal to k_1. The reverse process is unimportant until the reaction is nearly at equilibrium, because $k_{-1}[B]$ remains small. At equilibrium, the forward and reverse reactions, given by the following expressions, are equal:

$$\text{Rate}_f = k_1[A] = \text{Rate}_r = k_{-1}[B]$$

Both rates are small since at equilibrium $[A]$ and k_{-1} are both small. Similar considerations hold when $k_1 << k_{-1}$. In this case, the measured rate constant is essentially k_{-1}. Let's examine a particular example of a reversible first order process.

Example:. The kinetics of the reaction below were monitored by following the change in absorbance with time of R.

$$R \Leftrightarrow P$$

The following data were obtained.

Time, min	*Absorbance due to R*
1	.789
2	.648
4	.452
6	.333
8	.261
10	.217
12	.191
60	.150

Solution: Determine the rate constant for approach to equilibrium. The absorbance value at 60 minutes represents the equilibrium value, since enough time has elapsed for the reaction to have reached equilibrium. Thus $A_e = 0.15$. A value of $A - A_e$ is calculated for each time and a plot of $\ln(A - A_e)$ constructed. The linearity of the plot substantiates a first order approach to equilibrium. The slope of the plot gives the rate constant, 0.27 min^{-1}. This is actually the sum of the forward and reverse rate constants for the reaction.More complicated processes can be made to conform to equation by imposing pseudo-order conditions. For example, reaction involves forward and reverse processes that are second order:

$A + B \Leftrightarrow C + \text{D}\ [k_1, k_{-1}]$

Suppose that initially $[B]_0$ and $[\text{D}]_0$ are both made much larger than $[A]_0$, and $[C]_0 = 0$. Then the concentrations of B and D will remain constant during the reaction, and A will be converted to C in a pseudo-first-order process in which the forward rate constant, k_f, is equal to $k_1[B]$; and the reverse rate constant, k_r, is equal to $k_{-1}[\text{D}]$. We write the rate law for approach to equilibrium:

$$\text{Rate} = -\mathrm{D}([A]-[A]_e)/dt = (k_f + k_r)([A]-[A]_e) = k_{\text{tot}}([A]-[A]_e)$$

k_{tot} is studied first as a function of $[B]$ with [D] held constant:

$$k_{\text{tot}} = k_1[B] + k_r$$

k_1 can be obtained from a plot of k_{tot} versus $[B]$. Then k_{tot} is studied as a function of [D] with $[B]$ held constant :

$$k_{\text{tot}} = k_f + k_{-1}[\mathrm{D}]$$

k_{-1} is obtained from a plot of k_{tot} versus [D].

Bibliography

K.K. Sahani: *Chemical Equilibrium and Chemical Kinetics*, Cyber Tech, 2009.

K Sarn: *Chemical Kinetics* Rajat, 2005.

Morris Sylvin: *Problems in Chemical Kinetics*, Sarup and Sons, 2001.

Syed Aftab Iqbal: *Problems in Chemical Kinetics and Solutions*, Discovery Pub, 2011.

S.A. Khan: *Text Book of Chemical Kinetics*, Sonali Pub, 2011.

E. H. El-mossalamy: *Textbook Of Chemical Kinetics*, Discovery Publishing, 2011.

Pankaj Sethi: *Textbook Of Chemical Kinetics*, Campus Books, 2012.

Raphael D. Levine: *Molecular Reaction Dynamics*, Cambridge Univ Press, 2006.

Octave Levenspiel: *Chemical Reaction Engineering*, Wiley, 2006.

Lanny D. Schmidt: *The Engineering of Chemical Reactions*, Oxford University Press, 2007.

James Keeler and Peter Wothers: *Why Chemical Reactions Happen*, Oxford University Press, 2003.

Mohsen A. Hedaya: *Basic Pharmacokinetics*, Taylor and Francis Group, 2010.

Milo Gibaldi: *Biopharmaceutics and Clinical Pharmacokinetics*, Pharma Book Syndicate, 2005.

Milo Gibaldi: *Biopharmaceutics and Clinical Pharmacokinetics*, Pharma Med Press, 2008.

Gayathri V. Patil and Harpal Singh: *Biopharmaceutics and Pharmacokinetics*, Shree Pub, 2009.

V. Venkateswarlu: *Biopharmaceutics and Pharmacokinetics* : , PharmaMed Press, 2010.

K.K. Sahani: *Chemical Equilibrium and Chemical Kinetics*, Cyber Tech, 2009.

K Sarn: Chemical Kinetics, Rajat, 2005.

Stephen H. Curry: *Drug Disposition and Pharmacokinetics With a Consideration of Pharmacological and Clinical Relationships*, PharmaMed Press, 2008.

Taylor Springer: *Enzyme Kinetics and Mechanism*, Paperback, 2002

Cahn , Springer: *A Guide to Physics Problems, Part 2: Thermodynamics, Statistical Physics, and Quantum Mechanics*, Paperback, 2003.

G. C. M. Ray and Sharma: *A Textbook On Dynamics*, S. Chand Publisher , 2005.

Marx: *Ab Initio Molecular Dynamics*, Cambridge Univ Press, 2002.

Dominik Marx and Jurg Hutter: *Ab Inito Molecular Dynamics*, Cambridge University Press, 2009.

M. Amin: *Advanced Bioenergetics and Biodynamics*, Capital Publishing Company, 2011.

Bharat Bhushan Prasad: *Advanced Soil Dynamics And Earthquake Engineering*, PHI Learning, 2002.

V K Selvaraj: *Advanced Thermodynamics*, Campus Books Intl, 2008.

S K Aggarwal: *Advances In Fluid Dynamics*, ALP Books, 2009.

Barnes W. McCormick: *Aerodynamics, Aeronautics, and Flight Mechanics*, Wiley India, 2010.

Jagmohan Meher: *Afghanistan : Dynamics of Survival*, Kalpaz, 2008.

Diwakar: *Agriculture and Industry: Dynamics of Imbalances in India*, Manak Publication, 2002

R P Rastogi and R R Misra: *An Introduction To Chemical Thermodynamics*, Vikas Publishing House, 2008.

H Versteeg and W Malalsakra: *An Introduction To Computational Fluid Dynamics : The Finite Volume Method*, Dorling Kindersley, 2012.

V.K. Shrivastava: *An Introduction to Dynamics*, ABD, 2006.

Batchelor: *An Introduction To Fluid Dynamics*, Cambridge Univ Press, 2002.

Ardema: *Analytical Dynamics: Theory and Applications*, Springer, 2002.

B.K. Venkanna and B V. Swati: *Applied Thermodynamics*, PHI Learning, 2011.

Raghavendra Reddy and Narayan Goswami: *Applied Thermodynamics*, Pacific Books International, 2012.

Ashish Tewari: *Atmospheric And Space Flight Dynamics* , Springer, 2009.

Abul Kalam: *Bangladesh : Internal Dynamics and External Linkages*, The University Press, 1996.

Manish Rathi: *Basic Concept of Chemical Thermodynamics*, Cyber Tech Pub, 2011.

Index

A

Acetals 288, 289
Acids 283, 284
Acylals 288, 289
Alcohols 272, 277, 279, 280
Aldehydes 280, 282, 283, 285
Apparatus 58
Approximation 159, 160, 229, 236
Azines 291

B

Bases 295, 296
Bimolecular 11, 130, 133, 134
Boiling 271, 272, 283

C

Calculations 173
Carbohydrates 277
Carboxylic Acids 280, 283, 284
Catalyst Function 101
Catalytic 202, 226, 227, 228, 229, 234
Catalytic Converters 104

D

Decomposition 51, 66
Determination 7, 9, 75, 111, 139
Dilution 172

E

Electron 263, 265, 266, 267, 268
Electronegative 267, 268
Electrophile 265, 267, 268
Elementary 11, 12, 129, 130, 132
Energy 294
Ethylene 103
Eukaryotic 244
Explosions 88

F

Finite 142, 143
Free Energy 166
Functional Groups 277, 279, 281

G

Gases 182
Generalising 81
Graphical 9
Gravitational 248

H

Halides 269, 270, 271
Homogeneous 14
Hydrates 200
Hydrochloric 86
Hydrogen Peroxide 51, 66, 67
Hydrolysis 284

I

Imines 290
Initial 9, 64, 72, 83, 161, 168
Initial Rate Method 139
Instantaneous 32, 74
Integral 114, 116, 119, 120
Integrated 65, 66, 69

K

Ketones 282,
Kinetics 3, 7, 13, 17, 63

L

Ligand
203, 234, 235, 237, 242, 243, 245
Lipids 277

M

Macromolecule 234, 235, 236
Melting point 271, 272
Microgram 277
Molecular Collisions 88
Molecularity 11, 130, 132, 152
Molecule 165

N

Nitrogen 162
Nitrogenous 290
Numerical Analysis 142

O

Orbital 266
Organic Chemistry 271, 277
Organic Compounds 279
Organic Families 279
Orientation 127, 219

P

Parallel Processes 297
Parallel Reactions 298
Parameters 21
Phosphorylation 234
Polar 281
Polar Reactants 281
Polymer 271, 272
Potential Energy 92, 93
Pressure 162, 163, 164, 165, 166, 167
Proportionality 72
Protein 277
Pseudo-Order Methods 144

R

Radical 262, 263, 264
Reaction Mechanism 148
Reaction Rate 134
Reversible Reactions 300

S

Salts 185
Simulations 242
Solid State 272
Sulfuric Acid 104
Supplement Applications 166

T

Termolecular 11, 130, 134
Termolecular
11, 13, 124, 129, 131, 134
Ternary 217
Thermodynamics 2, 293, 294
Thioaldehydes 285
Toxicity 277
Trivial 285, 287

U

Unimolecular 11, 130, 133
Uniqueness 274, 278

V

Vapour Pressure 182, 191, 293, 296
Vitamin 277

W

Water 272, 275, 284